Governing Molecules

Inside Technology

edited by Wiebe E. Bijker, W. Bernard Carlson, and Trevor Pinch

Marc Berg, *Rationalizing Medical Work: Decision-Support Techniques and Medical Practices*

Wiebe E. Bijker, *Of Bicycles, Bakelites, and Bulbs: Toward a Theory of Sociotechnical Change*

Wiebe E. Bijker and John Law, editors, *Shaping Technology/Building Society: Studies in Sociotechnical Change*

Stuart S. Blume, *Insight and Industry: On the Dynamics of Technological Change in Medicine*

Geoffrey C. Bowker, *Science on the Run: Information Management and Industrial Geophysics at Schlumberger, 1920–1940*

Louis L. Bucciarelli, *Designing Engineers*

H. M. Collins, *Artificial Experts: Social Knowledge and Intelligent Machines*

Paul N. Edwards, *The Closed World: Computers and the Politics of Discourse in Cold War America*

Herbert Gottweis, *Governing Molecules: The Discursive Politics of Genetic Engineering in Europe and the United States*

Gabrielle Hecht, *The Radiance of France: Nuclear Power and National Identity after World War II*

Eda Kranakis, *Constructing a Bridge: An Exploration of Engineering Culture, Design, and Research in Nineteenth-Century France and America*

Pamela E. Mack, *Viewing the Earth: The Social Construction of the Landsat Satellite System*

Donald MacKenzie, *Inventing Accuracy: A Historical Sociology of Nuclear Missile Guidance*

Donald MacKenzie, *Knowing Machines: Essays on Technical Change*

Susanne K. Schmidt and Raymund Werle, *Coordinating Technology: Studies in the International Standardization of Telecommunications*

Governing Molecules

The Discursive Politics of Genetic Engineering in Europe and the United States

Herbert Gottweis

The MIT Press
Cambridge, Massachusetts
London, England

Set in New Baskerville using Ventura Publisher under Windows 95 by Wellington Graphics.
Printed and bound in the United States of America.

Library of Congress Cataloging-in-Publication Data

Gottweis, Herbert, 1958–
Governing molecules : the discursive politics of genetic engineering in Europe and the United States / Herbert Gottweis.
p. cm. — (Inside technology)
Includes bibliographical references and index.
ISBN 0-262-07189-4 (hardcover : alk. paper)
1. Genetic engineering—Government policy—Europe. 2. Genetic engineering—Government policy—United States. I. Title. II. Series.
TP248.6.G68 1998
338.9′26—dc21 98-25670
CIP

Contents

Acknowledgments

The idea of writing this book began to take shape in 1988. As an avid reader of the British *Financial Times* and of the German *Tageszeitung*, I noticed in 1987 that both newspapers had greatly extended their coverage of topics related to genetic engineering. While the *Financial Times* was excited about the commercial prospects of biotechnology and warned about too much regulation, the alternative-green *Tageszeitung* emphasized genetic engineering's risks and its unforeseeable environmental and social impacts and demanded strict laws to control the hazards of recombinant DNA. A fascinating conflict seemed to be in the making. I felt that I should follow the events and try to write a book on the unfolding political controversy about biotechnology.

Interviews with more than eighty individuals in science, government, business, and social movements in four countries were my central method of exploring the genetic engineering conflict, its history, its dynamics, and its actors. The interviews provided me with fascinating firsthand accounts of the everyday life of biotechnology policy-making. Furthermore, most of my interviews had the important side effect of helping me to get access to written documents, including internal correspondence, memos, and minutes. Without those documents many portions of this book could not have been written. I want to thank my interview partners in Britain, France, Germany, and Belgium for their time, their patience, and their interpretations of biotechnology policy.

Invaluable help came from Bernhard Zechendorf, the person in charge of BIODOC, the biotechnology documentation center located at the European Commission's Directorate General for Science, Research, and Development. BIODOCs unique collection of documents, scientific papers, newspapers, and journal reports related to biotechnology helped me to reconstruct some important stories.

The Gen-ethische Netzwerk in Berlin was an important source of information and helped me get my orientation in the jungle of biotechnology.

During the academic year 1989–90, the Minda de Gunzburg Center for European Studies at Harvard University was a perfect place to get some distance from my first round of fieldwork and to think about the historical dimensions of the politics of molecular biology. Discussions at the de Gunzburg Center with Peter Hall and Pnina Abir-am were inspiring and gave me new insights. Financial support for my stay at the de Gunzburg Center came in form of an Erwin Schrödinger Fellowship from the Austrian Ministry of Science. During my time in Cambridge I also met Charles Weiner from MITs Program in Science, Technology, and Society. His brilliant understanding of the tension between biotechnology and society and his friendship helped to make this book possible. I also benefited greatly from his earlier work on the beginnings of the genetic engineering controversy in the early 1970s, and particularly from his collection of interviews and documents in the Recombinant DNA History Collection at MITs Institute Archives and Special Collections. In 1992–93, having finished my interview work in Europe, I wrote the first draft of the book at MIT under the auspices of the Program in Science, Technology, and Society. Support during that year came from Andrew Mellon Foundation. Between 1993 and 1995, Cornell's Department of Science and Technology Studies provided a perfect setting for putting together a second draft of this book. My colleagues Sheila Jasanoff and Peter Taylor were creative commentators on my work. The undergraduate and graduate students in my courses helped me to test and elaborate many of my ideas and interpretations. During my time at MIT and Cornell, Lily Kay, whose pathbreaking work on the history of molecular biology had helped me to understand today's genetic engineering controversy, was a constant source of support, encouragement, and inspiration. Volkmar Lauber of the political science department in Salzburg was an always open but critical reader of my manuscript. Hans-Jörg Rheinberger of the genetics department helped me to develop a deconstructivist reading of the science of genetic engineering. Rogier Holla, Bernhard Gill, Les Levidow, and Timothy Young commented on the whole manuscript. The final draft was written in 1997 at the Hong Kong University of Science and Technology, where my stay was organized by Govindan Parayil.

Governing Molecules

NOCH MEHR WEISS BITT
EG KUNSTMILCH FABRIK
KON-SERVIER-UNGS-MITTEL
GESCHMACKS-VERSTÄRKER
VITAMIN-LAGER
FARB-STOFF -WEISS-

Introduction

In its June 1989 issue, the German farmers' magazine *Bauernstimme* published a feature on rBST, a new hormone drug that was soon to be launched on the market by the US-based multinational corporation Monsanto. Bovine somatotropin (BST) is a natural protein produced by the pituitary gland of cattle; rBST is a recombinant form of the protein developed with the help of genetic engineering. Added to feed, rBST promised a substantial increase in milk output per cow. Nevertheless, *Bauernstimme* was not enthusiastic about the drug. It reported considerable concern among farmers (especially small farmers) that rBST might have a negative effect on the quality of their cows' milk and on their economic future. A survey showed that more than 80 percent of Germany's dairies were not prepared to use rBST. These concerns were summarized in a cartoon showing an oversize cow that seems to be part of a factory. The cow is identified as a "European Community artificial milk factory." Pipes connect its body to containers labeled "preservatives," "vitamins," and "artificial color." A factory worker shouts "More white please!"

In the eyes of Monsanto, Germany's Bundesministerium für Forschung und Technologie (Ministry of Research and Technology) and other actors, the notion that rBST collapsed the boundaries between animals and technology constituted a gross exaggeration that could not be justified on scientific and technical grounds. In Monsanto's or the Bundesministerium's representation, rBST—a substance readily available in a cow's body—was simply being "enhanced" by technical means, and rBST milk was as natural as any other milk.

Thus, while Monsanto and the Bundesministerium saw nature when they looked at rBST, *Bauernstimme* and the farmers saw technology. From the perspective of the chemical company or the

Bundesministerium, rBST was a safe and useful product; from the perspective of many farmers, it was a potentially unsafe substance that might have unpredictable socio-economic consequences. Such differences in describing what might seem to be an objective artifact can render the consensual selection of policy options difficult indeed.

The importance of the multiplicity of meanings attributed to phenomena in the policy process becomes even clearer when the rBST debate is contextualized as part of a larger controversy over genetic engineering in the United States and in Europe. In the late 1950s and the early 1960s, the new biology that would eventually lead to the development of recombinant DNA techniques had already become an object of systematic state support. Shortly after genetic engineering became available, comprehensive policies for research in biotechnology and for the biotechnology industry were developed in Germany, in France, and in Britain, and also at the level of the European Community. In the United States, support for molecular biology focused mainly on providing a research base for the emerging biotechnology industry. On both sides of the Atlantic, scientists, administrators, and company heads hailed biotechnology as something that would significantly affect the future competitiveness of nations and trading blocs.

At the same time, critics pointed to potential risks for humans, animals, and the environment posed by work with recombinant DNA. While some perceived genetic engineering as a technology of the future, others saw it as a source of potential disaster. From the mid 1970s until the mid 1980s, the public policies developed in Germany, France, Britain, and the United States attempted to combine state support of genetic technology with regulation. In the United States the recombinant DNA controversy had peaked in the mid 1970s. It Europe it was in the mid 1980s that the conflict culminated in demands for more stringent regulations of genetic engineering, for a broader consideration of the ethical, ecological, economic, and political implications of genetic engineering, and even for the banning of certain products (including rBST) in the European Community. Far from being settled, the debate over genetic engineering continues today in various fields, from genetic testing to the cloning of sheep. The goal of this book is to explain how genetic engineering became a controversial technology that some sought to promote by any means and others sought to block entirely.

The Need for a Poststructuralist Approach

In this book I outline a conceptual framework for policy analysis inspired by poststructuralism and apply it to the study of policy-making—in particular, to the comparative analysis of regulatory and technology policies. For social scientists the most important analytic message of poststructuralism may be the need to pay careful attention to the complicated ways in which language and discourse are used to constitute social, economic, scientific, or political phenomena, to endow them with meaning, and to influence their operation. Accordingly, I interpret the genetic engineering controversy as a process that was inseparable from the mapping—the social construction—of the political, economic, and scientific worlds. I emphasize the importance of interpretations, framings, and definitions in the construction of reality, subjectivity, and identity in the realms of science and politics. That is, I argue that there is a need to examine how discourses and narratives—stories that create meaning and orientation—constitute the policy field of genetic engineering. What are the parameters of state regulation? What counts as a rationale for state support? Who is constructed as a legitimate actor in a policy field? Which institutions are defined as central in a policy field? How is the boundary between state and civil society defined and regulated? Which strategies demarcate science from nonscience, and how does scientific knowledge contribute to the shaping of social identity? While my approach underscores the importance of language and knowledge for political analysis, it does not lose sight of actors and institutions; I argue, however, that actors and institutions must always be analyzed in close association with the discourses in which they are constituted and with the policy stories that define the logic, the actors, and the institutions that matter in a policy field.

Poststructuralism has had a considerable impact on a number of disciplines, including history, science studies, and cultural anthropology (Frank 1989; Lenoir 1994; Rosenau 1992). In political science, poststructuralist voices have become particularly important in the subfields of international studies and political theory (Gooding and Klingemann 1996, pp. 21–22; George 1994). Other fields, including comparative political studies, have remained relatively untouched by this new approach and continue to look at politics through a lens narrowed by realist epistemology and positivist methodology.

Conventional political science views policymaking as a struggle between competing groups, as a game played by rational actors, or as determined by institutional structures. The approach developed in this book is critical of the assumption that we can simply assume the "existence" of actors and structures in politics. Furthermore, I question the narrow understanding of "the political" underlying traditional political science approaches. Following a broader understanding of "the political," I suggest conceptualizing the process of policymaking as situated at the intersection between forces and institutions deemed "political" and those apparatuses that shape and manage individual conduct in relation to norms and objectives but are deemed "nonpolitical," such as science or education (Rose 1996, pp. 37–38). Instead of assuming stable boundaries between sectors such as politics, the economy, and science, this approach moves the creation of such boundaries—the micropolitics of boundary drawing—to the center of analytical interest.

This theoretical perspective has a number of important implications for my conceptualization of the main research problems of this book. The "politics of genetic engineering" cannot be reduced, for example, to citizens' pressing for the regulation of recombinant DNA technology. A "politics of genetic engineering" is also taking place when, for example, scientists carry out certain experiments in genetic engineering and subsequently make claims with respect to the significance of this work for the understanding of nature and of human behavior. Arguing that such different articulations of the "politics of genetic engineering" are interrelated, I describe the working of a more complex, multiple regime of governability—a system of fields and sites ranging from laboratories to parliaments where strategies focused on the manipulation of genetic material were deployed and negotiated. Guided by this understanding of "the political," I analyze policymaking against the backdrop of a constellation of locations where different strategies have been pursued to turn genes into objects of technological and political intervention—to make them "governable."

While this book's understanding of politics is broad, its analytical focus is on policymaking. I show how, through the mutually reinforcing relationships between various "political" and "nonpolitical" forces, genes were construed not only as objects for technological manipulation but also as legitimate concerns for regulatory efforts and for state support. The following questions are addressed: How did molecular biology and genetic engineering become objects of state intervention?

How and with what impact did risk emerge as a major topic in the politics of genetic engineering? Which arguments framed and constructed genetic engineering as a technology of the future that would have a deep impact on socio-economic development? What is the relationship between eugenics' and molecular biology's conceptualization of properties of life and genetic factors? How can we explain the strong resistance to genetic engineering in Europe in the 1990s? What lessons can we draw from the study of genetic engineering politics concerning how national political systems and the newly emerging political system of Europe handle the deep and pervasive transformations and challenges introduced by modern science and technology? Are these processes of technological transformation guided by principles of negotiation and compromise? Or has decisionmaking on scientific development and its economic exploitation become a form of "subpolitics" (Beck 1993) that is removed from the scrutiny of democratic institutions and suffers under a chronic deficit of political legitimization?

In dealing with these questions we are constantly confronted with problems of meaning, interpretation, and definition. For instance, the importance of potential hazards associated with genetic engineering for the development of regulatory policies cannot be denied—but what exactly constitutes such a hazard, and how is its interpretation socially constructed? In a similar way, it seems to be obvious that international competitiveness is a major reason for state support of biotechnology—but what exactly does it mean to state that there is a technology race going on between nations? How is this image of competition being created, and what is its role in policy rationales? These questions address political issues and refer at the same time to broader contexts of culture and society. What matters are definitions of the "truth" of a problem, a challenge, or a solution—in other words, the drawing of the always flexible and contingent borders and territories of politics, science, and society. Such boundary constructions constitute important demarcations of power—for example, by separating science from processes of political shaping, or by attributing to experts (as opposed to representatives of the public) the exclusive authority to evaluate the environmental risks of genetic engineering.

This book's interest in boundaries and in the construction of meaning has much to do with an important transformation in contemporary life. Today's social conditions seem to be characterized by an overall sense of uncertainty, reflecting the absence of analytical and political guarantees, the dissolution of privileged political positions and agents, and the coexistence of a plurality of political spaces and social logics through which various forms of social and political identity are constituted. One important expression of this "postmodern" constellation that has enormous implications for policymaking is the blurring of boundaries of phenomena in a variety of social and political sectors (Smart 1992, p. 219; Latour 1993). Examples include the increasingly unclear boundaries between experts and non-experts in scientific decisionmaking, the collapsing boundaries between nature and technology, and the great difficulty of drawing clear borders around the economy of the nation-state. Scientific expertise does not enjoy as much legitimacy as it did some decades ago. Frozen embryos seem to be a form of life but also a form of technology. Globalization undermines traditional functions of the nation-state. How is policy made under such new conditions? Does democracy still provide the space in which to negotiate the new demarcations of power? In this book I attempt to answer these questions by looking at the "politics of genetic engineering" in France, Germany, the United Kingdom, and the European Community. Although the focus is on Europe, I also examine a number of important developments in the United States. The European "politics of genetic engineering" is part of a larger political texture interwoven with images, developments, and myths of American biotechnology.

Outline

Rather than give a comprehensive overview of genetic engineering policies in the United States and Europe, I focus on episodes that help to explain how genetic engineering and biotechnology gradually emerged as objects of political intervention, how they became controversial, and how policymakers dealt with the various challenges they posed.

In chapter 1, I locate the approach I will use in the rest of the book within the current discussion about the comparative study of policymaking. I discuss several models that are important for the study of comparative politics, assessing their strengths and their limitations. Then I outline an alternative, poststructuralist framework—inspired by the work of Michel Foucault, Ernesto Laclau, and Chantal Mouffe and by recent work in science and technology studies—for the analysis of technology policymaking. This chapter underscores the importance of discourse, knowledge, and narratives in both scientific and political practice, and it clarifies my understanding of policymaking. The argument that language and stories matter in politics also has important normative implications, as I will argue in the concluding chapter.

In chapter 2, I discuss how molecular biology policy came into existence as a new policy area. A number of mutually reinforcing developments explain molecular biology's rise as a topic of political interest in the United States and in Europe. I emphasize that molecular biology was always seen as more than just an interesting new field of basic science. Since the late 1930s the supporters and the critics of the "new" biology have emphasized that molecular biology is also a project of population politics. There are a number of interesting links between the "old" pseudo-science of eugenics and the "new" field of molecular biology. Eugenics focused on the "gene pool" as a site of intervention; molecular biology shifted its interest to individual genes. I discuss the relationship between eugenics and the new biology by looking at important transformations in US biology during the 1930s.

The focus of the book then shifts to Europe, where, from the late 1950s on, political discourse interpreted molecular biology primarily as a new scientific field of central importance for medicine, for industry, and for general social well-being. I show that in Europe the identification of molecular biology as an object of policymaking was predicated on the deployment of narratives that articulated molecular biology as an element in broader policies of modernization. Thus, a discursive link was created between ongoing processes of economic restructuring and discursive shifts in the discipline of biology. These policies of molecular biology set in motion a dynamic of technological development whose socio-political implications and relationship to eugenics would again become objects of public scrutiny and discussion only a few years later.

In chapter 3, I move on to the 1970s and to the semiotic construction of genetic engineering's risks as a field for regulatory policymaking. Immediately after recombinant DNA technologies were developed, they became objects of multiple ascriptions of meanings. Genetic engineering had raised highly difficult boundary questions about the relationship between nature and technology. What for some was the scientific basis of a new industry appeared to others a dangerous technological intervention that might even bring back eugenics in a new form. The evolving regulatory policies did not address the many ethical, social, and environmental concerns raised with respect to genetic engineering; it focused mainly on the creation of a system of risk regulation. The new regulations also implied a specific and contested mapping of the boundaries among society, science, and politics that affected the procedures of regulatory decisionmaking. I argue that these regulations inscribed the risks of genetic engineering in a way that contributed to (rather than undermined) the framing of modern biotechnology as a socially acceptable technology of the future.

Chapter 4 shows how the interpretation of genetic engineering as a controllable technology and the creation of the myth of the US biotechnology industry became essential elements in the shaping of biotechnology policies in Britain, Germany, France, and the European Community during the 1980s. I argue that the dominant narratives of biotechnology contained a number of assumptions (about such things as the structure of international research and development in the pharmaceutical and chemical industry, the working of financial markets, and the current state of art in genetic engineering) that together gave meaning to the notion of biotechnology as a high-technology industry of the future. These assumptions soon turned out to be highly problematic and made implementation of the adopted policies difficult. At the same time, these policies also reflected significant ideas about how the new genetic technologies could help to bring about social transformation—the creation of a "Bio-Society." This "bio-politics" and its relationship to older forms of bio-politics, such as eugenics, soon moved to the center of social conflict.

In chapter 5, which focuses on the many voices that have been strongly critical of genetic engineering, I argue that the European genetic engineering controversy of the 1980s must be understood in the context of a number of important contextual changes. The articulation of a new discourse of modernization oriented to ideas of ecology and sustainable development the emergence of a social movement sector organized around the topic of genetic engineering had created a new discursive constellation for policymaking. The critics compared molecular biology to eugenics and construed it as a strategy of cultural hegemony that could change and normalize the parameters of human self-recognition and the understanding of nature. Furthermore, deconstructive readings of the biotechnology policy narratives rejected the politics of connecting the collective identity and future of Britain, France, Germany, and Europe with the project of rewriting life through the new genetic technologies. Most important, this critique was also directed against a partitioning of the political space of biotechnology and against the construction of policy boundaries that separated regulatory topics from the discussion of research policies and from considerations of the ethical, social, economic, and political implications of biotechnology.

Chapter 6 shows how the discursive struggle over the boundaries separating genetic engineering, politics, and society defined the political dynamics of one of the most critical episodes in European biotechnology policy in the 1980s: the drafting of new risk regulations for the contained use and the deliberate release of genetically modified organisms in Germany, Britain, and France. These policies, which involved the establishment of a precautionary regulatory approach toward genetic engineering's risks, introduced significant elements of stabilization in the discursive field of genetic engineering. Although the environmental and health risks of genetic engineering could be negotiated to some extent in the context of the newly established regulatory structures, other topics of concern were still without a place where they could become objects of institutionalized contestation and deliberation.

Despite numerous attempts to end the debate, genetic engineering continues to be a highly controversial topic. In my concluding analysis of the relationship among genetic engineering, democracy, and identity construction, I emphasize the importance of creating a political space for the democratic negotiation of biotechnology. Failure to do so might lead to consumer boycotts against the biotechnology industry and to other forms of resistance. Finally, I emphasize that policymaking with regard to genetic engineering is a case in point which seems to indicate that, under postmodern conditions, policymaking can be successful only if it understands the need to mobilize existing institutional mechanisms and to set up new institutional mechanisms that will ensure tolerance of and respect for the multiplicity of socially available policy narratives and reality interpretations in a policy field.

1

What Is Poststructuralist Science and Technology Policy Analysis?

In this chapter I outline the conceptual framework of this study and my approach to the relationship among science, politics, and power. An important theoretical inspiration for my framework comes from the emerging field of science and technology studies (STS), which has emphasized the socially constructed and contingent nature of scientific knowledge and technological practice. The field of STS has, however, paid only scant attention to science and technology policymaking. In contrast, political science has devoted considerable attention to the relationship between the state and science and technology, though most political science studies of topics related to science and technology remain informed by a view that conceptualizes science as a unique truth-seeking activity and technology as some sort of application of the truth (Bimber and Guston 1995, p. 555; Elzinga and Jamison 1995). The complicated mutually constitutive and mutually reinforcing relationships between the state and science and technology have rarely been explored in this line of research (Jasanoff et al. 1995, p. 527). In addition, the epistemological realism informing most political science studies excludes important dimensions of analysis that could help explain the dynamic interactions between politics and science and technology. The poststructuralist approach disagrees with central concepts and underlying realist assumptions of the science and polity models used in the current literature of comparative politics. These models have serious shortcomings, many of which result from the neglect of the critical relationships among knowledge, language, meaning, and power. In contrast, the poststructuralist critique highlights a number of new and previously ignored dimensions of political analysis, such as the importance of discourses and narratives for the construction of political reality, the constructed nature of actors in politics and society, and the phenomenon of the competing, conflicting, and often contradictory structures of meaning and expression in social and political life.

Typically, comparative political studies try to explain phenomena by mobilizing concepts of actor and structure. Why did political party A adopt a particular technology policy? Was it the world economic situation, or perhaps just a strategic game between feuding groups within the party in which the group that came to dominate used the technology policy as part of its game plan? What explains why one country decides to build a welfare state while another abstains from adopting comprehensive social policies? Might different patterns of class relationship explain the variations in policy choice? Or might the alternative approaches reflect diverging strategies of the labor movements in the two countries? Different schools of political science will give different answers. Ultimately, these answers will depend on the conceptualization of the relationship between structure and agency underlying the approach.[1] Whereas conventional schools of political science privilege either actors or structures in their accounts, poststructuralist political analysis avoids such a dichotomization by offering a language or a discourse-analytical perspective that acknowledges the importance of structural phenomena and contexts for the understanding of politics without reducing actors to "outcomes of structures." I will develop this framework for the study of science and technology policy in the following discussion, but first I want to offer a brief look at the most important approaches to comparative politics in order to show the sorts of problems raised by the conventional conceptualization of the actor-structure framework.

Discourses of Comparative Politics

Historically, the Anglo-American discourses of politics, and comparative politics in particular, have identified institutions as a central analytical concept. The "old" institutionalism focused on detailed configurative studies of different administrative, legal, and political structures (Bill and Hardgrave 1973, pp. 2–6; Thelen and Steinmo 1992, p. 3). During the "behavioral revolution" in political science in the 1950s and the 1960s, this old institutionalism, with its emphasis on formal rules and administrative structures, was thoroughly rejected as incapable of explaining actual political outcomes or behavior. The behaviorists argued that individuals or groups of individuals constituted the fundamental building blocks and that political results were simply aggregations of individual actions. In this framing, institutions were viewed as empty shells to be filled by individual roles, statutes, and values. Following the methodological suggestions of logical positivism, the behaviorists emphasized the importance of systematic empirical research, conceptualization at various levels of abstraction, and hypothesizing and testing by empirical data (Chilcote 1981, p. 56; Shepsle 1989, p. 133). The progress of political science was closely associated with the development of "models" and conceptual frameworks. Central to these efforts was the idea of conceptualizing politics as a "system" (Gunnell 1983, pp. 19–20). From the 1960s on, behaviorism in comparative politics found itself challenged by the rational-choice approach. While behaviorism conceptualized social actors as reactive and passive, rational choice framed human agency as purposive, proactive, and oriented toward the maximization of privately held values. In this view, actors are not connected to the social structure in which they are embedded. Espousing an atomistic conception of political life, such a view champions the application of social-choice theory, game theory, and decision theory to politics. With individuals rather than whole systems as the principal units of analysis, politics is conceptualized as a competition among individuals whose goals are access to power or scarce resources and whose means are rationally calculated to achieve those ends most efficiently (Shepsle 1989, p. 134; Hall 1986, pp. 10–12; Riker and Ordeshook 1973).

The 1950s and the 1960s saw the development of still another direction in political science: group theory. For the group theorists, whose most prominent younger representatives were active in the "behaviorist revolution," the essential unit of analysis was neither the political system as a whole nor the individuals within it, but the social groups or classes that come into conflict within the polity. In its pluralist version, group theory asserts that democracy is premised on diverse interests and on the dispersion of power. In the work of Bentley, Truman, and Dahl, the central assumption is that political life consists of a bargaining process in which various groups interact in negotiation and compromise (Bentley 1908; Truman 1951; Dahl 1961; Hall 1986, p. 13). On the left end of political science's spectrum, the class approach took as its basic unit of analysis aggregates of individuals, not unified by a common goal or shared interest as in the group approach, but sharing rough equality in one of the fundamental or central distributive values (power, wealth, or prestige). In this reading, society-wide stratification is the fundamental reality of social and political life and the basic determinant of conflict and change. But in class analysis as well as in group analysis, the basic unit of analysis—the aggregate of individuals—is considered to be crucial in explaining the operation of the political system (Bill and Hardgrave 1973, pp. 140, 191).

From the late 1970s on, models of politics have begun to refocus on the importance of institutions in the policy process. Clearly, this reorientation constituted in part a rebellion against positivist premises and methodological formalism such as are expressed in rational choice's atomistic model of politics, which too often had yielded only limited insights into the political process (Rockman 1990, p. 27). One strand of theorizing and empirical work, "state-centric" theories, contended against pluralist and Marxist models of politics that policy is not primarily a reaction to pressure from social groups or classes. The "statists" argued that the state should be seen as far more independent from societal pressures than group theory suggests and that it is capable of formulating and implementing its preferences.[2] Another strand, historical institutionalism, developed a more elaborate model that entailed a reconsideration of the state's relationship with society or, more broadly, with its environment.[3] Institutions are defined—following Peter Hall's widely accepted definition—as the "formal rules, compliance procedures, both formal and conventional, that bind the components of the state together and structure its relationship with society" (Hall 1986, p. 19). According to the historical institutionalists there are two key ways in which institutions affect political outcomes. First, they tend to distribute power in the political system in such a way as to give actors representing some social interests more power than others. Second, they affect the actors' perceptions of their own interests by making some political outcomes more feasible than others (Hall and Taylor 1994, p. 3). Within the discourse of historical institutionalism there are varying efforts to define what counts as an institutions. Whereas Ellen Immergut's study of health politics in France, Sweden, and Switzerland emphasized a rather limited theory of institutions (Immergut 1992, pp. 231–232), thereby focusing on the importance of traditional political institutions such as the legislature and government, Peter Hall's study of economic policymaking in Britain and France espoused a much broader understanding of institutions that included the organization of labor, capital, the state, the political system, and the structural position of the country within the international economy (Hall 1986, p. 232).

"Rational-choice institutionalism" was developed around the same time as historical institutionalism. The atomistic versions of rational choice had run into a dilemma: under the rules of majority voting they had predicted unstable and paradoxical decisions with rapid "cycling" from one bill to another as new majorities appear to overturn any bill that is passed. But the reality of political life in the United States did not seem to be constant flux; rather, stability seemed to be pervasive. The new explanation given by rational-choice institutionalism was that much of the instability inherent in pure majority voting systems is eliminated by legislative rules. According to Kenneth Shepsle, the "old" preference-driven explanation of equilibrium had given way to an institutionally enriched, structure-induced equilibrium. Rational-choice institutionalism viewed institutional arrangements as responses to problems related to collective action—as arrangements that arose because the transaction costs of political exchange were high (Shepsle 1989, pp. 134–138; Shepsle and Weingast 1994; DiMaggio and Powell 1991, p. 5).

Analytical intentions similar to those driving the shaping of the new institutionalism were behind the discourse of the "policy network" or "policy community" approach. At the core of the policy community approach was the development of institutionalized relations between governmental and nongovernmental bodies, which facilitate the making and the implementation of policy. The goal of the network approach was to shift the study of politics away from an emphasis on institutions such as the legislature and the cabinet and toward a focus on the complex and intricate configurations of group, departmental, and non-legislative-based policymaking arrangements—a goal that clearly was commensurate with the new institutionalism's interest in the organizational aspects of the state-society interaction (Jordan 1990; Rhodes 1988; Wilks and Wright 1987; Jordan and Schubert 1992; Atkinson and Coleman 1989). In fact, most network approaches in political science attempt to enrich the conceptual metaphor of the network by explicitly incorporating elements of rational-choice theory and institutionalist approaches.[4]

Both the various strands of the new institutionalism and the political community and network approaches are characterized by a number of shortcomings. One difficulty involves the methodological realism these approaches espouse. Roughly speaking, for these approaches, institutions and "laws" guiding human behavior (as in rational choice) have the same ontological status as elephants. They are conceptualized as existing "out there" and waiting for the political scientist to represent them in a "truthful" way.[5] Despite repeated assurances that, for example, historical institutionalism offers a solution to the micro/macro problem in political analysis,[6] these approaches end up privileging structures (conceptualized as observable entities) over actors. And the structures seem to exist independently of the actors, which in turn receive little conceptual attention. Rather, the strategies and goals of the actors are seen as shaped by the institutional context. Whereas the historical institutionalists see goals *and* strategies as shaped by institutional context, the rational-choice models of the new institutionalism see primarily the *strategies* as imposed by institutional contexts (Thelen and Steinmo 1992, p. 8). Rational choice theoretically brackets the formation of preferences (goals) by assuming that political actors are rational and will maximize their interests. The rational economic actor is conceptualized as a unified subject who in a bargaining game encounters decisionmaking with an identity already formulated in terms of his or her preferences. These preferences are understood to determine the behavior of the utility-maximizing actor. Furthermore, rational-choice theory assumes the existence of determinate laws of human action that, once known, can be used to predict rational action without reference to the author's framing of reality (Metha 1993, pp. 88–89).

Hence, the rational-choice version and the historical version of institutionalism (and, as a result of borrowing from theses approaches, the network models) avoid such issues as the ascription of meanings and the construction of identity and subject positions involved in any process of policymaking. The methodological "realism" of the approaches discussed above is closely connected to a systematic neglect in this work of the phenomena of language and knowledge. This is not to say, however, that realism fails to comprehend the importance of ideas in political life. There has in fact been a surge of interest in this topic. Recent contributions to the study of comparative politics and innovation politics have begun to point toward the importance of frameworks of knowledge or paradigms and their evolution for the process of policymaking. For comparative politics, Hall's 1989 work on economic paradigms has been most significant in reopening the debate on the relevance of ideas and discourse in the study of politics.[7] But Hall sees the influence of Keynesian ideas as determined by a variety of institutional factors, including the structure of the state and state-society interactions. In their edited volume of essays, Judith Goldstein and Robert Keohane (1993a) stress the importance of ideas in the political process and focus on variations in beliefs and on the effect the institutionalization of ideas in societies has on political action. In the same volume, Nina Halpern examines the adoption of Stalinist economic policy ideas in China, a process that occurred against the will of the Soviet Union. As Goldstein and Keohane (1993b, p. 14) sum it up, "Soviet ideas, not Soviet power, explain the choice of economic policy." The "realist" reading of ideas that is expressed in the notion that Soviet power somehow has nothing to do with ideas presupposes a particular reading of power as contingent on and independent from practices of signification.

This brief discussion of some important schools in comparative politics shows that the conceptualization of the actor-structure problem is at the core of the analytical frameworks used in these models. Political science has certainly made important progress in overcoming the "atomizing" versions of political action developed in the early days of rational-choice theory. Network approaches and the new institutionalism have arrived at sophisticated conceptualizations of the actor-structure problem. However, these approaches continue to be limited by their epistemological realism, which creates a natural barrier to a nondichotomizing understanding of the relationships connecting human agency, institutions, and structures. As a result, these approaches view such central elements of the political as discourse, ideas, and knowledge as epiphenomena. It is precisely an emphasis on language as *constitutive* of politics that is at the center of the poststructuralist framework of analysis.

Postpositivism, Science Studies, and Political Science

Poststructuralist political analysis is part of a larger movement in the social sciences that has grown out of a critique of the positivist presuppositions that inform most empirical political science. Neo-positivist social science describes scientific practice as a process of hypothesis testing with the goal of finding causal generalizations that are quantitatively testable. Language is viewed as a neutral medium used to record observations. This conceptualization of language and of the idea of a search for universally generalizable findings is rejected by postpositivism (Fischer 1996).

The intellectual roots of the postpositivist movement in political science can be traced back to German sociology and social theory of the late nineteenth century and the early twentieth century, and in particular to the work of Max Weber, Edmund Husserl, and Alfred Schutz. The work of these men constituted an important influence on the evolving sociological schools of symbolic interactionism, critical phenomenology, and ethnomethodology. Whereas positivist social science was searching for universal laws or empirical regularities of human behavior, interpretative approaches were more interested in the human capacity for creating, communicating, and understanding meaning, which was seen as located in social practices and texts (Yanow 1996, pp. 4–9). Another important influence on the emergence of postpositivism was the later work of Ludwig Wittgenstein. His reasoning undermined the empiricist epistemology of logical positivism, which took as given the atomistic nature of the relationship between the objects of "the world" and their meanings as expressed in elementary linguistic expressions. Wittgenstein contended, against that view, that to understand reality through language is to engage in complex social practices that defy the atomized logic. According to Wittgenstein, one should concentrate not on the logical independence of things but on the systemic relationship among them, which invests them with social meaning. For Wittgenstein, language was not an exclusively descriptive medium but something active that could be used to give meaning to things, which, in turn, was to constitute social reality (George and Campell 1990, p. 273).

Interpretative approaches in line with Wittgenstein's philosophy have had enormous influence on the relatively new academic discipline of science and technology studies (STS). In the past 20 years, STS has launched a full-fledged attack on the concepts of science espoused by logical positivism. The new STS approach marked a distinctive break with earlier work in the sociology of science, which was concerned with science as an institution and with the study of scientists' norms, career patterns, and reward patterns. In contrast, STS took as its subject the actual content of scientific ideas, theories, and experiments. The new sociology of science explored science "from the inside," as a body of knowledge and as a social system inextricably linked to society and to politics (Pinch and Bijker 1987, pp. 18–21). Scholars "opened up the black box" of science. What many of them found were meanings, social constructions, inscriptions, and discourses. According to Michel Callon and Bruno Latour, black boxes are much used in scientific practice. A black box, which contains what no longer must be considered, stabilizes situations of meaning (Callon and Latour 1981, pp. 284–285). The new social studies of science came to the following important conclusions: We should avoid "black boxing" (that is, assuming as "given objectively") the facts and entities of science and technology. Rather, we have to view scientific and technological phenomena as resulting from complex social constructions. Thus, a judgment as to whether or not a hypothesis has been verified (or falsified) is the culmination of complex social processes within a particular environment. Furthermore, scientific knowledge does not arise from the application of preexisting decision rules to particular hypotheses or generalizations, nor is language a neutral medium of communication. With this move, the focus of science and technology studies shifted to the social processes behind science and technology, but also to language as the constitutive medium through which knowledge, truth, meaning, scientific and technological objects, and (ultimately) social identity are made possible (Lenoir 1994; Gieryn 1995). One important way in which science studies challenged the social sciences was by asking this question: If the actual practice of the natural sciences does not obey the ideal model of neo-positivism, why should sociology or political science structure their ways of research and reasoning along what increasingly looks like an outdated fantasy about the way science works?

Political Science after Structuralism and Metaphysics

Just as neo-positivist epistemology had given rise to a number of quite different approaches in political science (such as behaviorism and rational choice), postpositivist influences have led to the gradual development of new intellectual tendencies (including interpretative studies and discourse analysis). These tendencies share a focus on language, but they differ significantly in their conceptualization of language and in the way they relate discursive to nondiscursive phenomena. In this book I introduce a poststructuralist mode of political analysis that draws on the work of Michel Foucault, Jacques Derrida, Ernesto Laclau, Chantal Mouffe, and Nikolas Rose and from the current sociology and philosophy of science, in particular the work of Ian Hacking, Michel Callon, Bruno Latour, and Hans-Jörg Rheinberger. The unifying theme holding the work of these diverse scholars together is that of creating a thinking space after structuralism and metaphysics. A new theory of language is at the center of this project.

Classical structuralism, as paradigmatically developed in the work of Ferdinand de Saussure, constitutes the point of departure of poststructuralism's critique of language and meaning. Saussure developed his notion of structure through the analysis of language.[8] For him, structure refers to the ordering principle according to which the lexicon of a language is articulated. This is done in such a way that it can be recognized and mastered as the lexicon of one and the same national language. This process happens in acts of differentiation and connection; that is, all elements of expression that make a sign audible and legible must be clearly distinguishable from one another—one must be able to distinguish between signs. According to Saussure, this is not immediately possible on the basis of the meaning of signs alone; it requires an understanding of their expression. Meaning itself is amorphous, lacking a clear profile. Therefore, one can distinguish among meanings only by differentiating between sound images or written images—that is, between the signifiers (acoustic or written images) rather than the signified (the concept). Hence, a structure is a system of pairs—meaning and expression, or signified and signifier—such that one and only one signified is assigned to every signifier. The individual signs are exact applications of an invariable law to which they are related; they cannot proliferate and become uncontrollable.

But it is exactly the concept of uncontrollability that poststructuralism brings into the debate. For Jacques Derrida, the idea of a closed structure reflects metaphysical thinking and a desire for control. Structuralism searches for general ordering principles and universal regularities that make the world amenable to technological and scientific mastery—that give clear orientation in a world that would otherwise seem to be out of control. According to Derrida, Saussure took an important step in his critique of language by rejecting the idea that words are somehow images of preexisting ideas and noting that meaning is constituted solely by differences between signifiers within a system of interrelations—that one can think about the redness of an apple only in relation to yellow apples and green pears. But Saussure insisted on the existence of something like a structuring principle of language. This, argues Derrida, constituted a move in the language game of metaphysics. Metaphysics not only believes in a transsensual world, it also offers orientation in the form of control and domination, for example by searching for general ordering principles and universal regularities. Derrida argues that structures can decompose, and that we are entangled in structures and have no possibility of getting beyond our "Being-inside-structures." He says that "everything is structure," but he does not mean that "everything" is taxonomy; rather, every meaning, every signification, every view of the world is in flux. Nothing escapes the play of the differences, and thus nothing can be tracked down and fixed in its meaning. How can this poststructuralist critique of language instruct a reconceptualization of the actor-structure problematic? Let me outline some of the central concepts of the poststructuralist framework of analysis for political science.

Representation

First, the poststructuralist perspective argues that there is no such thing as a theory-neutral observation language. Whereas the (neo-positivist) representationalist or correspondence theory of truth believes that there "is a truth out there" and that it can be represented through the neutral medium of language, the anti-representationalist view of poststructuralism rejects such a "picture theory" of language (according to which the physical properties of the world are considered fixed while language is in the business of meeting the needs of their description). Rorty (1989, p. 5) puts it this way: "Truth cannot be out there—cannot exist independently of the human mind—because sentences cannot so exist, or be out there. The world is out there, but descriptions of the world are not. Only descriptions of the world can be true or false. The world on its own—unaided by the describing activities of human being—cannot." What does that mean for the analysis of a phenomenon such as the politics of science? It means that we must conceptualize articulations of science and technology as the outcomes of complicated processes of inscription, of re-presentation. Neither the truth of "genetic engineering," the policy problem "hazards of genetic engineering," nor the "high-technology gap" (to give examples from this study) is simply "out there," needing only to be discovered or studied. Rather, what constitutes a "risk" or a "high technology" is a result of historically specific and ongoing constitutive practices of writing and inscription. Consequently, the "truth" of an event, a situation, or an artifact will always be the contingent outcome of a struggle between competing language games or discourses that transform "what is out there" into a socially and politically relevant signified (Daly 1994, pp. 167–177; Rorty 1989).[9]

Writing

Precisely because the "nature" of reality is inherently undecidable, there is the human drive to organize or structure the world. Structure is never a given. According to the theory of poststructuralism, writing is the process by which human agents inscribe order into their world; it is a way of fixing the flux and flow of the world in spatial and temporal terms. Government programs, political speeches, and party platforms are excellent examples of such attempts to create order. They strive to give sense and coherence to the otherwise confusing or contradictory realities of political life. The determination of entities such as the state or the economy and the delineation of their boundaries cannot count on any prediscursive settlements but are based on processes of writing or articulation.[10] For Derrida, writing has material potency. It is described as the act of inscribing notations, marks, or signs on a surface that bring representation into being, be it a sheet of paper or a DNA sequence. Writing is concerned not with the meaning and content of messages but with the structure and organization of representations. Accordingly, ministries, political parties, and social movements can be regarded as symbolic productions that emerge from being "written" and constituted by those who observe and participate in them (Cooper 1989; Linstead and Grafton-Small 1992, p. 341).

The writing of a book also constitutes a material practice. This book is the result of a complex process of attending to others' experience, transcribing and analyzing that experience, and telling about it by putting it into the form of a book. The textual activity of analysis, as in this book, inevitably constitutes a process of rewriting and transforming; in this respect, reading must be understood as a process of writing. My reading of the "social movements" critique of genetic engineering is, in a way, a practice in the production of the meaning and the implications of controversy in the context of policymaking. Hence, the practice of writing political analysis participates in the production of the political, rather than only representing it (Game 1991, p. 5; Riessman 1993, pp. 8–15).

Decentered Subjects

Derrida's theory of writing helps us to gravitate in the direction of a nondichotomizing conceptualization of the actor-structure relationship. From a poststructuralist perspective, subjects or actors cannot be viewed as origins of social relations, because they depend on specific discursive conditions of possibility. For example, those who "write" and create organizations or policy programs should not be conceptualized as autonomous, rational actors. In what Derrida criticizes as the "logocentric tradition" of thinking, self-conscious, rational minds produce thoughts. These thoughts provide meaning, which is articulated by speech. Speech itself is inscribed by writing. The unmediated presence of consciousness is privileged over speech (logocentrism), and speech in turn is privileged over writing. This means that meaning is more authentic the closer it is to the origin: the immediate consciousness. Derrida overturns this logocentric conception of writing and argues that speech would not be possible without writing. He thus directs our intention to the specific and multilayered textual processes by which subjects come into being. He shows that writing determines consciousness, which is not immediate and unreflective but rather is the result of a relationship with what has already been inscribed: the trace of previous individual or collective experience (Linstead and Grafton-Small 1992, pp. 341–342).

In this perspective, the self-conscious modern subject begins to be replaced by the idea of an intertwined and fragmented texture as the place for the emergence of subjectivity. For example, gender, though constructed, is not necessarily constituted by an "I" or a "we" who precedes that construction in any spatial or temporal sense. Judith Butler (1993, p. 7) writes: "Subjected to gender, but subjectivated by gender, the 'I' neither precedes nor follows the process of this gendering, but emerges only within and as the matrix of gender relations themselves." Therefore, instead of talking about subjects it is preferable to talk about subject positions, which are contingent and strategic locations within a specific discursive domain. Actors do not have stable subject identities; they constantly develop their subjectivity in a discursive exchange (Laclau and Mouffe 1985, p. 115; Hajer 1994, p. 5). From a logocentric view, the human agent represents a holistic and clearly bounded universe. But self-consciousness is never pure, authentic, unmediated experience, as rational-choice theory believes. Social and historical traces enter into the structuring of consciousness. In this way, individual subjects or actors are constituted through symbolic systems that fix and differentiate them in place while remaining outside of their control. This is not a "cultural dope" conception of human agency; it is a perspective that points out that human agency is produced reality (Linstead and Grafton-Small 1992, p. 343; Smart 1982, p. 135). Also, this perspective does not imply that "there are no human actors" in politics or that we have to stop to talk about agency as we analyze political processes. There is no question, for example, that a particular high-level administrator in the Commission of the European Community is in a powerful position and can take important steps to mobilize support for his goal of imposing strict environmental regulations on genetic engineering, or that we can and should analytically follow his actions. But we have to understand that this administrator does not act independently from the discourse of European environmental policy, which in many ways influences how this administrator views the world, defines his goals, and structures his actions.

The Social Boundaries of Science and Knowledge

As was mentioned above, an important inspiration for postpositivist policy analysis comes from the field of social studies of science. For decades political scientists have tried to model their research practice on those of the natural sciences. Gunnell (1975) described convincingly how his fellow political scientists were basing their work on an ideal model conjured up by the philosophers of logical positivism. Since the early 1980s, however, social scientific studies of science practice have accumulated a picture of the actual practice of research showing that the ideal model had little to do with the everyday life of scientific practice in fields such as biology or instrumental physics. This work is significant for political science for two reasons: it forces upon policy analysis a reconceptualization of the notions of scientific knowledge and technology, which play a role in many types of policy; more fundamentally, it supports postpositivist tendencies by emphasizing the socially and linguistically constructed character of scientific activities.

Although in the standard view science is a process by which researchers struggle to discover the truth about objects existing independently in the natural world, contemporary science studies conceptualize science as a textual as much as a material practice. A common approach in recent sociology of science has been to focus analytically on representations of scientific practice and knowledge in social situations where science is being demarcated from nonscience (Gieryn 1983, 1995). Simply put, the focus is on what Gieryn (1983, p. 782) has called *boundary work:* "the attribution of selected characteristics to the institution of science (i.e., to its practitioners, methods, stock of knowledge, values and work organization) for purposes of constructing a social boundary that distinguishes some intellectual activity as nonscience." The distinction between science and nonscience is a critical element in boundary work, but on a more general level boundary work encompasses those acts and processes that create, maintain, and break down boundaries between knowledge units, such as disciplines or subdisciplines. Thus, boundary work incorporates the processes whereby legitimacy and cognitive authority are accorded to knowledge units. In this reading, science is a space that acquires its authority from and through episodic negotiations of its flexible and contextually contingent borders and territories (Fisher 1990, p. 98; Gieryn 1995, p. 405). In this context, boundary objects give definition to realities across sites (Star and Griesemer 1989, pp. 388–393). They inhabit several intersecting worlds, such as those of politics and of science.

A boundary object such as "gene" is an object that is plastic enough to adapt to local needs and constraints of the several parties employing it yet robust enough to maintain a common identity across sites. Boundary objects separate groups of people (the scientist has a "scientific" concept of the gene, the politician "a layperson's view") but also link them. "Gene" means different things to a politician and to a scientist, but both share some basic and similar views of its meaning. They can talk to each other. This is exactly why boundary objects such as "gene" help the communication between different social worlds. But this link or communication is also characterized by the maintenance of a boundary between the scientist and the politician. As these two individuals talk about genes, the politician remains the layperson and the scientist remains the expert. This means that a specific partitioning of the political space takes place and that a hierarchy of credibility is being established.

But boundary work in science is not only a project of creating meaning and authority but also a material practice of transformation. As Hacking has pointed out, the birth of the Baconian Program in the seventeenth century signaled the ascendance of a science whose prime aim was to manipulate and control nature for the utility of man, thereby collapsing the distinction between representation and intervention. This suggests that the analysis of science should focus on the inseparable relationship between representing and intervening (Hacking 1984). In this sense, scientists who do research in biology write texts on life that at the same time shape life. We can conceptualize the way scientists write texts on life as using experimental systems—the smallest functional units of research, designed to give answers to questions that we are not yet clearly able to ask. Rheinberger (1992) has suggested that we interpret experimental systems as comprising two different yet inseparable structures or components: the scientific object (or the epistemic thing) and the technological object. The scientific object is a physical structure, a chemical reaction, or a biological function whose elucidation is at the center of the investigative effort. It cannot be fixed, and we can say that it is something like a questioning machine. The technological object, in contrast, can be compared to an answering machine: it is determined and performs within and according to known regularities. The technological object contains scientific objects, embeds them, restricts them, and controls the proliferation of their meanings. Technological objects are (to stay with molecular biology) gels, ultracentrifuges, or electron microscopes, which determine the mode of representation of the epistemic thing. At the same time, this very operation of the control of meaning, the representation, constitutes the framework for intervening into life. In this sense, representation *is* intervention. Genetically modified plants, hybrid animals, and recommendations based on prenatal screening are expressions of such interventions.

The "semiotic character" of scientific practice becomes clear when we begin to understand that an experimental system creates a space of representation for things that otherwise cannot be grasped as scientific objects. Usually it is said that such a representation presents a model of what is going on "out there in nature." Thus "in vitro systems" are models for "in vivo systems." But one cannot know what goes on out there unless one has a model for it. Consequently, the reference point of any model can be nothing else but another model. It is impossible to go behind a signifying chain, for representations do not merely disclose some underlying reality but actually constitute it. Hence, a scientific object realizes itself as a kind of writing, an *écriture,* as Rheinberger argues. Through measuring devices and technical arrangements, the scientist creates the basis for a material textual structure. In other words, the realized representation is an epistemic thing that relies on a particular technical and instrumental background.

If the scientific object does not immediately produce visible traces, "tracers" are introduced into it: radioactive or fluorescent markers, pigments, etc. A particular sequence of DNA is "read" as a ladder of black bars in four adjacent columns on the autoradiograph of a sequencing gel. Temporally and spatially, the object is an *inscription.* This is an important point. The experimenter uses such elements as a chromatogram or a DNA sequencing gel to arrange these graphemes and to compose a model. It is this graphematic reality in which the scientist is immersed. "Nature" is an object of experimentation only insofar as it is already representation—insofar as it itself is an element, however marginal, of the game. Hence, what goes on is neither a "reading" of the "book of nature," a depiction of reality, nor a deliberate construction of reality. Scientific representation is not expressing something intrinsic; instead, what occurs is graphematic activity: the shaping and arranging of material entities along partitions that are neither revealed, imposed by the existing world, nor prescribed through our actions. In short, science is a form of writing.

Boundaries of the Political

The poststructuralist conceptualization of representation and writing and the specification of the boundary concept in science studies have a number of implications for our understanding of politics. Politics and political phenomena do not simply exist. Their borders are always drawn (Gieryn 1995, p. 405), and thus we can view politics as an "empty space" until it is demarcated and partitioned through struggles involving boundary drawing.[11] Here the boundaries separate the "political" from the "nonpolitical." It is through practices of articulation and inscription that nodal points are constructed that partially fix meaning and construct political spaces as different from other spaces, such as the space of science. There are no entities (such as "society," "politics," or "the state") that constitute a totality and thus can be linked or "sutured" to become capable of "interactions," as in "state-society interactions." No single underlying principle (e.g., Marx's mode of production or Weber's process of rationalization) is identified as responsible for connections and relations among groups and, in that way, as constituting the whole field of differences. Rather, social, economic, and political "reality" are constructions made possible through articulatory or writing practices.[12] These practices consist of the construction of nodal points, "bridges of meaning," that partially fix meaning. In the process of policymaking, the unification of a political space through the institution of nodal points constitutes a successful attempt to define a political reality, subject identities, and modes of action.

Boundaries of the political can be rigid or flexible, and can be permeable or impenetrable. But political boundaries are always a temporary phenomenon, and the relational system defining the identities of a given political space can undergo processes of weakening. This weakening may result in the proliferation of floating elements of discourse and in a crisis of signification. States, for example, never achieve full closure or complete separation from society; indeed, their precise boundaries are usually in doubt and contested.[13] Many political conflicts evolve around definitions of these boundaries (Alonso 1994, p. 381). The debate about the desirability of regulating the Internet raises questions concerning the boundaries between legitimate scrutiny and surveillance that undermines core liberties of civil society. Such boundary disputes often lead to redefinitions of what constitutes the field of the state and what constitutes the field of society. Expertise is another case that demonstrates the importance of boundary drawing. In science policy the boundaries drawn around what constitutes expert knowledge are of special importance. In the most extreme case, experts are capable of constructing *expert enclosures*—relatively bounded locales or types of judgment, or elements of a regime of governability, within which the power and the authority of a particular group of experts are concentrated. The arguments and assessments of these experts become obligatory modes for the operation of the policy network as a whole (Rose and Miller 1992, p. 188).

Government

My conceptualization of politics and knowledge as boundary phenomena affects another central concept of policy analysis: government. In the usage of conventional political science, government is associated with the activities of political authorities (such as a cabinet or a ministry). Recently, a concept of governance has been introduced to emphasize that, besides the state, there are other important (private) mechanisms of governance (such as the community and the market, with their guiding principles of spontaneous solidarity and dispersed competition).[14]

In the context of this study, government focuses on practices: mechanisms and techniques that, in the name of truth and the public good, aspire to inform and adjust social and economic activities. Under democratic conditions, government implicates truth and rational problem solving. In this way it is intrinsically linked to knowledge—to scientific theories, technological practices, experiments, or economic forecasts.[15] This "knowledge dependence" of government has important implications for policymaking. Because the legitimacy of policymaking relies increasingly on technical and scientific arguments, power becomes intertwined with knowledge: the exercise of power is predicated upon the deployment of knowledge. At the same time, knowledge is always underdetermined, in the sense that, faced with choices, researchers tend to adopt interpretations, theories, or lines of research that enable them to monopolize areas of investigation, dismiss the work of competitors, or answer to the social demands of a particular class (such as the state, a political party, or a church). This underdetermination is overcome by nonscientific power interests that relate power internally and essentially to scientific knowledge. This phenomenon has been described by Michel Foucault as the power-knowledge nexus (Kusch 1991, pp. 150 and 162–163).

In this perspective, the state is one important site of government among others. There are sites of government that influence policymaking but have nothing to do with the state and its institutions. In this conceptualization, the focus on repressive modes of the exercise of power as visualized by the Hobbesian state is supplanted by the perspective of the "multiple regime of governability" relativizing the notional boundary between the state and society (Gordon 1991, p. 36). Markets, research institutes, biotechnology companies, and financial institutions (in other words, knowledge-producing institutions of civil society whose practices define and structure social and economic realities but which are usually constituted as being "nonpolitical") are not necessarily "objects" of state intervention but may coexist in a relationship of intimate symbiosis with state strategies and tactics (Hunt 1992, p. 27). In short, they govern, contribute to making things governable, and are parts of a larger structure. Thus, we can say that policymaking is situated at the intersection of the various sites of a regime of governability. Regimes of governability are systems where individuals, nature, and artifacts are transformed into objects of interventions and become "governable." The objects of government (such as pregnancy, genes, the economy, or global warming) co-emerge at a number of locations that are not necessarily considered "political" (such as in hospitals or research institutes) but where nevertheless significant influence on the logic and the rationale of a policy field can be exerted.

Faces of Power

Underlying the concept of governability is a relational model of power. At the core of this model is an assumption that there are forms of power that make possible certain global effects of domination that emanate not from centralized locations such as the state or the economy (as in top-down models of power) but from places such as laboratories or scientific journals.

The debate about power that has occupied political scientists for more than 30 years started with a discussion of "the two faces of power" by Peter Bachrach and Morton Baratz (1963). They argued that Robert Dahl's (1957) Weberian understanding of power—as a situation where A can get B to do something that B would not otherwise do—reflected only one face of power. What this understanding does not encompass is the important phenomenon of "non-decision-making"—the practice of limiting the scope of actual decisionmaking to "safe" issues by manipulating the dominant community's values, myths, and political institutions and procedures. In such a situation, A doesn't interact directly with B but might nevertheless effectively prevent B from doing what B wants to do. Steven Lukes (1974) added that power can also be exerted by A through a manipulation of B's very desires and wants, to a point that B wants to do what A desires despite the fact that it goes against B's interests.

Despite the substantial differences, the theories of Dahl, Bachrach and Baratz, and Lukes all focus on the repressive side of the power phenomenon: power is about making B do something, keeping B from what he or she might otherwise do, or getting B to act against his or her own interests. In contrast, the relational concept of power (or the "fourth face of power") emphasizes the productive side of power. This concept does not displace the others but provides a different level of analysis. Neither A nor B nor institutions with their biases and political ideologies (the first three faces of power) "simply exist" as power phenomena. They come into existence through discursive processes that transcend the traditional boundaries of the political (Digeser 1992, p. 978). In this way regimes of governability can be understood as sites where individuals are shaped and constructed. Neo-liberalism, for example, not only describes principles for regulating and limiting governmental activity but is also closely tied to particular forms of rational self-conduct of the governed, such as the entrepreneurial and competitive conduct of economic-rational individuals. This suggests that the process of governing involves individuals' adopting particular practical relationships to themselves in the exercise of their freedom. Such practices of shaping individuals often encounter resistance (Connolly 1991, p. 198; Burchell 1993, p. 268; Robson 1993, p. 462; Digeser 1992).

Narrating the Political

Government always involves the deployment of policy narratives, which are special forms of discourse. These policy narratives serve to integrate the various sites that constitute a regime of governability. Narrative and discourse are structural phenomena that further help to specify the poststructuralist project of overcoming a dichotomizing perspective of the actor-structure problematic. They underscore the idea that the boundaries of politics, science, and technology are always drawn within the larger semiotic context of the various stories that give a society its identity and hold it together.

"The narratives of the world are numberless" is how Roland Barthes begins his famous essay on narrative theory. Narrative is present in myth, legend, fiction, cinema, weather reports, government documents, political speeches, and articles about molecular biology. News of the world comes to us in the form of "stories," our dreams narrate about the subconscious, and for each of us there is a personal history, the narrative of our life, which enables us to construe who we are and where we are going (O'Neill 1994, p. 11). However, only recently has the theory and study of narratives emerged as a field of interest in the social and human sciences, in disciplines ranging from literary theory to philosophy, economics, ethnography, and, more currently, political science.

This "narrativist turn" must be contextualized as a part of the broader anti-representationalist movement identified above as deriving from the breakdown of logico-deductive models of reason and knowledge and of transcendental truth claims. The anti-representationalist perspective asserts that there is no objective reality that can be universally represented through a neutral medium of language. The methods of natural science as interpreted by logical positivism are thus inadequate for understanding society and culture. This development opened the way for new approaches, such as narrative analysis. Martin (1986, p. 7) states that "mimesis and narration have returned from their marginal status as aspects of 'fiction' to inhabit the very center of other disciplines as modes of explanation necessary for an understanding of life."

The exact nature of narratives and their relationship to politics require explication. On the most general level, narrativity is the representation of real events that arises out of the desire to have real events display the coherence, integrity, fullness, and closure of an image of life that is and can only be imaginary (White 1981, p. 23). Narratives bring elements of clarity, stability, and order into what usually tends to be the complicated and contradictory world of politics. This power to create order is an attractive quality that makes narratives essential for the shaping of policies, the settling of conflicts, or the securing of legitimacy for political action.

Narratives also interconnect the various sites of a regime of governability, such as laboratories, scientific journals, and research ministries. In other words, they function as "networks of meaning," establishing relationships among a variety of entities. They problematize, enroll, and mobilize persons, procedures, artifacts, and representations in the pursuit of a particular policy goal. *Problematization* determines the sets of agents or artifacts that might be involved in a policy field and defines their identities. For example, a policymaker in a country's Ministry of Industry might argue that agriculture needs to change and industrialize because of dramatic transformations in world markets, and that the chemical industry should play an important role in this process. *Enrollment* then defines sets of interrelated roles and attempts to attribute them to certain actors. At this stage, a policymaker might argue that the chemical industry should invest in applications of genetic engineering in the oil seeds business, where genetic technologies may offer superior strategies. The chemical industry might then take this new direction on its own, or it may accept subsidies from the government for research and development. Finally, *mobilization* describes the transformation of enrollment into active support. Actors have accepted their roles and thus become part of policy schemes or models of political reality. In a budget meeting, the Minister of Industry might argue that the chemical industry is an important ally of the government in its attempt to restructure national agriculture by supporting biotechnology, and that resources should be reallocated from the Ministry of Agriculture to the Ministry of Industry to ensure the future of the government's newly developed strategy for national agriculture. In this way the chemical industry and genetic technologies have become important nodal points in an emerging network of meaning that, under the leadership of the Ministry of Agriculture, will attempt to restructure the agricultural system (Callon 1986).

The most important features of a narrative require further clarification. First, events of some sort, such as the election of a new government or the invention of a new laboratory technique, are a central necessity if we are to speak of a narrative. These events constitute the raw materials of narratives, which arrange them in a temporal and causal order. Times, places, and characters interact in a complex fashion in the narrative transaction. Narrative events are a function of time, setting (place), and characters of both time and place (Martin 1986, p. 124; O'Neill 1994, p. 17). Narratives depend on texts. Text is a concrete and unchanging product, words upon a page. It is only through the text that we an acquire knowledge of the narration. At the same time, it is through narrative that a text manages its articulation in time and orchestrates the relations of its writers and readers. The sentence "Bangkok is one of the most polluted cities in the world" constitutes a text or statement. Set into the context of a report on new strategies for pollution control drafted by the Thai government, the sentence becomes part of a narrative; it assumes narrative character. As soon as any story becomes a story by being told, it gives rise to a proliferation of possible intersecting meanings and enters the realm of textuality. Textuality refers to the voluntary and involuntary interactivity of authors, readers, and texts, creating the phenomenon of *intertextuality* (O'Neill 1994, pp. 23–25).[16] Two articulations of narrativity in political discourse are *political metanarratives* and *policy narratives*.

Political metanarratives describe general concepts and values of the social order. Metanarratives offer a conceptual framework that provides a polity and its subjects with an imagined collective political identity situated in historical time. In this way, the discourse of modernization, dominant state projects such as the welfare state or neoliberalism, and the idea of democracy can all be construed as elements of a political metanarrative. Though broad and general in content, metanarratives are also specific. For example, though it might seem universal, the idea of modernization has different meanings in China and in France. This is a function of the intertextual nature of metanarratives, which are always linked to specificities of historical experience and culture. Metanarratives are not simply "out there"; they are performative practices, they do things with words, and they are always written, rewritten, read, and reinterpreted. Thus, there cannot be any "general theory" of political metanarratives. The study of political metanarratives is always the study of interwoven practices taking place in contexts of time, space, and sociality. Therefore, analyzing political metanarratives requires careful empirical examination of specific settings, such as modernization narratives in France in the 1950s or in China in the 1990s. The ensemble of the metanarratives dominant in a particular society constitutes the *political imaginary*—a cognitive mapping and its accompanying values through which individuals see and experience the world. The political imaginary is a social construct that is part of the available repertoire of political visions and identifications in one's social situation. It consists of the configurations of the most common representations, stories, and ideals defining a particular social space and its boundaries (Kellner 1989, p. 188; Baudrillard 1976).

Policy narratives are more specific and describe the frames or plots used in the social construction of the fields of action for policymaking.[17] These frames or plots are principles of organization that govern events. They describe a structure of relationships that endows the events contained in the account with meaning by integrating them into a narrative. Policy narratives are constructed and used to make sense out of events. Their raw materials are dispersed events, such as a particular experiment in laboratory X, a branch of industry Y, the world economic situation, and the financial resources of Ministry Z. Constituting such a frame or narrative would be the explanation of a policy that underlies Ministry Z's distribution of resources to laboratory X in order to bolster the competitiveness of industry Y in a time of heightened international competition (for example, laboratory X might be producing knowledge highly relevant to industry Y, which is overseen by Ministry Z). The construction of such a narrative also will typically involve references to certain metanarratives, such as statements about the importance of industrial policy for capitalist development or about the general character of capitalist development. Political metanarratives endow policy narratives with a higher authority and legitimacy as they tie policies to the traditional systems of value and identity codes of a community. For example, in the US, state support for molecular biology and genetics was justified by arguments not only that this work would contribute to scientific progress and better health but also that it would help to create new high-tech industries that were essential for the economic future of the nation. In this way, the (medical) science policy narrative established a relationship between the human genome project and the metanarrative about the industrial future of the United States.

Discourse

But where do political metanarratives and policy narratives come from? Who is behind these narratives? Are political narratives just vehicles and instruments for skillful actors and interests in the game of power? Intimately tied to the concept of narrative is the notion of discourse. In the wake of structuralism, discourse has been characterized by "subjectivity," the explicit or implicit presence of an "ego" as the person who maintains the discourse. By contrast, narrative is defined by "objectivity," the absence of all reference to the narrator. However, discourse (particularly historical and political discourse) is endowed with "narrative structure." These forms are "narrativizing discourses," as Hayden White (1981, p. 3) puts it—discourses without a narrator. No one seems to speak, and the events seem to speak for themselves. Narrativizing discourse refers to the narratives that every culture makes available for those of its members (such as administrators) who might wish to draw upon them for the encoding and transmission of messages (for example, about pollution control). In other words, there are individuals and interests behind narratives that bring narratives into the world. But the individuals give birth to narratives only within the confines of the available discursive possibilities. Actors cannot freely choose the narratives they deploy. The given discursive possibilities describe the large reservoir of narratives that can be mobilized for political purposes. For example, whereas it had certainly been for decades within the realm of discursive possibilities to craft neo-liberalism into a new political metanarrative, not before the 1980s did neo-liberalism actually achieve the status of a new political metanarrative in many countries. In this context, Foucault has emphasized the importance of the *archive*—the general system of forming and transforming statements existing at a given period within a particular society. This archive determines what may be spoken of in a discourse, which statements survive or are repressed, and what individuals or groups have access to particular kinds of discourses (Foucault 1972; Laclau 1990, p. 100; Smart 1985, p. 48). However, it is important to see that the regulated and systematic character of a discourse does not imply a sense of closure. The sayable and the thinkable are delimited not only by the rules internal to discourse but also by the rules of combination with other discourses. In the European Community, for example, the rise of neo-liberalism to the status of a political metanarrative was closely connected to a discourse of "Europessimism"—of a Europe increasingly unable to compete with Japan and the United States. In fact, the systemic character of a discourse implies its articulation with other discourses.

In this context, it could be stated that a discourse is always caught in a web of materiality that is a historical product. A discourse is not just a system of words; it is an entity that operates in a social, an economic, or a political context. But discourses are not simply reflections of these contexts; rather, they are complex mediations between various codes by which reality is to be assigned possible meaning (White 1981, p. 202). For instance, a political discourse on industrial development that discusses different strategies for supporting an industry cannot be separated from practices in the companies to which it refers. At the same time, the objects of this discourse do not coincide with the range of activities and phenomena that they systematize as central to specific discourses, though they bear relations to them. For example, a discourse of industrial development is not equivalent to the investment and product development decisions of companies, though there is a specific relationship between these two domains of the theoretical and the material. To some extent, it can be said, context is already in discourse, discourse bears the traces of content, and the study of discourse involves investigation of the extent to which discourses are able to come to terms with contexts and vice versa. Discourse should be seen as a practice of production that can be characterized as material, intertextual, and complex, always inscribed in relation to other discourses and thus part of a "discursive economy" (Henriques et al. 1984, pp. 105–106).

Industrial production, for example, is governed by a proliferation of discourses and cannot exist prior to or outside of its articulation with a set of other social and political discourses in concrete historical conjunctures (Daly 1991, p. 100). Discursive formations such as biotechnology are intertwined with disparate other factors, such as institutions or economic forces, which establish flexible relationships and interactions that merge into a single apparatus (Dreyfus and Rabinow 1983, p. 121). The fact that one discursive formation gains influence over another, that it becomes *hegemonic,* is related to the degree of congruence and complementarily that this discursive formation has within a given discursive constellation and within an apparatus or a historical bloc. Rising unemployment rates, lack of success in high-technology fields such as information electronics, and the discourse of Europessimism are examples of the constellation in Europe in which neo-liberal narratives began to gain political currency (Laclau and Mouffe 1985).

We have now arrived at a critical point in our poststructuralist reconceptualization of the actor-structure problematic. When we talk about structures in the context of politics, we refer to something that is not given, but is the result of a writing process and boundary work. The semantic structure of politics, the political imaginary, and the many existing policy narratives and political metanarratives constitute a matrix that informs policymaking—the emergence of its actors and institutions—but at the same time is produced and reproduced through the articulatory practices of policymaking. Consider the statistical methods used to measure economic performance. Description of an economic situation by reference to the inflation rate and to unemployment figures is a process based on inscriptions that gain meaning within the context of a particular narrative of what constitutes economic performance. But this narrative of economic performance, in its turn, produces prescriptions for policymaking that are based on the statistical data collected on inflation and unemployment. Or consider the statistical methods used to measure risks related to certain experiments in genetic engineering. The truth of these measurements is constituted in the context of a certain practice of statistical calculations that presupposes a particular representation of risk. On the basis of this representation, policy narratives then suggest certain regulatory strategies (Latour 1990, pp. 19–68).[18]

Hence, policymaking can be described as an attempt to manage a field of discursivity, to establish a situation of stability and predictability within a field of differences, to maintain a specific system of boundaries (such as those separating the state, science, and industry), and to construct a center that fixes and regularizes the dispersion of a multitude of combinable elements. The elements that are stabilized by the process of policymaking originate in the different sites of regimes of governability. In the case of science and technology policy, laboratories or scientific journals are typical sites for the production and interpretation of scientific and technological facts and artifacts that, far from being stable in their meaning, continue to be reframed and transformed in the process of policymaking.

Ultimately, what is of importance in policymaking is the intermediation between policy narrative, discursive constellation (or discursive economy), and discursive context. The success of policymaking depends as much on the inscription of a policy narrative into the given discursive constellation as on its ability to mediate between competing codes by which economic, scientific, or political reality (context) is assigned meaning. However, this attempt to mobilize actors, interpretations, meanings, and artifacts and to stabilize a political space by means of a dominating narrative is usually only temporarily successful: government is a fundamentally unstable and conflict-ridden operation. Policy narratives are constant objects of interpretation and reinterpretation, as different readings are brought to bear upon the text of politics. Though policy discourse is often successful in defining the actors in a policy field, it may also fail to do so. Environmental activists, for example, might challenge a narrative that construes them as a group of actors that should be excluded from the formulation of environmental regulations. The process of writing the political intertext (the determination of the structure and organization of representations of "the political") is grounded in the multiplicity of meanings of any text—in the dissemination and the dispersions or instabilities of meanings as an inherent feature of intertextuality. As much as semiotic structures may influence action, they remain effects of precarious stabilizations and thus inherently uncontrollable. In this respect, narratives about the boundaries of the biotechnology industry are as much objects of systematic discursive struggles as is the political identity of a society. The analytical tools of poststructuralist policy analysis focus especially on the attempts of policymaking to create order and structure under conditions of instability that would seem to undermine such efforts.

2

Molecular Biology and the Rewriting of Life: Origins of American and European Genetic Engineering Policies

In this chapter I will discuss the early stage of molecular biology politics in the United States and Europe and show how a policy field comes into existence. Typically, such a field has a number of central elements, such as actors, institutions, and sets of goals that orient behavior. It is important to understand that these elements are usually products of complex, long, and contingent histories of discursive construction that take place in a number of different locations, do not always articulate a single logic or intention, and constitute a moment of stabilization in a multilayered discursive struggle. As a result of this struggle, the initially "clean slate" of politics becomes demarcated with more or less defined boundaries separating groups of actors and defining the institutions, objects, and contexts that matter and the goals and strategies that actors pursue.

I will trace the politics of molecular biology to its origins in the 1930s and will show how, during the late 1950s and the early 1960s in the United States, Britain, France, and Germany, the state became increasingly involved in what I shall call the governing of molecules: the creation of interrelated policies, strategies, and institutions to facilitate research on and technological intervention in genes at the subcellular level. I will develop three interrelated themes: the relationship between genetic discourses and processes of subjectification, the emergence of genes as a socio-political issue, and the articulation of the state into the politics of biology. I will show how, from the 1940s on, molecular biology acted as a subtle, individualizing "politics of life," and how, despite strong efforts to dissociate itself from previous traditions of genetic research, the field remained in many ways linked to the population-oriented science of eugenics, pursuing some of eugenics' central goals by new means. In this reading, molecular biology became an important site from which power operated. Beginning in the 1930s, programs of molecular biology evolved as a result of complex alliances among scientists, universities, medical leaders, and philanthropic organizations, notably the Rockefeller Foundation. Only gradually, in the late 1950s and the early 1960s, were these actors and institutions linked to the state apparatus.

Foucault (1980) has pointed to a significant historical transition, contemporaneous with the shaping of industrial capitalism, in which emphasis shifted from the primacy of sovereignty, law, and coercion or force to "take life" to the development of new forms of power constitutive of life. Such processes of subjectification can occur in the form of the subjection of individuals to techniques of domination and discipline, or through subtler techniques of the self. From the perspective of poststructuralist analysis, power in regimes of governability originates in a number of different sites, some (e.g. state agencies) considered political and others (e.g. scientific disciplines) considered nonpolitical. In our story, the new scientific discipline of molecular biology became a central site from which fundamental parameters of human self-understanding were redefined and circulated as knowledge to other social fields.

At the same time, the emergence of genes as an object for state politics is inseparable from the operation of newly shaped policy narratives and discourses of modernization. By the 1960s, molecular biology had acquired the status of a "cutting-edge" discipline and was increasingly represented as a scientific field that would eventually deliver substantial benefits and serve economic development and medical progress. In the early 1970s, the technology of genetic engineering became available and was immediately portrayed as a scientific breakthrough with significant economic implications. Soon afterward, the support of genetic engineering and its various applications moved to the center of biotechnology policies. But the goals and rationales of these new biotechnology policies of the 1970s and the 1980s were, to a considerable extent, influenced by a discourse on life that had originated decades earlier. Recombinant DNA technology seems to be a phenomenon of the 1970s. But, as this chapter will show, strategies to govern molecules, such as the identification of social and scientific problems to be solved by biological research, the definition of the locus of life at the submicroscopic level, or programs of research and institution building—in other words, the identification of a terrain for government—emerged much earlier.

On the Government of Molecules and Populations

The 1930s saw the emergence of a new field of biological research—molecular biology—that subsequently became the dominant disciplinary trend in the biological sciences (Kay 1993, p. 3). Molecular biology offered a new narrative of life phenomena. Until the 1930s, virtually all genetic research had been done with plants and animals. This research approach had some disadvantages: the number of units that could be observed was relatively small, and the life cycles of even short-lived organisms were relatively long. Molecular biology turned to bacteria and viruses—microscopic organisms whose life cycle lasts for less than an hour and which can be raised in the billions overnight. The fundamental chemical composition of bacteria is no different from that of higher cells, but their organization differs in several important aspects from that of eukaryotes.[1] Most important, the volume of the prokaryotic cell is smaller than that of the typical eukaryotic cell by a factor ranging from 10 to 10,000. Hence, not until the electronic microscope became available did it become possible to construct a detailed picture of the anatomy of the bacterial cell. Of the vast number of bacterial species, *Escherichia coli* (the name of which was derived from its discoverer, Theodor Escherich) became the best-understood cell in biology research.

The new biology stressed the unity of life phenomena and began to study fundamental life phenomena on their smallest level in bacteria and viruses. In this process of defining life in terms of fundamental physiological processes, molecular biology ultimately narrowed its focus to the level of macromolecules. Emergent properties and interactive processes occurring within organisms, between organisms, and between organisms and their environments ceased to be objects of interest for this kind of biological research. In the narrative of molecular biology, the locus of life phenomena was conceptualized at the submicroscopic level, between 10^6 and 10^7 centimeter.

This reconstruction of the scientific object set in motion a representational dynamics driven by the fact that the submicroscopic level of life could be investigated only with complex and sophisticated instruments and techniques, such as electron microscopes, ultracentrifuges, electrophoresis, and x-ray diffraction. This had an enormous impact on molecular biology: its program was defined and conceptualized in terms of technical possibilities and capabilities (Kay 1993, p. 5). It was only through the aid of advanced technologies that the fundamental processes of life became visible, legible, and changeable or transformable. Technological representations provided the means of intervening into life processes. Or, to put it differently, the representation of life came to be predicated on modes of technological intervention: life had become a technology.

The discourse of molecular biology after the 1930s attempted to explain biological functions exclusively through the study of physical processes and chemical structures. Biological phenomena were conceptualized as the consequences of the interaction of large assemblies of biologically significant molecules, and it was claimed that the constitutive chemical components determined, and ultimately explained, the foundation of all biological phenomena. A new space of representation had developed, providing a new language and a new system of signification to represent life processes.

This new science of molecular biology and its representations of life processes were hardly attributable to autonomous developments in laboratories. The rise of molecular biology must be seen in the context of a complex process of boundary drawing in the micropolitics and the macropolitics of science—a process involving diverse and mutually reinforcing developments: the significant changes in the discourse of biology since the end of the nineteenth century, the decline of eugenics as a scientific discipline, systematic institutional support by the Rockefeller Foundation, and the shaping of a new hegemonic international order.

Critical financial and logistic support for molecular biology came from one of the major institutional actors in pre-World War II American science policy, the Rockefeller Foundation.[2] In 1933, when the Rockefeller Foundation inaugurated a support program for biology, science policy was not yet a well-defined policy field; it depended on largely uncoordinated support by industry, by the state, or by philanthropic organizations.

The Rockefeller Foundation outlined the rationale for its support as follows: For the previous 100 years, physics and chemistry had reigned supreme, and questions of human behavior had been neglected. The new goal was to accomplish social control through understanding and knowledge of the very basic elements of the human body. The trustees of the foundation stated:

Science has made significant progress in the analysis and control of inanimate forces, but science has not made equal advances in the more delicate, more difficult and more important problem of the analysis and control of animate forces. This indicates the desirability of greatly increasing emphasis on biology and psychology, and upon those special development in mathematics, physics and chemistry which are themselves fundamental to biology and psychology. . . . The challenge of this situation is obvious. Can man gain an intelligent control of his own power? Can we develop so sound and extensive a genetics that we can hope to breed, in the future, superior men? Can we obtain enough knowledge of the physiology and psychobiology of sex so than man can bring this pervasive, highly important, and dangerous aspect of life under rational control?[3]

In 1938, Warren Weaver, the man in charge of the Rockefeller biology program, unveiled the name "molecular biology." Weaver compared the divisibility of the cell (the "older" unit of analysis) into subcellular units to the divisibility of atoms into subatomic units. Apparently, according to Kay (1993, p. 49), the Rockefeller Foundation's interest in molecular biology was not entirely scientific in motivation: "This molecularization of life was to be applied to the study of genetic and epigenetic aspects of human behavior. Based on the faith in the power of upward causation to explain life, Weaver and his colleagues saw the program as the surest foundation for a fundamental understanding of the human soma and psyche—and ultimately the path to rational social control." But such reasoning did not play an important role in the public representations of the new biology, which focused mainly on the cutting-edge scientific aspects of molecular biology.

The new subfields of biology were targeted by the Rockefeller Foundation for enormous grants. The influence of the foundation's New Biology Program was not confined to the United States; it also had a major impact on biological work in Europe. This influence lasted until the late 1950s, when the foundation turned its attention to other areas of research.

The Rockefeller Foundation's support for molecular biology had certainly not built the new approach to studying life *ex nihilo*. A number of researchers in biology and genetics, including the highly influential T. H. Morgan, had espoused a mechanistic view of life and saw biology as headed in the direction of intensified cooperation with physics and chemistry (Kay 1993, p. 79). A critical element in the creation of molecular biology as a new scientific discipline was the gradual linking of a number of initially only loosely interacting experimental systems that provided physical, chemical, and functional characterizations of living organisms in terms of macromolecules (Rheinberger 1995, p. 3). But by privileging the molecular biology approach over the evolutionary, ecological, and organismic traditions, the Rockefeller

Foundation emerged as a critical coordinating agency. Its financial muscle reinforced the ongoing disciplinary tendencies, and the foundation became an integral part of a system of power relations. The foundation's importance was the result of a multifaceted process of enrollment and boundary work that defined particular individuals, structures, representations and institutions as elements in a new configuration and linked them together through processes of legitimization and distribution of cognitive authority (Kay 1993, p. 17). In the mid and late 1930s, for example, all the avant-garde work in protein studies, which at the time was the focus of interest in molecular biology, was carrier out by Rockefeller grantees (Abir-Am 1982, p. 369).

The Rockefeller Foundation's promotion of the new biology must also be seen in the context of the decline of old-style eugenics.[4] Until the 1920s, eugenics was generally viewed as a useful explanatory framework illuminating important social and biological phenomena. But soon many geneticists in the United States began to realize that the efficacy of applying Mendelian genetics to human breeding in order to obtain precise and lasting modifications within a few generations was grossly overrated. By then studies had shown that some genes were pleiotropic (simultaneously influencing several traits) and that many traits were polygenetic (determined by the action of several genes). As evidence accumulated challenging the idea that a single gene determined a phenotypic character (a core idea of eugenics), inbreeding or outbreeding complete traits such as character or temperament became increasingly viewed as goals unlikely to be reached with the methodologies proposed by and available to eugenic thinking. Not only was eugenics more and more viewed as a pseudo-science; furthermore, the rise of eugenics in Nazi Germany attached to it a stigma that made funding increasingly problematic. Eugenics' fall from grace within the scientific community did not, however, imply a rejection of its fundamental premise: the desirability and eventually practicability of selectively controlling human reproduction. These goals, which continued to influence scientists, found articulation in high school and college textbooks well into the 1930s (Kay 1993, p. 37). Neither did the gradual demise of eugenics as a scientific doctrine bring the disappearance of another of its core principles of genetic determinism: the idea that genetics lies at the root of most human talents and disabilities.[5]

But the fundamental unit of genetics, the gene, had remained largely devoid of material content. The fundamental unit of classical genetics was an indivisible and abstract gene. Yet the gene's physical nature remained unclear, for it was not known how the gene managed to preside over specific cellular processes or how it achieved its own faithful replication in the cellular reproductive cycle (Stent and Calendar 1978, p. 23). This lack of knowledge necessitated more basic research, which eventually led to the transition from classical genetics to molecular genetics. In the early 1950s, the fundamental unit of molecular genetics became a concrete chemical module, the DNA double helix, while the gene was relegated to the role of a second-order unit aggregate comprising hundreds or thousands of nucleotides. At a time when the "old," "crude," "population-oriented" eugenics was represented as scientifically outdated and politically unacceptable, the much subtler, subcellular method of understanding and controlling life became "the technology of choice" (Kay 1993, p. 9). Geneticization—the framing of central dimensions of humans in terms of biological factors (genes)—provided continuity behind the surface appearance of discontinuity between the "old" eugenics and the "new" biology. The object of intervention had changed from the population (gene pool) to the subcellular level, but the vision of social control and the shaping of a particular political order remained inscribed as a goal of biological research. With these developments, the biological determinism of the "old" eugenics had moved from the cellular to the subcellular mode of reasoning. Properties of life, health, and disease continued to be defined in biological terms, only with the important difference that now they were now construed not in terms of the abstract gene, but in terms of genetically directed macromolecules.

The Rockefeller Foundation envisioned molecular biology as a science that could yield a means for social control and that could tacitly serve as the successor science to eugenics. At the same time molecular biology was translated into the approach that would constitute the future of biological research. In the early 1930s the trustees of the Rockefeller Foundation had talked about the new biology's potential to yield a means of social control. However, such concerns were not in the forefront of public reasoning. The meaning of molecular biology was stabilized as the new and cutting-edge way to go about research on life phenomena which would ultimately lead to pathbreaking progress in the science of life processes and, eventually, yield substantial benefits for humankind. This politics of meaning was a critical step in the forging of associations and cooperation between the Rockefeller grant recipients in the scientific community and universities, the state, and the foundation.

Furthermore, according to the Rockefeller Foundation's science policy narrative there would also be an important international dimension to its project. Not only American institutions but also a wide range of actors in other countries should play important parts in the development of molecular biology. The Rockefeller Foundation's engagement in the "New Science of Man" came during a time of dramatic changes in the global political economy. The Eurocentric nineteenth-century order had come to an end with World War I. The international configuration of forces took the form of spheres of influence and rival imperialist blocs (S. Gill 1991, p. 282). In October 1929, after the Wall Street crash, a general economic decline took hold. By 1932 many currencies were floating and international finance had virtually collapsed. It was at this time of crisis that a new historical bloc emerged in the United States. Capital-intensive industries, investment banks, and internationally oriented commercial banks became the key elements in this new configuration. This new hegemonic bloc, able to accommodate millions of mobilized workers amidst the depression, was at the core of the New Deal. Capital-intensive firms such as the Rockefeller-owned Standard Oil, which used less human labor, felt less threatened by labor turbulence. They had the flexibility to accommodate workers' demands. In addition, as world and domestic leaders in their fields, these firms only could gain from free trade (Ferguson 1989, p. 7).

The Rockefeller Foundation's engagement in social reform was congruent with the newly emerging system of international relations. Projects in the social and biological sciences were intended to increase economic productivity and social stability worldwide; at the same time, motives of social control and surveillance directed the foundation's interest in the human body on the individual level as well as on the collective level (Kay 1993, p. 26). In other words, from its inception the new hegemonic project of multinational liberalism involved not only a reorganization of political economy but also a project to increase social control by means of knowledge gained from research in the new biology. In the context of this socio-economic configuration, the Rockefeller Foundation's science policy narrative construed support for molecular biology not only as a project that should take place in the United States but also as an international strategy to stimulate the development of molecular biology abroad. The more actors and institutions shared and exchanged knowledge, experiences, and technologies, the more molecular biology's narrative on life consolidated, disseminated, and legitimized.

The Rockefeller Foundation's financial resources, new technological modes of representing life, transformations of world capitalism, and dispersed scientific research in laboratories became critical elements in the discourse of the foundation and contributed to the construction of the new scientific field of molecular biology through processes of problematization, enrollment, and mobilization. Molecular biology, thus, developed in mutual reinforcement with the rise of multinational liberalism, which, in the form of the Pax Americana, would soon become the dominant ordering principle of the postwar economic order. Thus, molecular biology was not simply a strategy of social control pushed by the Rockefeller Foundation; nor was it solely the result of the work of ingenious scientists and researchers. Rather, the gradual rise of molecular biology to hegemonic status in the life sciences operated within a complex discursive constellation in which the Rockefeller Foundation's influence on society was as much tied to the authority of the knowledge of molecular biology as the conditions for scientific work in molecular biology were connected to the power and resources of the Rockefeller Foundation and its embeddedness in the international capitalist order. It is in the nature of scientific disciplines that over the course of time such complex intertexualities tend to move into the background, while scientific practice seems to be mainly determined by the inner logic and workings of everyday laboratory life.

Toward Policies of Molecular Biology after World War II

It was in the postwar period that molecular biology was gradually articulated as a topic of concern for state intervention in the United States and in Europe. Important changes in the discursive economies of the United States and Europe led to the emergence of new actors and institutions in "life politics." The Rockefeller Foundation continued to play an important role during the 1950s,[6] but the foundation started to withdraw its support. Soon new actors were to emerge.

The International Politics of EMBO

In the postwar Europe, the practice of molecular biology was characterized by an ongoing stream of meetings, a preponderance of collaborative efforts, and a gradual shaping of scientific networks across countries (Abir-Am 1993, p. 153). In particular, after World War II the sudden absence of war-related travel restrictions contributed to the increase in international meetings and collaboration. After Sputnik and until the late 1960s, a variety of bilateral and multilateral scientific cooperation schemes were developed, such as those in the field of molecular biology between the National Institutes of Health in the United States and institutions in the United Kingdom and in France (Abir-Am 1993, p. 170). The prominent French biologist François Jacob (1988, p. 286) describes the developing of nctworks among molecular biologists as of 1956 this way: "This biology was performed by a very small, very exclusive club; a sort of secret society to which belonged perhaps a dozen laboratories throughout the world. Some forty researchers who wrote to each other, telephoned, visited, exchanged strains and information, traveled to each other's labs to do experiments, met periodically here and there to keep abreast of the current state of the field."

Systematic efforts to support molecular biology on the European level date back to the 1950s and the 1960s. This support was not limited to the national level. Relatively early on there emerged a discourse on the need to coordinate and conduct research in molecular biology on the European level. The threatening migration of European scientists to the United States and the possible cessation of American funds for European science were important elements in this evolving policy narrative. The European Molecular Biology Organization (EMBO), created in 1963, was an institutional response to this debate.

A model for EMBO was set in up in 1954 by CERN, the European Center for Nuclear Research, which had been created to give European scientists access to the costly facilities required for high-energy physics. Initiated by Leo Szilard, CERN director Victor Weiskopf, James Watson, and John Kendrew, EMBO went on to nominate as members 140 European molecular biologists from twelve countries (Abir-Am 1993, pp. 171–172). EMBO's mission was defined as including the creation of a European Molecular Biology Laboratory, the establishment of a clearing house for biological information, the creation of a network of European laboratories, and the institution of a European fund that would provide grants for laboratories. Supported by Germany's Stiftung Volkswagenwerk (Volkswagen Foundation) with seed money of 2.76 million marks,[7] and later by the member states, EMBO became an important actor in the shaping of a European space of molecular biology by organizing international workshops, fostering collaboration, and facilitating mobility among young scientists.

Driven by a desire to operate on an international scale, but also by a mission to reorient the direction of European biology, the Stiftung Volkswagenwerk and its scientific advisers contributed decisively not only to the creation of German molecular biology but also to the initial efforts to create a European space governing molecular biology. A representative of the Stiftung Volkswagenwerk put it as follows: "Measurement, this was a term the Stiftung Volkswagenwerk always liked to use. We did not want contemplative observation, but analysis with the hard methods of chemistry and physics."[8] The advancement of "hard methods" was very much the program of EMBO, which did not limit its politics to facilitating scientific networks, but whose mission was very much built around assuming an avant-garde position of a technology-driven molecular biology with the EMBO infrastructure as its core. In 1967 EMBO began to discuss plans for an European Molecular Biology Laboratory, which came into operation in 1977. The narrative behind EMBL's formation emphasized expensive and complex technologies for research projects that could not be carried out nationally because they either were too expensive or required too many scientists. Apparently, the similarities between the origins of molecular biology and the Rockefeller Foundation and the Stiftung Volkswagenwerk and EMBO were due not only to the fact of philanthropic influence on the shaping of the discipline but also to the focus on the most advanced technology as the defining feature for disciplinary advancement in the creation of an international space for molecular biology (Yoxen 1981, pp. 99–100).

But the vision of EMBO and EMBL as the engines of European molecular biology was flawed from the beginning. As the justification for the creation of EMBL began to be questioned, the black box of European molecular biology began to open up. According to Michel Callon and Bruno Latour, black boxes are critical devices for activities such as scientific practice or policymaking. A black box contains that which no longer needs to be considered. Black-boxing allows growth. An actor grows with the number of relations he or she can put away into a black box. Black-boxing stabilizes situations of meaning and allows for further action (Callon and Latour 1981, pp. 284–285). In many respects, molecular biology was such a black box. It was an arrangement of assumptions which fulfilled certain functions without necessarily living up to the test of reality.

In contrast to representations from the early 1960s, molecular biology, unlike high-energy physics, was not dependent on enormously expensive research facilities, such as those provided by CERN. This problem was already anticipated by critics of EMBO—particularly in the 1960s, when EMBO went ahead with its plan to create a central European Molecular Biology Laboratory (which eventually went into operation in 1977). In the 1960s, critics argued that specialized facilities and large laboratories would funnel molecular biology and its practitioners into an elitist group that would have little influence on the development of science in Europe (Newmark 1989).

Hence, resistance within the scientific community as well as the logic of disciplinary development (which evolved around smaller research operations, rather than CERN-like centralized operations) reduced the politics of EMBO to network building rather than core building. However, backed by the molecular biology elite and by scientists such as Kendrew, Watson, Max Perutz, and Jacques Monod, EMBO gave high visibility to the practice of a highly specialized, technology-intensive biology. This modeling was a strategic move to keep European science competitive at a time when most European countries still had to design their national policies of molecular biology (Yoxen 1981, p. 99). EMBO, hence, played an important role in the shaping of a European research-and-development space by creating European networks of science and research and by giving legitimacy to the evolving national policies of molecular biology.

Discursive Economies of Science: Molecular Biology and Modernization Narratives

The state's increasing involvement in the shaping of molecular biology in the United States and in Europe must be seen in relation to a number of political and economic events and developments, and in relation to their coding in national political discourse. To be understood, the development of a policy such as that of molecular biology must be situated within the dominant discursive economy—the broader context of tropes, concepts, and social projects summarized in the political metanarratives specific to a society.

In the United States, World War II was a watershed for science. The war had created new institutions and reconfigured older relations. Wartime military research was conducted through the Office of Scientific Research and Development, which was headed by Vannevar Bush. In his vastly influential 1945 book *Science: The Endless Frontier*, Bush had called for a strong federally funded science geared toward the nation's political and economic goals. The National Institutes of Health grew dramatically and became the main state agency supporting molecular biology. Beginning in the 1950s, the NIH heralded the rise of state-dominated biomedicine in the United States, thus completing the Rockefeller Foundation's never fully successful attempts at comprehensive support of the biomedical sciences (Pauly 1993, p. 145). What was accomplished over the years was the creation of a highly specialized cadre of experts grouped together in universities, in medical schools, in the various branches of the NIH, and in the field of clinical medicine (Yoxen 1981, p. 95).

The major US investments in molecular biology and the resulting developments in the field gave rise in Europe to a discourse on the "delay" and "backwardness" of European research. In the case of EMBO, a similar framing of the United States as the leader in molecular biology research became an important element in a more complex policy narrative on the need to create national support policies for molecular biology. Other important raw materials for this narrative came from new "theories" on the nature of international competition, the changing role of the state, and the importance of science and technology for economic development. Political narratives arrange disparate events and developments in a temporal and causal order and create order and stability in the otherwise contradictory world of politics. Which events, developments, and discursive changes entered the construction of the evolving policy narratives of molecular biology?

First, a number of developments in the international political economy were important in the shaping of the new policy narratives. The new US-centered postwar order had developed along several interrelated dimensions: the political, economic, and socio-cultural reconstruction of the defeated Axis powers; the economic rebuilding of Western Europe under the Marshall Plan; the militarization of relations between the United States and Europe through NATO; and the Cold War (S. Gill 1991, p. 283). The central premise of this new hegemonic order was to establish the unrestricted globalization of the structure of accumulation that had emerged at the end of the nineteenth century (Cox 1987, p. 212). The corporate state operated according to a system that coordinated government, business, and labor. The social side of the liberal international system was supported by the "internationalization of the New Deal," which promoted social welfare and state protection of vulnerable social groups. But the state also accumulated significantly more responsibilities on the economic side. The state projects—the metanarratives concerning the proper role of state in society—underwent deep change. One important field that was redefined as critical for state intervention was science and technology policy.[9]

Central to the emergence of science and technology policy as a legitimate field of state activism was the idea that scientific development, political stability, and economic growth were intrinsically linked to one another. The rise of this idea was related to the experience of the postwar economic boom. Western Europe, Japan, and the United States had rebuilt their economies after 20 years of depressed consumer demand. Between 1945 and 1973, favorable economic conditions in the United States stimulated growth throughout the world economy. The expansion of the US economy helped the European countries and Japan to reconstruct their own war-devastated economies. This reconstruction, based on American mass-assembly principles, was designed to achieve the "economic miracles" of the late 1950s. The unleashing of pent-up consumer demand for homes, household appliances, and other consumer goods stimulated production. Technological innovations in electric equipment, machinery, scientific instruments, transportation, and the chemical and aerospace industries caused rapid growth. All these conditions increased total demand. The world economy seemed to be geared toward a never-ending boom, which seemed to be linked to scientific progress.

During this period, worldwide governmental spending on research and development had grown to unprecedented levels. Soon, the dramatic postwar increase in R&D spending and the period of unprecedented economic growth were interpreted as causally related. A new political-economic discourse defined scientific development and economic growth as intrinsically related to each other and concluded that, because of this, the state ought to support industry by promoting research. During the 1960s, the "science push theory," which argued that the key to innovation was the "push" of scientific development, gained in importance. According to this theory, the state had to push science. Especially during the 1960s and the 1970s, science and technology policies were being systematically developed in the industrialized countries. Scientific and technological progress became represented as a central condition for a nation's viability as expressed by economic growth, agricultural expansion, medical advances, and competitiveness in the international economy. In this increasingly dominant discourse, the societal goals of economic growth and scientific progress were integrated under the trope of "modernization," which conceptually collapsed social and economic development with accelerated industrialization.[10] Modernization became a political metanarrative, a narrativizing discourse that began to be used in a number of policy contexts to legitimize, justify, and explain the rationales of policies.

The increase in R&D expenditures was also influenced by the Cold War discourse and by the related escalation in defense expenditures in the United States and in Europe. R&D had become a core element of the evolving politics of security. What had emerged was a politics of security that focused on the "ensemble of the population," in its outward as well as its inward orientation. The discourse of foreign policy, oriented toward the Communist menace to the collective, was part of a governmental rationality of which the promotion of the well-being, health, and prosperity of the population were other crucial elements (Gordon 1991, p. 20; Campbell 1992). An important aspect of the Pax Americana was the creation of a particular kind of individual and collective identity based on a form of social cohesion that was underwritten by such interrelated practices and methods as military policy, research policy, industrial policy, and welfare provisions. Support for molecular biology in the postwar period must be located within this discursive political economy.

Also important for the emergence of molecular biology policy narratives were changes in the discourse of molecular biology. During the 1950s and the 1960s, molecular biology underwent a dramatic shift from the protein paradigm to the DNA paradigm. Genes were now interpreted as consisting of DNA, which was seen as "encoding" information that determined the processes of replication and protein synthesis. Linguistic tropes such as "code" and "information" became naturalized with the scientific and cultural discourses of the postwar era to the point that it became virtually impossible to think of genetic mechanisms and organisms outside the discursive framework of information. Molecular biologists came to view organisms and molecules as information-storage-and-retrieval systems. Biology's narratives of heredity and life were rewritten in experimental systems that aligned molecular biology with the new knowledge-power nexus of the Cold War.[11] In the postwar discourse, molecular biology acquired the meaning of a science that not only could contribute to medical progress, nutrition, and health but also could help a nation's competitiveness in the international economy. This interdiscursivity—this simultaneity of relations between discourses on industrial development, the meaning of modernity and the international economy—was central to the emerging cultural dominance of molecular biology.

In European discourse, American molecular biology gradually turned into a mythical entity. Roland Barthes (1972, pp. 131–132) characterized myth as stolen language, as depoliticized speech. Myths do not deny things; they talk about them. But usually talk about things is open to many interpretations. In contrast, mythical language makes contingency appear eternal and creates a natural image of reality (ibid.). European representations increasingly framed American molecular biology as the unquestioned world leader, and this "fact" emphasized the importance of European initiatives in this field. At the same time (during the late 1950s), molecular biology was inscribed into the larger context of the then-emerging technology and industrial policies which were aimed at modernizing the European economies. Whereas the idea of "modernization" had acquired the status of a metanarrative in Europe, specificities of political discourse varied from country to country, playing a crucial role in the final shaping of national public policies and in the identification and construction of its actors and institutions. Because discourse and narratives always operate within a given discursive constellation, in order to scrutinize similarities and differences in the construction of a political space of molecular biology in Britain, France, and Germany we must now turn to the specific political discourses in those three countries and to their importance in the shaping of national research policies supporting the new biology.

France: Molecular Biology and "Developmentalism"

The shaping of a policy narrative on molecular biology in France in the late 1950s must be understood in relationship to the development of a new political discourse that framed the transition of French capitalism to Fordism[12] and related changes in the political-institutional framework from the Fourth to the Fifth Republic, and, on the political-strategic level, to the emergence of Charles de Gaulle's politics of modernization, which introduced a new model of socio-economic transformation.

After World War II, high civil servants who had worked in the Resistance and American Marshall Plan advisers operated in a France whose former hegemonic bloc had been defeated. During the Third Republic, industrial growth had not been a central goal of political strategy. This hegemonic bloc, composed of industrialists, merchants, peasants, and savers, had limited the state's role to defending property. After the war, liberal capitalism was held responsible for the economic problems of the interwar period and for what was perceived as national decline. Distinctive and different from the situation in Britain or in the United States was the public officials' reading of economic backwardness as the central problem of France. This framing contributed to the reorganization of the French state project around the goals of modernization, efficiency, productivity, and competitiveness.

Under this new project, the state assumed the tasks of managing and stimulating the economy and sustaining a constant flow of investment (Kuisel 1981, pp. 272–279). This "developmentalism" was, to a considerable extent, made possible by the creation of a credit-based, price-administered financial system, which opened up the way for administrative influence and discretion in the allocation of capital (Zysman 1983, p. 168). This strategy was coupled with a gradual exposure of the French economy to foreign competition. Military weakness during World War II and involuntary decolonization forced French international policy to focus away from the Third World and toward the First, a process underlined in 1957 by France's co-founding of the European Economic Community (Adams 1989). Modernization efforts after 1944 came exclusively from high civil servants supported by social groups (consisting mainly of wage earners) that had been pushed aside until the defeat of 1940. Their project of modernization equated social progress with enough growth in popular purchasing power to expand demand and thus guarantee full employment. Supplementing this consumption-driven economy was a system of social security, which mitigated the wage earners' risks of unemployment, illness, and old age (Lipietz 1991, p. 23). This model encountered resistance from the elites. The Centre National des Indépendent et Paysans and the Mouvement Républicain Populaire tried to block the emerging mode of regulation. In response, the state began to displace the propertied elites. It undertook to create a centralized system of wage regulation, social security, and price administration, and to shape a productive apparatus through subsidies, nationalization, and planning (ibid., p. 24; Zysman 1983, p. 105).

This process was advanced by the founding of the Fifth Republic, which brought a significant shift of power to the executive, and by the transformation of the French political system from a parliamentary to a presidential one. During the Third and Fourth Republics, the parliament had held a decisive role in the administration of the state; the executive had emanated from parliament and had very few controls over it. And the nature of the French party system had made stable coalitions unlikely, thus weakening the executive further (Hoffmann 1991, p. 43). Charles de Gaulle and Michel Debré drafted a new constitution, which went into effect in 1959 and which was eventually to rescue the executive from impotence in times of crisis (as in the matter of Algeria) and in periods when France needed major economic and social transformations (Hoffmann 1991, p. 44). With these shifts in the discursive economy of postwar France, the rewriting of the institutional framework provided the organizational opportunities for governmental intervention in science and technology.

The emerging discourse of modernization lent itself to the reformulation of postwar science policy. Representations critical of the state of French science go back to 1952, when the Commission on Education of the Planning Commissariat declared that the French economy's demand for scientists and engineers would soon exceed the supply. This assessment, along with other reports which deplored the state of science in France, led in 1954 to the first appointment of an Under Secretary for Scientific Research and to the establishment of a Supreme Council of Scientific and Technical Progress. The council and its committees received the task of drafting the report on science for the third plan of social and economic modernization (1957–1961) (Rouban 1988, pp. 94–95).

In 1956 a parliamentary committee on the expansion of scientific research, in collaboration with a number of scientists, called for a national conference to discuss the problems of French science and to develop a comprehensive plan of action. The Colloque de Caen (November 1–3, 1956) brought together France's most important scientists, politicians, and industrialists. They drafted a twelve-point national plan for the renovation of French science and technology, which became the cornerstone of all subsequent reforms and actions in the Fifth Republic (Gilpin 1968, pp. 188–191). Crucial for the politics of molecular biology was the fact that the prominent biologist Jacques Monod served as one of the principal organizers of the Colloque de Caen. This put Monod and other prominent molecular biologists in a good position to launch the French politics of molecular biology.

The rise of the new biology in France was closely connected with the Institut Pasteur and the work of Jacques Monod, André Lwoff,

François Gros, François Jacob, and Elie Wollman. In the interwar period, research in biochemistry, genetics, and embryology was less developed in France than in Germany. However, Monod's biochemical research and Jacob's genetic work in conjunction with other scientific work at the Institut Pasteur led to important French contributions to the emerging field of molecular biology, as evidenced by the development of the operon model and the elucidation of ribosomal DNA (Gaudillière 1991). This important intellectual tradition, however, was strongly enhanced by systematic state support starting in the 1960s.

It was at Caen that a broad-ranging critique of the state of biology in France was first articulated on the political level. As was pointed out in this critique, nearly all of biology, genetics, biochemistry, general physiology, embryology, and microbiology was absent from university science faculties, which were dominated by zoologists and botanists. Where these new fields were represented (for example, in medical and pharmaceutical education), they only had marginal status (Gaudillière 1991).

With the third plan (1957–1961), science policy became a field accorded special attention by the state. For de Gaulle, in power since 1958, science became a means to achieve national sovereignty. It came to be considered essential to break with past traditions and to establish an alliance on the highest level between the state and the scientific community—an alliance empowered by a set of new organizational devices (Rouban 1988, p. 151). The Délégation Generale de la Recherche Scientifique et Technique (DGRST), le Comité Interministeriel de la Recherche Scientifique et Technique (CIRST), and the Comité Consultatif de la Recherche Scientifique et Technique (CCRST) were all set up in 1958, respectively constituting a "political section," an "administrative section," and a "parliament" (also known as the "Comité des Sages"). In this discourse, science was politically redefined as a high-priority area for state intervention in the interest of national grandeur (ibid., p. 110).

In one of its first meetings, the CCRST discussed the case of molecular biology. The arguments made for a special policy supporting molecular biology were as numerous as they were connected. The narrative underlying the arguments highlighted scenarios such as the French lag in the field in comparison with developments in the Anglo-Saxon countries, the lack of flexibility within French universities, and the French "potential" in this sector as it was reflected by the international recognition of French research in this area—a recognition that was not reflected by the position of the research field within the French university system.

To single out modern biology as a field for special state support was not an obvious thing to do in the early 1960s. The new French science and technology policy was strongly geared toward economic restructuring, and it was by no means clear what the emerging field of molecular biology could contribute to this endeavor. But, as Jean Monnet, the architect of French postwar planning said, "La modernization n'est pas un état de chose, mais un état d'esprit,"[13] and in the Pasteurian scientists' successful framing molecular biology succeeded in becoming an element in the articulation of this spirit.

The translation of molecular biology as a contribution to French socio-economic development created a link between two discourses that was instrumental in shaping an alliance between scientists (mostly from the Institut Pasteur) and state officials from the Délégation Generale and in creating a flexible structure composed of a few personalities with control over considerable means. Molecular biology had become an important boundary object. It played a crucial role in spanning the initially separated worlds of science and politics and in shaping a new network of meaning. Though administrators and scientists had different understandings of molecular biology's meaning and of its various theories, instruments, and artifacts, they shared a basic understanding of the main issues and concepts involved, and they were able to interact productively and to cooperate. Thus there evolved a new network of meaning that identified the actors and structures that, it was agreed, should figure in the strategies by which molecular biology would be supported. What began to take shape was a new regime of governability that articulated the state and its institutions as important and legitimate actors in the game of molecular biology. As one of its first steps, the scientific committee responsible for molecular biology drafted a report on the prospects for the first "action concertée" (1960–1965) in molecular biology. According to the "action concertée," three key arguments supported developing the field of molecular biology: that molecular biology was a new field deserving attention; that this new field constituted a revolution in biological knowledge; and that, in the face of important developments in the field abroad, France could not afford to delay. Further materials that entered this narrative for the French strategy in molecular biology were found in institutional developments in other countries, such as the establishment of a new institute in Cambridge around Francis Crick, John Kendrew, Max Perutz, and Sydney Brenner and the founding of an institute of molecular biology at the University of Cologne under Max Delbrück. According to the scientific committee led by Jacques Monod, in past decades molecular biology had been responsible for a succession of spectacular discoveries which had resulted in a new theoretical and methodological body of knowledge covering biochemistry, genetics, immunology, biophysics and other disciplines (Gaudillière 1991, p. 82).

What made this narrative powerful and persuasive was not the "rightness" or "wrongness" of "facts" and "evidence" assembled. It was the narrative's organization of "reality" into the reasonably coherent form of a story that not only gave an orientation but also helped to create an agenda for state action. The scientific committee recommended new research institutes, systematic training of young scientists, and travel money for research congresses. These recommendations far surpassed the initial ideas of the DGRST concerning molecular biology. In particular, the suggestions to create research institutes and to train young scientists had the implication of bypassing the authority of the established research organizations, such as the Centre Nationale de la Recherche Scientifique, and of displacing the university as the primary training ground for young scientists. Nonetheless, the suggestions were largely adopted by the DGRST. Molecular biology received some 30 millions francs, a sum comparable to the funding for the established field of biochemistry in the fourth plan (1962–1965) (Gaudillière 1991, pp. 82–86).

In 1965, Lwoff, Monod, and Jacob received the Nobel Prize. This confirmation of the success of French molecular biology led to an extension of the action concertée from 1966 to 1970, endowed with the same funds as the first one (30 million francs). Monod developed the rationale for the second phase of the program. According to Monod's narrative, molecular biology was no longer a source of methodologies and technologies; it was a new doctrine that allowed for the rational interpretation of all fundamental properties of living beings in terms of the structures and functions of the biological macromolecules. The text took as elements of its doctrine the elucidation of the mechanisms of the genetic code, the discovery of messenger and transfer DNA, and the characterization of the control mechanism of the synthesis of protein—the latter two elements being closely connected to the work of Monod, Jacob, and Gros (Gaudillière 1991, pp. 97–99).

Thus, the narrative of the new molecular biology skillfully integrated the state's promotion of scientific research and the push for modernization with a new discourse on life—that is, a new regime of truth explaining life as a manifestation of macromolecules determined by the structure of nucleic acids, the units that contain and transmit the information necessary to build proteins.

French modernization discourse followed tacitly the "science push" idea: the hope that the support of research would eventually translate into application and innovation on an industrial level. But the science and technology offensive of the 1960s generally did not affect technical research; many manufacturers regarded technology as a secondary factor in their growth strategies, and much of the money spent in the 1960s went into basic science.[14] Similarly, in Britain there was no shortage of political statements pointing to the need for an integration of R&D policies and applications. In fact, the fourth plan already stressed the need to combine scientific planification with the general plan (Salomon 1991, p. 41). Seen against the background of a fragile system of political regulation, with strong state steering displacing consensus between the social groups, the channeling of resources into the research system and its restructuring did not necessarily lead to the desired pace of technological innovation. But the inscription of molecular biology into the "developmentalism" narrative constituted an important semiotic resource that put the French state into a position to create links between research and industry—links that in the 1970s and the 1980s would be specified in the form of regulatory and industrial policies that supported the new biology as a modernizing effort. A new architecture of meaning had been created which later on became instrumental for the explanation of biotechnology policies and the enrollment of its actors.

Britain: Molecular Biology and the "White Heat of Science"

In France' s political discourse, the scientific discipline and material practice of biology was socially constructed mainly as a contribution to the enhancement of the development the economy. Whereas in the French policy narrative molecular biology was framed as part of an overarching strategy of "catching up" and modernizing, in Britain molecular biology found support within a constellation of political and economic discourses focused on reversing what was perceived as the country's long industrial decline. In contrast to Germany and France, Britain never moved successfully into a Fordist trajectory. The political character of the postwar period in Britain was defined by the "settlement" of the 1940s. An unwritten social contract had emerged, with the right accepting the idea of the welfare state, comprehensive education, and Keynesian management of the economy as well as a commitment to full employment as the terms of a peaceful compromise with labor and the left agreeing to work within the framework of a modified capitalism and within the Western bloc's sphere of influence. After the "restoration" of capitalist imperatives during the 1950s within the framework of the United States' perceived hegemony, there followed in the 1960s and the 1970s a period during which Britain was dominated by the Labour Party. Labour attempted to manage the new big-state-and-big-capital corporatist arrangements that had partially developed as the basis of economic policy and planning by harnessing the working classes to corporatist bargains through the trade unions and by disciplining them through Labour's traditional alliance with the unions.

But British Fordism was flawed at crucial points: productivity did not increase to the same extent as in other countries; the Fordist wage relation was defective, owing to a voluntaristic collective bargaining system; and mass consumption was financed through demand management and the social wage as well as through productivity increases. This flawed Fordism was reinforced by Britain's approach to the international political economy: British firms tended to look toward imperial markets in Africa and Asia at a time when fast growth and integrated markets were found in North America, Japan, and Western Europe. Britain proceeded cautiously with its integration into the European Community, and its membership was delayed until 1973. Furthermore, the development of the British economy was hampered by the dominance of financial capital in the market hierarchy. State intervention in the economy was based on resources from the budget, but not through management of the allocation of credit through the financial system. This approach limited the scope of the government's influence by forcing it to enter the market from the outside rather than enabling it to act through the financial system itself. As a result, the financial system restricted rather than facilitated interventionist public polices. This was the rather unsettling institutional background against which the British government attempted to reverse the nation's long decline from industrial preeminence (Jessop 1988, pp. 13–16; Zysman 1983, pp. 171–232; Overbeck 1990, pp. 100–104). In contrast to France, Britain's political discourse was not framed by a strategy of modernizing the country from agriculture to industry. Rather, this discourse focused on a politics of change within an entrenched industrial economy. But the major political parties and social actors—Labour, the Conservatives, the unions, and industry—were divided among themselves in rewriting the British state project, and that situation limited the chances for settling on a project for state intervention in economic development (Zysman 1983, pp. 206–210). This situation was exacerbated by the organization of the British state: the vertical links between central officials and the leading economic and social groups were weak, thus making systematic and sustained intervention in economic affairs rare.

It was within this interdiscursive constellation that science and technology policy was articulated as a strategy contributing to the reversal of the decline of British industry. At the annual conference of the Labour Party in 1963, the principal address was delivered by the party's leader, Harold Wilson, on the subject of "Labor and the Scientific Revolution." The unprecedented speed of technical change and the possibility of mass redundancy due to automation, Wilson noted, had created a case for socialism. Instead of leaving scientific and technological development to free enterprise and an unregulated economy, such development should be integrated into a comprehensive, purposeful social and economic plan. A new Ministry of Science should be created which would ensure that scientific resources would be deployed in "productive" sectors rather than on prestige projects, that new state industries would be developed on the basis of government-sponsored research, and that these industries would be located in high-unemployment areas (Vig 1968, p. 34).

Wilson's narrative of "the white heat of science" that was to forge a new Britain was an expression of Labour's attempt to reconfigure its political ideology around the idea of scientific and technological change. Labour's position was a reaction to the Conservative Party's new theme of "modernization," which the latter party adopted shortly after the 1959 elections. This response acknowledged Labour's need for a new and progressive image after ten years in power. In fact, as appropriated by Labour, the idea of modernization, along with the emergence of new problems, issues, and voter preferences, became the central theme of the 1964 election (Vig 1968, pp. 37–38). But unlike in France, where modernization was semantically linked to a social struggle over "developmentalism," in Britain modernization was coded as a social struggle over "decline reversal." The postwar constellation in France, the old hegemonic bloc being defeated, had allowed a redefinition and a discursive stabilization of the state's project that took the form of a new institutional framework defined by the idea of "developmentalism." By contrast, in Britain, where the war did not disrupt class relationships as it did in France and where relatively favorable economic conditions obtained in the years after the war (such as the absence of Germany, France, and Japan as serious trade competitors), the struggle for re-articulating the state into the management of the economy became a "war of movement" (Gramsci)—a continuous hegemonic struggle with frequent institutional changes.

Despite the new commitment of the Tories to modernization, the ruling government remained committed to the assumption that science is not a proper field for government control and planning. But in 1962, pressured to display a new outlook on the integration of R&D with economic and social development, the Tory government set up a committee under the chairmanship of Sir Burke Trend to inquire into the organization of civilian science. The Trend committee found that the various agencies concerned with the promotion of civil science did not cooperate, and that there was a lack of state coordination of scientific research and education (Gummet 1991, p. 19).

When Labour came to power in 1964, it continued its ideological transformation into a party associated with modernization and focused its political strategy on the reconstruction of Britain it had begun in 1945 (Jessop 1992, p. 14). The new government followed most of the suggestions that the Trend committee's report suggested might improve the coordination of research; in addition, it set up a Ministry of Technology, which eventually became the main agency for the promotion of technology and evolved into a ministry of industry. In 1963 the University Grants Committee (UGC), which administered the large block grants that supported universities, was transferred out of the Treasury and placed under the Minister of Science in a move that enabled the UGC not only to finance the universities but also to shape priorities for science and education. As the universities expanded in the 1960s, the UGC promoted development selectively through Memoranda of General Guidance that emphasized certain areas for development. The new Labour government completed the reorganization of the machinery of science policy by dividing up jurisdiction over science and technology. The Department of Education and Science became responsible for institutions of lower and higher education, including the University Grants Committee and the Research

Councils. Technology, on the other hand, came under the purview of a new Ministry of Technology. The responsibilities of the former Department of Scientific and Industrial Research were redistributed between the Department of Education and Science and the Ministry of Technology, the former acquiring two new research councils: a Science Research Council responsible for research in the natural sciences and a Natural Environmental Research Council responsible for research in such fields as geology and ecology (Shattock 1991).

What had happened was that Labour had stabilized modernization discourse by erecting a substantial apparatus to pursue its original "science push" philosophy. During these years, this philosophy was substantially influenced by the economic discourse on innovation that was dominated by the "science push" theory (Coombs et al. 1987, pp. 223–229). The basic assumption of this narrative was that the more R&D is "pushed" into the system, the greater will be the flow of finished innovation coming from the system. At the time, demand-side factors received only scant attention (Rothwell and Dodgson 1992, pp. 226–227).

But this discursive stabilization of the new British science policy had the effect of institutionalizing a separation between "pure" science and its "application." Work at the universities and the research councils was defined so as to isolate it from direct political control, whereas industrially related science was written into the responsibilities of the Minister of Technology. The research councils became a minor agenda of the Secretary of State for Education and science, whereas applied science came under immediate jurisdiction of the cabinet-level Minister of Technology (Wilkie 1991, p. 74). Under the leadership of Anthony Wedgwood Benn, the focus of the new department changed from the promotion of advanced technology in industry to highly controversial and often not very successful large-scale interventions in industry (ibid., p. 75).

In addition, soon after Labour took power it became evident that the underlying economic and political conditions necessary for the Wilson government's politics of planning and modernization did not exist. Labour, unable to realize its program, soon retreated to short-term economic crisis management (Jessop 1992, p. 14). The contradictions inherent in the postwar settlement between capital and labor turned out to be too strong. Simultaneous commitments to full employment, the welfare state, an international role for the pound sterling, the City of London, and costly defense undertakings prompted efforts to cut back or abandon one or more of those commitments (ibid., p. 16). This resulted in a stop-and-go cycle of policymaking during the 1950s, and in vacillations between planning and laissez faire in the 1960s (ibid., p. 21). Labour's new course of social intervention and indicative planning was bound to fail due as a result of the inability of the available institutional mechanisms to carry out such a program (Budge and McKay 1993, pp. 211–212). It is against the background of this particular discursive struggle to define the nature of the problem and the strategies for reversing industrial decline through modernization that we have to understand the articulation of the British politics of molecular biology.

The expansion of molecular biology in Britain was mainly the work of physicists, who, drawing on the new social standing they had acquired through their contributions to the war effort and on the advances of new physical technologies, secured state funds for what was called biophysical research: the application of physical technologies to biological problems (de Chadarevian 1994, p. 5). In the postwar era, a newly formed network of scientists engaged in biophysical research saw their influence and their advisory functions growing in a situation of a general expansion of R&D budgets (ibid., p. 6). In particular, the Unit for the Study of Molecular Structures, set up by the Medical Research Council (MRC) in Cambridge in 1947, gradually emerged as the center of molecular biology in Britain. A number of important breakthroughs soon followed. In 1953 Crick and Watson presented their model of the double helix. In 1957 Kendrew introduced his "sausage model" of myoglobin, which for the first time showed the conformation of a polypeptide chain in protein. Perutz's work on the more complex structure of hemoglobin was gaining pace, and Crick and Brenner worked with phage mutants to find out how the genetic code works. All these achievements are celebrated today as hallmarks of molecular biology (ibid., p. 12).

As in France, in Britain the creation of a more coherent molecular biology policy was preceded and provoked by a process of problematization, with scientists playing a key role in the "definition of the problem." Again we observe that the support for molecular biology was hardly an "obvious" act; rather, it was the result of the operation of a narrative that construed molecular biology as the future of research in biology and medicine. As early as 1962 the Royal Society's ad hoc Biological Research Committee pointed out that "major changes in the organization of biology departments" were necessary if

Britain was to keep its place in international research—a point of view that was not accepted and not acted upon by the more traditionally minded University Grants Committees (Yoxen 1981, p. 98). This move toward a more systematic politics of molecular biology received a further push with Labour's ascent to power in 1964 and with the deployment of the new politics of science. The 1966 Annual Report of the Council of Scientific Policy, founded by the new Labour government, cited molecular biology as an example in order to draw attention to the problem of utilizing the country's scientific resources in the country (de Chadarevian 1994, p. 16). John Kendrew, who at the same time played a crucial role in the creation of EMBO, was a member of the council.

The same year the report was published, the Council of Scientific Policy set up a committee or "working group" chaired by John Kendrew and composed of eight leading scientists, mostly from the MRC's research units in Cambridge, Edinburgh, and London, to "inquire into the present conditions of, and future plans for, teaching, recruitment and research in molecular biology in the United Kingdom." Two years later the report was published. It concluded that most progress in biology took place in the area of molecular biology. The impact of research in molecular biology, so the story went, not only had "deep intellectual significance" but also was likely to yield "social and economic dividends of inestimable value" through its biomedical and agricultural applications. However, the report diagnosed a decline in British leadership in molecular biology, where the United States was making great strides. The continuation of a high level of support for molecular biology was seen as crucial to ensure that molecular biology would be a source of prestige to British science, to protect a relatively inexpensive investment that might pay high socio-economic dividends, and to maintain molecular biology' s role as the engine of progress in biology as a whole (Abir-Am 1992, pp. 167–168).

On the institutional level, the MRC played a significant role in supporting molecular biology. This was demonstrated by the establishment of the Laboratory of Molecular Biology in Cambridge in 1962. Furthermore, King's College and the Microbial Genetics Unit in London put much emphasis on research in molecular biology. In 1968–69 in the United Kingdom, 7 million pounds out of a total government expenditure on biological research of 70 million pounds was spent on biology at the molecular level.

In the evolving policy narrative, molecular biology became a reflection of Harold Wilson's "white heat of science" and a functional element in the new project of social and economic modernization. The inevitable success of molecular biology and its capacity to transform the life sciences had effectively become a black box—a translation device that established a network of meaning among molecular biologists, administrators, the MRC, and the Labour Party and ultimately linked the practice of molecular biology research with the state apparatus. Black boxes stabilize meaning, and molecular biology' s meaning as cutting-edge science increasingly enjoyed broad acceptance. There were also dissenting voices. Biochemists, for example, were afraid of a colonization of biology as a whole by the new field of molecular biology. In fact, a report of the British Biochemical Society, published in response to the Kendrew report in 1969, argued that there was no need to establish new departments of molecular biology, but that there was an urgent need to expand biochemistry. At the same time, the biochemists rejected governmental intervention on behalf of molecular biology as a violation of academic freedom (Abir-Am 1992, pp. 173–177). But it was exactly molecular biology's "openness" to such intervention and the field's intention to shape an alliance with the state that made it such an attractive candidate for the new interventionism of Labour.

However, as was noted above, Labour's discourse on the logic of technological innovation had been framed within a more pressing policy imperative: that government intervention in molecular biology would eventually yield tangible economic and social benefits. However, it was not only in the case of molecular biology that this technology policy discourse had failed to materialize into an institutional apparatus enabling the state to intervene effectively in the process of economic and industrial management.

Germany: Molecular Biology as Industrial Strategy

In Germany the field of molecular biology was discovered relatively late by policymakers. The German national policy narrative established a relationship between molecular biology and modernization which had a different meaning than that of France and Britain. Unlike these other two countries, Germany was under no pressure in the postwar period to make dramatic adjustments in the sectoral base of the industrial economy, but only to reestablish and expand existing industries.

Widespread rationalization before the war had laid the groundwork for postwar competitiveness (Zysman 1983, p. 254; Reich 1990). The labor movement gained codetermination and worker participation but was also obliged to work within the limits of a strong market rationality in the so-called *soziale Marktwirtschaft* (social market economy). This economic formation was characterized by the dominance of private-sector capital, a key coordinating role for banking capital, only limited direct and open state intervention, and a welfare state organized along corporatist rather than liberal lines. The postwar expansion was based less on the production of consumer durables than on a strong export-oriented capital goods sector (Jessop 1988, pp. 17–19). Cut off from its former markets in Eastern and Southern Europe and without historically significant colonial ties, West Germany sought its trading partners almost exclusively among the advanced capitalist countries of Western Europe and North America. The country became an early member of the European Economic Community in 1958. Economic recovery was further characterized by a re-establishment of the prewar patterns of industrial concentration, cartel-like arrangements, and centralized semi-official trade associations. Rationalization, competitiveness, and integration within the world market became essential to the continuing reproduction of the German social formation and subordinated the main social groups to the "logic" of the export-oriented political economy (Graf 1992, pp. 14–20).

After World War II the left and the right in Germany shaped a growth-oriented social alliance aimed at the reconstruction and consolidation of the capitalist market economy. High taxes and low wages were part of a policy of strengthening the position of higher-income employees, property owners, and the traditional middle class. The situation changed in the late 1950s and the early 1960s, when, with the erection of the Berlin Wall, a sharp drop in the available supply of labor put pressure on the low-wage structure and trade surpluses forced a revaluation of the Deutschmark. The past regime of production, which had relied on existing technology, cheap labor, and little state intervention, became obsolete. The imported (dollar) inflation exacerbated the situation. In 1966 the German economy no longer grew. It was in this situation that political discourse increasingly defined economic planning and a reconfiguration of available organizational resources as central to the future of the "German model" (Katzenstein 1987, p. 14; Hirsch 1970, pp. 79–81). The idea that modernization was to be achieved only by the planning of scientific and technological development was increasingly gaining support. One outcome of this new narrative was the substitution of general support policies, such as those for universities, with directed programs in important scientific and technological fields (Fleck 1990, p. 50). It was in the context of this changed discourse on the politics of science, which gave new momentum to economic growth, that molecular biology became earmarked as a new field deserving systematic support. Thus, the framing of molecular biology as an element of modernization politics took the form of a discourse of strategic industrial adjustment, a discourse which must be understood in relationship to the highly internationalized character of the German economy. This shows again an important characteristic of discourse: discourse is material in the sense that it is always caught in the practices to which it refers without coinciding with the activities and phenomena that are systematized as objects of discourse. In other words, the evolving German discourse of molecular biology can hardly be understood apart from the complex political economy described above. However, this political-economic context does not offer an "explanation" for the emergence of molecular biology discourse. Rather, it constitutes context and material for this discourse.

Key institutional resource in this transition to a more interventionist mode of governance were the parapublic institutions linking private and public, such as the Deutsche Forschungsgemeinschaft (DFG) and the consulting commissions of the Bundesministerium für Wissenschaft und Forschung (BMWF). The main impact of the new discourse on R&D organization was a certain organizational tightening and efforts toward better coordination within the already-established system of institutions interlinking science, the state, and industry.

Molecular biology entered the scene of policymaking in the now-familiar forms of problematization, assessment, and evaluation. It is through problematization that problems—as well as sets of actors, their identities, and future strategies—are defined. The initiative to reconsider the state and the direction of German biological research did not come from individual "great scholars" or famous institutions in the field; it originated in a rather unspectacular way in the DFG and the Stiftung Volkswagenwerk. This might be related to the fact that, at the time, there were no "great German scholars" in the field of molecular biology with the reputation and standing of a Monod or a Kendrew. In addition to reflecting a lack of any significant research tradition in the emerging field of molecular biology and a more decentralized network of scientists and state institutions, this absence was also linked to organizational features of the German research system.

As early as 1958, a report of the DFG came to the rather gloomy assessment that the dramatic changes in the new biology were hardly reflected in the practices of German biology:

> Whereas in the last twenty years abroad, in particular in the United States, chemists, physicists and geneticists have understood very quickly that there is a new field of problems emerging which only waits to be dealt with by means of new methods and quantitative thinking, very little of this development has been taking place in Germany. Only a very few steps in this direction were taken in our universities and there exists no textbook which deals with the new problematic in a conceptual and methodological way which is even of minimal satisfaction.[15]

Six years later, in another report on "Development and Under-development of the Sciences in Germany," the DFG came to the following conclusion:

> All in all we see that those fields of biology are difficulties in which impulses have to come from neighboring disciplines and cooperation with chemistry and physics and also by adopting mathematical statistics.[16]

The report, taking into account that a country of the size of Germany cannot support research in all fields of science, goes on to argue that the creation of points of emphasis for the practice of research policy is a necessity.[17] The narratives underlying these reports were important factors in a discursive field where research in the new biology, as financed by the DFG and conducted by the Max-Planck-Gesellschaft, grew steadily during the 1960s. In 1964, for example, the DFG created a new support emphasis on molecular biology research with roughly 2.5 million marks of annual research funding (Zarnitz 1968, p. 76).

In 1962, an important new patron of molecular biology, the Stiftung Volkswagenwerk, entered the scene of research support. Founded in 1961, the Stiftung Volkswagenwerk, with 1.1 billion marks in capital, immediately became the largest research-supporting foundation in Europe. The financial muscle of the foundation was particularly impressive during the early 1960s. During this time, its available funding was comparable to that of the DFG. This situation eventually changed during the 1970s, when the expansion of the DFG surpassed the gradual increase of the Stiftung Volkswagenwerk's capital.[18]

Early on, the Stiftung Volkswagenwerk identified potential areas of support. After consulting the Hungarian-American physicist Leo

Szilard, the foundation targeted molecular biology and physical biology as major fields of support activity. Later, in 1965, approached by the prominent German scientists Hans Inghoffen, Hans Brockmann, and Manfred Eigen, the Stiftung Volkswagenwerk played a key role in the foundation and funding of the Institute for Molecular Biology, Biochemistry and Biophysics in Stöckheim/Braunschweig, which in 1968 became the Gesellschaft für molekularbiologische Forschung (GMBF).[19] Initially set up to become a non-university elite institute of molecular biology, the GMBF turned into one of the largest biotechnology R&D institutions in Germany in the following years. Another important activity of the Stiftung Volkswagenwerk involved the funding of Max Delbrück's years at the University of Cologne as a tactic to introduce the new biology into Germany: "We wanted input from abroad to build up modern biology as quickly as possible. . . . What the Rockefeller Foundation had anticipated and stimulated in America we could now see there and we wanted it for Germany."[20] Between 1963 and 1974 the foundation gave 45 million marks in support of research and a similar amount for the creation of the GMBF.[21]

Together with EMBO, the Stiftung Volkswagenwerk emerged as one of the largest institutions supporting molecular biology in Europe. During this time, the DFG, the Max-Planck-Gesellschaft, and the Stiftung Volkswagenwerk began to increase their funding for molecular biology. A working group within the Bundesministerium für Wissenschaft und Forschung began to consider the establishment of support policies for "new and exiting developments in biology." The Bundesministerium für Wissenschaft und Forschung began to support biological research in 1964. This was done within the framework of its nuclear power and space exploration programs, its main areas of direct support (Zarnitz 1968, p. 79).

The German politics of molecular biology was remarkable for its systematic character and for its coordination of various participating research funding institutions, which were attributable to the institutional resources that could be mobilized for the purpose of molecular biology politics. The peculiar organization of the German R&D system can be traced back to the nineteenth century, when, under the considerable influence of industry, important non-university state research institutions such as the Kaiser-Wilhelm-Gesellschaft (succeeded after World War II by the Max-Planck-Gesellschaft) were founded with the intention of establishing research institutes reflecting the interests of industry, state, and science (Hirsch 1970, p. 129).

In fact, despite the Stiftung Volkswagenwerk's impact on the shaping of molecular biology in Germany, one would be mistaken to compare its tactics to those of the Rockefeller Foundation. Whereas the Rockefeller Foundation was guided by the will of its independent board and trustees (drawn primarily from the business sector), the Stiftung Volkswagenwerk was part of the highly complex organizational structure of the German R&D system, which linked together the state, self-governed research institutions, and business interests. The basic structure of the board of the Stiftung Volkswagenwerk, composed of representatives from science, business and the state, is mirrored by most of the other key institutions of the German R&D system.

This remarkable level of coordination becomes clear when one looks at the two pillars of German R&D, the Deutsche Forschungsgemeinschaft and the Max-Planck-Gesellschaft. In terms of their agendas the two institutions are quite different. The DFG, responsible for the general support of science, is financed by the central government and by the *Länder* (states) and does not have its own research institutes. The Max-Planck-Gesellschaft, financed mainly by the states, is engaged in the conduct of basic research and applied projects in close cooperation with industry and has its own institutes. But both organizations have a board structure that secures the considerable influence of state officials and industry representatives in the decisionmaking process (Hirsch 1970, pp. 214–215). The Bundesministerium für Wissenschaft und Forschung has a similar structure. The philosophy of keeping the staff of the BMWF small led to the establishment of a myriad of consulting commissions to advise the ministry on the various research areas. In 1969 some 155 civil servants were advised by a total of 373 consulting committee members, who determined the ministry's policies to a considerable degree. Hence, the BMWF became part of the larger structure of the German R&D system, composed of a highly complex network which created a continuous stream of interaction between science, industry, and the state, thereby establishing—widely removed from public scrutiny—in a quasi-autonomous manner the dominant patterns of science and technology planning and policy (Hirsch 1970, pp. 196–197).

In fact, the BMFT soon began to assume a leading role in exploring the commercial potentials of molecular biology. In 1969 the BMFT launched its New Technologies program, which was explicitly designed to support "high technology" and which selected biological and medical technology as one of the sponsored fields. As a civil servant put it:

We thought that in the long run the bringing together of modern biochemistry, genetics, microbiology and process technologies should result in new opportunities to produce new products. We thought it is impossible that the theoretical study of physiological and morphological processes on the molecular level will not result in the discovery of new metabolism and the identification of new products which eventually will be of practical interest. . . . We knew from practice that there are enough biological processes which are useful. If this assumption is right, we thought we only had to ask German scientists and industrialists which new scientific and technological developments and related products are possible.[22]

This statement highlights an important development in the postwar era: the formation of a new network of meaning in the politics of molecular biology. The creation of this network benefited from certain organizational features of German science policy, such as institutions that facilitated the cooperation between science and industry. But an institutionalist explanation of the emergence of German molecular biology policy alone is not sufficient. It was only through the ordering impact of narrative that these organizational features were mobilized, began to matter, and came to be alive; it was only through the impact of a network of meaning and translation that connections were established between such disparate elements the German science system, industry, theories of national science development, reconfigurations of the gene concept in the discourse of molecular biology, and the blackboxing of molecular biology as the future of biology. Business, private foundations, and research were linked to the state apparatus, and the state was positioned to exert influence over the shaping of molecular biology in the direction of industrial applications.

Conclusions

In this chapter I have discussed the interrelated scientific and political practices that helped to set up the political space where the objects, actors, structures, and strategies of European and American molecular biology policymaking emerged. Although various actors in various policy fields pursued various goals, and although strategies and political institutions influenced them in doing so, in each case it was necessary to ask what explained the very presence and the specific relationships of certain actors and institutions in a particular policy field. Rather than assume the existence of actors, goals, and structures of policymaking, I have focused on the importance of discursive processes for the initial construction of the central elements in a policy field. I have discussed the gradual shaping of a regime of governability which at its core identified genes and (by implication) their human carriers as objects for a new type of socio-political intervention. Molecular biology policymaking came into existence when policy narratives were developed that connected the various sites of the emerging governability regime by relating the new understanding of life phenomena in the discourse of molecular biology to central national political goals, institutional structures, and socio-economic contexts. Historically, this construction of the political space of molecular biology pertains to a number of varied but interrelated factors, such as the rise of multinational liberalism, the involvement of the Rockefeller Foundation in social and scientific policy, and the evolution of a discourse linking scientific progress with international economic expansion. Thus, this chapter has shown that policy narratives originate within larger discursive economies—that is, in complex systems of signification which are sources of specific political codings of contextual factors, such as developments in the political economy. In other words, contexts such as economic formations or regimes of accumulation are highly important for policymaking, but their "reality" is never fixed and is always intermediated by language.

This chapter has also demonstrated the importance of "bottom-up" flows of power, which often originate in sites considered to be non-political. The science of molecular biology was such a site, and the considerable power that emanated from it began to shape society. The local knowledge of molecular biology extended through an array of experimental systems, interacting projects, practices, and capabilities from dispersed laboratory sites to the surrounding world and finally began to display a pattern, a direction, and the features of the development of a new scientific discipline. The Rockefeller Foundation's "New Science of Man" emerged with the prestige associated with being at the "cutting edge" of science. But the power of molecular biology was a subtle one (Lenoir 1992, p. 8). Whereas the "old" eugenics needed to coerce bodies, the "new" biology operated by delineating new opportunities for human conduct, such as in medical knowledge.

In postwar Europe, the representation of molecular biology as a site of governmental intervention was articulated with discourses on modernization. The framing of molecular biology as an element in the politics of modernization provided the crucial link between the inscription of modernity qua industrialization and the emergence of a sectoral policy of molecular biology. The broad economic and technological diffusion of molecular biology became synonymous with successful modernization. To a considerable degree modern society was represented as a society whose future was determined by the successful application of molecular biology. This new science deserved systematic support by the state. Unlike the earlier discourse on eugenics, the language of social control that had been present in the Rockefeller Foundation's initial justification of its support of the new biology had by and large vanished from narratives of justifying the new biology by the early 1950s. The population-control-oriented eugenic "politics of life" had been replaced by the individualizing strategies of subjectification of molecular biology. The site of intervention had shifted from the "gene pool" to the "gene." Practices of technological intervention on the subcellular level of life moved into the core of molecular biology's approach to the study of life processes. In particular, a group of scientists operating on the international level (e.g., through their involvement in the creation of EMBO) as well as on the national level played a crucial role in mediating between the general socio-political imaginary of the 1950s and the 1960s and the emerging fields of science policy and molecular biology.

The process of translating molecular biology into a politics of modernization went remarkably similarly in France, Germany, and Britain. In the initial stage of problematization, reports were written in which the "reality" of molecular biology was assessed. The "state of the art" was scrutinized, and molecular biology was found be a science of the future, underdeveloped in the home country, that required substantial government support. Developments in molecular biology in other countries, particularly the United States, were said to be a model. They were also a warning and a further argument for mobilizing domestic resources and actors for molecular biology. By the late 1960s, American science had achieved mythical status in European discourse. But the argument that molecular biology was a cutting-edge science was not enough. In a further step, molecular biology was inscribed as a contribution to modernization. This allowed for a rather spectacular process of displacement. Simultaneously, relationships were established between the discipline of molecular biology and strategies of "industrial developmentalism" (in France), the reversal of industrial decline (in Britain), and international competition (in Germany). The narratives used in this "translation process" were crucial for the institutional stabilization of the discourse of molecular biology in the form of research grants and for the founding of new departments, institutes, and scientific organizations. It was also crucial for the definition of those actors and organizations legitimately engaging in the policymaking process, such as molecular biologists and research ministries. These stabilizations were not independent from the articulation of the state's role in the administration of science and technological development. Here we see the importance of political discourse varying from country to country for the organizing and shaping of policymaking. In France, the relatively early and forceful institutionalization of molecular biology was guided by a science policy narrative subscribing to the "science push" approach, which expected the benefits of science somehow to spill over into industry. In Britain, owing mainly to the later drafting of molecular biology policies, the representation of the impact of scientific and technological development was more complex and assumed the importance of intermediary institutional mechanisms. However, because of unresolved social struggles, this more complex image lacked institutionalization in the form of corresponding scientific and technological organizations. In contrast, the relatively late stabilization of the molecular biology discourse in Germany and a consensus-based institutional system for science and technology policymaking led to the inscription of molecular biology as a "high technology" that was to be explored and developed in close cooperation between science and industry.

These different constructions of the boundaries of molecular biology were important for the later framing of biotechnology policy. Despite differences among the three countries considered here and on the European Community level, molecular biology experienced institutional stabilization during the 1960s and began to transform the practices of biological research. At the same time, molecular biology policy became part of the larger politics of security that interrelated tropes of international competitiveness, economic growth, and national progress with images of medical advances and new insights into what constitutes human beings in particular and life in general. This micropolitics of molecular biology, with its promises of a "New Science of Man" and a new understanding of nature, soon became the center of a political storm: the recombinant DNA controversy.

3

Molecularizing Risk: The Asilomar Legacy in the United States and in Europe

Molecular Biology as a Risk

In the preceding chapter I traced the emergence of genes as an object for socio-political intervention from the 1930s until the early 1960s. State-led molecular biology policies began to take shape during the 1950s and remained relatively uncontroversial until the late 1960s. But in the 1970s, risk and concerns about the social and ethical implications of molecular biology emerged as major issues in the new policy field.

In accordance with the discourse-analytical strategy pursued in this book, I do not assume that the risks associated with molecular biology and concerns about its social and ethical consequences were phenomena that had come into existence mainly because certain scientific developments in biological research had occurred. By the late 1960s molecular biologists had begun to conduct experiments involving the controlled recombination of DNA from different sources, but this did not mean that these experiments carried specific risks that were obvious to everybody in one and the same way. Risks do not exist. They come into existence through complex and multiple processes of inscription, interpretation, and boundary work carried out by a variety of actors and informed by scientific and political discourses. Typically, different actors involved in a risk-regulation dispute tell different risk stories. Eventually one story begins to dominate the risk definitions in a policy field. We therefore need to conceptualize the risks of molecular biology as gradually emerging in a political space set up by interrelated discursive practices. These discursive contributions came from different authors, such as scientists, labor union officials, or administrators. As a result of the new risk discourse, the new policy field of molecular biology became even more complicated. Next to the support of the new biology, its regulation moved into the foreground of consideration. Definitions of the risks of genetic engineering were closely tied to arguments about which kind of actors and institutions should get involved in the making of regulatory policy for genetic engineering.

Risk became an issue of political concern after a number of significant changes in molecular biology during the 1950s and the 1960s. During those decades, the concept of the genetic code was deployed and definitions were provided for the processes responsible for the replication of DNA and protein synthesis, for the biochemical pathways involved in the replication, expression, and natural replication of DNA and protein synthesis, and for the enzymes that catalyze these processes. In the early 1970s this process seemed to culminate in the development of recombinant DNA technology. Genetic engineering was immediately portrayed as a significant scientific breakthrough with far-reaching social and economic implications (Wright 1994, pp. 67–70). This "success" of the new biology led to a further complexification of the system of linkages between the state apparatus and the practices and institutions of research and, later, production. Molecular biology was reconceptualized as the purveyor of technologies for a new scientific-industrial configuration: the "new biotechnology." At the same time, the risks connected with recombinant DNA work became a major topic in political discourse.

In this chapter I will show how in the early 1970s the debate about the boundaries of the risks of genetic engineering and the need to regulate molecular biology was by no means an obstacle to its expansion, but, to the contrary, was critical for the diffusion of genetic engineering into economy and society. During the 1970s the particular modes of inscribing the risks connected with recombinant DNA technologies that achieved hegemonic status contributed to the social construction of modern biotechnology as a "technology of the future" and as a terrain for governmental intervention, rather than undermining it. The stabilization of the meaning of "recombinant DNA's hazards," and its hegemonic status in the political discourse, were based on the crafting of a new regulatory policy narrative that created a semantic link between such diverse fields such as the protection of citizens from the hazards of genetic engineering, workplace safety, and economic modernization.

As we will see in this chapter, molecular biologists and experts from related fields played an important role in the definition of the risks of genetic engineering. But the determination of the "risks of genetic engineering" and the development of measures to "control these risks" was hardly a process in which a number of specialists simply had done their best to identify the "true nature of the risks of genetic engineering" and, after much research and soul-searching, provided the world with a good solution to the problem. As I will show, there is little evidence in the social history of the recombinant DNA controversy for this kind of "expert heroes" interpretation. However, we do not gain much if we shift the focus of our explanation from benevolent actors to utility-maximizing actors or contexts and interpret the recombinant DNA controversy as a process in which a number of social, political, and economic forces used the debate over technical issues to attain their own political or economic ends. My story is a more complicated one. It is a story of the tangled and layered political and economic histories that co-produced the signifiers of genetic engineering's risk and welded them into a signified, the meaning of "recombinant DNA hazard" (Lenoir 1994, p. 134). In this context, experts were important because they were specialized professionals who knew the language of genetic engineering. They also could draw on the scientific discourse of molecular biology, which by the early 1970s had assumed a dominant status in describing life processes. This constellation allowed the practitioners of molecular biology to position themselves as the key actors in the regulatory process. Furthermore, by contributing significantly to the crafting of regulatory policy narrative, scientists were active in linking genetic engineering to other policy discourses (such as workplace safety) and to different aspects of the larger political imaginary (such as the idea of modernization) (Rose 1994, p. 365). Thus they contributed to the articulation of other important actors (such as government officials or labor union officials) into the emerging regulatory system. Civil servants engaged in regulating genetic engineering demonstrated the state's intent to protect the citizenry and simultaneously to support the progress of science. In a similar way, labor union officials and business representatives played the role of the "participating public" in the regulatory decisionmaking process, which was to make sure that the risks of genetic engineering would be 'balanced against its benefits. Discourses of scientific expertise entered a mutually reinforcing relationship with narratives of modernization and workplace-safety philosophies. The regulatory state—the state in charge of overseeing the regulation of recombinant DNA—was one of the effects of this partitioning of the political space of genetic engineering. As we will see, this creation of policies for the regulation of genetic engineering was by no means consensual from the beginning, and it entailed the careful crafting of a complex chain of signification which by the late 1970s offered a widely uncontested, hegemonic interpretation of the nature of the risks of genetic engineering.

New Technologies of Writing Life

In order to understand the specific modes of the emergence of the recombinant DNA risk, we need to take a brief look at the development of a new language to talk about life, which was at the same time an instrument to modify it. By 1970 all the techniques necessary for genetic engineering with bacteria had become available. Plasmids and bacterial viruses served as vehicles to carry foreign DNA into living cells; techniques were developed that enabled bacteria cells to take up relatively large pieces of DNA without being killed in the process; methods for synthesizing DNA and for making DNA copies from messenger RNA provided a limited means of making pure sources of DNA; various enzymes discovered and isolated in the 1960s enabled scientists to join DNA from different sources; and the site-specific restriction enzymes that were becoming available in the late 1960s made it possible to cut DNA at exact locations (Wright 1994, p. 67).

A shift in molecular biology seemed imminent. Until the end of the 1960s the new biology had largely rested on techniques that could be classified within the traditional disciplines of biophysics, biochemistry, and genetics: x-ray crystallography, ultracentrifugation, radioactive tracing, chromatography, and gene mapping. Classical biophysical, biochemical, and genetic technologies are usually characterized as in vivo, in situ, or in vitro approaches, meaning that these technologies attempt to construct the milieu of the living cell by the creation of technical conditions under which metabolic structures and processes can be referred to as model structures or model processes. In short, these technologies furnished an extracellular representation of an intracellular configuration.[11] Recombinant DNA technology has inverted this situation. Its central tools, such as restriction, transcription, and replication enzymes, plasmids, and all sorts of vectors, as well as pieces and bits of DNA and RNA, are of the order of molecules themselves. Hence, with Hans-Jörg Rheinberger (1993, p. 8), we can argue that the very tools of manipulation of the molecular biological endeavor have become molecularized; they are themselves of the nature and dimensions of the processes into which they intervene:

> The scissors and the needles by which the genes get tailored and spliced are enzymes, and the carriers by which they are transported are giant molecules. This kit of enzymes and purified molecules constitutes a "soft" technology. . . . With gene technology, roughly speaking, informational molecules are constructed under biochemical conditions which then are implanted into the intracellular environment of the organism who transposes them, reproduces them and 'tests' their characteristics. With that, the organism itself advances to the status of a technical object, that is, to the status of a space of representation within which new epistemic things are becoming probed and articulated. . . . The molecular engineer constructs objects, that are, in essence, information bearing molecules, which no longer need to preexist within the organism, and in reproducing them, expressing them, and analyzing their effects in reproduction and expression, he uses the milieu of the cell as their proper technical embedding. The organism itself, in a rather strong sense, is turned into a laboratory. From now on, what is at stake is no longer the extracellular representation of intracellular structures and processes, that is the 'understanding' of life, but rather the intracellular representation of an extracellular project, that is the deliberate and quite literal "rewriting life." (ibid., pp. 8–10)

With this move, DNA technology had turned from a mode of discovery, or the benign illusion of simply constituting a mode of discovery, into a deliberate praxis of molecular writing, of bio-construction (ibid., p. 19). Molecular biology, from its inception predicated on interventions aimed at reshaping vital phenomena and social processes (Kay 1993, p. 280), acquired in its shift from the protein paradigm to the DNA paradigm the status of a discipline for which intervention was not only the mode of representation but also the explicit program. Intervention into subcellular processes of life became both the condition of molecular biology's intelligibility and representability and the goal that legitimized the discipline.

The Making of the Recombinant DNA Debate

What were the social, political, ethical, and economic meanings of the development of this new language of molecular biology? Different readings of the meaning of the "genetic revolution" began to proliferate during the 1960s. A small group of scientists working in molecular biology played a critical role in providing new theories of the meaning of their research practice. Many of them would soon become "experts" in the discussion of genetic engineering's risk. Whereas in the past the new biology was mainly coded as a cutting-edge science with momentous implications for the understanding of life processes and potential benefits for the economy, the re-presentation of molecular biology now focused increasingly on its "revolutionary" characteristics and "transformative" potentials.

This process can be traced back to the 1962 Ciba Foundation Conference on "Man and His Future," where members of the scientific elite in genetics, evolutionary theory, medicine, and biochemistry, including the Nobelists Herman Muller, Joshua Lederberg, and Francis Crick, came together (Wolstenholme 1963). Concerned about the diagnosed trend toward global overpopulation and the perceived increase in genetic diseases, the participants reflected on what the "new biology" could already offer and would soon be able to offer as a remedy. The participants expressed the hope that, with the "new biology," man could finally become the master of evolution. Among the strategies proposed were direct manipulation of germ lines and deliberate support of positive hereditary traits, both of which were elements of a broader strategy intended to establish a better future for man. The publication of the conference proceedings led immediately to broad public debate. In Europe and in the United States, a variety of voices expressed concerns over the dangerous eugenic territory the "new biology" was about to explore.

It was also in the 1960s that the term "genetic engineering," denoting the deliberate and controlled modification of genetic material, gained currency. Such leading scientists as Sidney Brenner, Arthur Kornberg, Joshua Lederberg, and James Watson devoted considerable energy and diligence to communicating the fruits of their research to a mass audience. Scientific research was in need of financial support, and scientists felt the urgency of communicating the relevance of their work to the public in order to justify their demands for more research money.[2]

In 1969 a group of young molecular biologists working at the Harvard Medical School published a paper in *Nature* announcing the first isolation of a bacterial gene (Shapiro et al. 1969). At a press conference, they emphasized the political significance of their work and pointed to the potential misuse of their research. In English newspapers, their presentation was reported under such headlines as "Genetic Bomb Fears Grow" (*Evening Standard*). The *Daily Mail* observed: "Scientists find the secret of human heredity and it scares them." (Yoxen 1978, p. 312) In 1970, the British Society for Social Responsibility organized a conference on "The Social Responsibility of Modern Biology." That conference brought together leading molecular biologists and critics, who again were divided over the question whether scientific research should be more thoroughly subjected to social scrutiny, or whether the direction of science should be left to scientists alone.[3]

In 1971, Paul Berg and his collaborators conducted the first successful experiment in genetic engineering at Stanford University. The Berg group used DNA from the bacterial virus lambda and from a monkey tumor virus known as SV40. In the first step of the process, each loop of DNA was cut with a restriction enzyme. Next, the ends of each piece were enzymatically modified. Then adenine nucleotides were added to the ends of one type of DNA and thymine nucleotides to the ends of the other type. When the two types of DNA molecules were mixed together, their ends joined, forming circular hybrid molecules containing the SV40 and lambda virus DNA. After this first successful experiment in gene splicing, the next step was the insertion of foreign DNA into bacteria in such a way that it would be replicable, accomplished by Herbert Boyer, Stanley Cohen, Robert Helling, and Annie Chang in 1973. They showed that DNA from the toad *Xenopus laevia,* which normally codes for a specific type of RNA, could be introduced into *E. coli* and replicated in the bacterium. This move to recombine genes from different species was widely regarded as a remarkable achievement (Wright 1994, pp. 72–75). But immediately the question arose whether this experiment did not go beyond what seemed to be ethically and politically acceptable. In 1975, Paul Berg recalled the response of a European audience of scientists to his work at the NATO molecular biology workshop in Sicily:

> I came to lecture on protein synthesis to this sort of a school for Europeans, and the lecture, as a sort of rap session, that I was asked to give, was about the work we were doing in the construction of this hybrid DNA molecule of SV40 and lambda virus. And I did give that, and there was a very, very strong response on the part of the young European students, that this was really sort of the beginning of an new era and potentially dangerous, and raising the spectrum of genetic engineering in humans, behavior control. . . . And they asked, could we have an informal session to discuss the political and social consequences of it. And this German fellow organized an evening session up in the ramparts of the old castle . . . and we sat up till about midnight, this whole crew drinking beer, about eighty people, back and fourth discussing the possible hazards, prospects for genetic engineering.[4]

The doubts concerning the implications of inserting SV40 DNA into *E. coli* intensified, and Berg and his collaborators postponed their experiment. Soon afterward, Berg and others decided to organize a series of conferences about the hazards of SV40, the first of which was held at the Asilomar Conference Center in Pacific Grove, California, in January 1973. Researchers from a broad range of disciplines within the biological and health sciences—all from the United States—were in attendance. They focused on viruses, especially on those that were capable of producing infection in humans or tumors in animals. No recommendations were made to limit research; rather, the emphasis was on getting more information (Krimsky 1982, pp. 58–69).

The debate continued and culminated in June 1973 at the Gordon Conference on Nucleic Acids, a conference with considerable participation of European scholars. Herbert Boyer reported on his pathbreaking work with Cohen, Helling, and Chang. After long debates, the two chairpersons of the conference, Maxine Singer of the National Institutes of Health and Dieter Soll of Yale University, sent a letter to *Science* discussing the potential impact of the new technological possibility of combining DNA molecules from different sources. The letter emphasized the potentially beneficial sides of the new technology, but it also pointed out the potential hazards for laboratory workers and the public. Singer and Soll proposed that the National Academy of Sciences establish a committee to examine the problem and develop specific actions or guidelines to meet that concern (Krimsky 1982, pp. 70–80).

The National Academy of Sciences asked Paul Berg to lead a study panel, later to be called the "Committee on Recombinant DNA Molecules, Assembly of Life Sciences." The committee, made up of James Watson, David Baltimore, and other leading American molecular biologists, introduced a new element to the construction of the recombinant DNA problematic when it identified three classes of experiments as warranting special attention: type I experiments (in which two types of genes, those coding for antibiotic resistance factors and those coding for toxins, were linked to bacterial plasmids), type II experiments (in which DNA segments from a cancer-causing virus or other animal viruses were linked together), and type III experiments (in which fragments of animal DNA were joined to bacterial plasmids in a random way and then the plasmid was implanted into the bacterium).

In its final statement, the Committee on Recombinant DNA Molecules requested that experiments of types I and II be deferred until hazards could be better evaluated. This recommendation reflected mounting concerns in the public health community about dangers inherent in the insertion of genes that would give antibiotic resistance to organisms that are normally sensitive to antibiotics, and about the release of these organisms into the environment. In a similar vein, the creation of bacteria with new toxins that might invade the ecosystem was seen as a menace. It was recommended that type II experiments, a category that described a much wider class of potential recombinant DNA experiments than the other two groups, be "carefully weighed"—a suggestion that was not further specified as to its implementation. In a further important step, the committee translated the risks of molecular biology into a problem for further research and, most critically in relation to the first proposal, into an object for governmental intervention: the Berg panel requested that the director of the National Institutes of Health establish an advisory committee to oversee an experimental program of risk assessment, to develop procedures to prevent the spread of recombinant molecules, and to develop guidelines for research. Overall, the Berg committee's deliberations were characterized by considerable uncertainty concerning the potential hazards stemming from recombinant DNA work. Published in 1974 as a letter in the *Proceedings of the National Academy of Sciences* and later in *Science* and in *Nature,* the final version of the committee's conclusions received wide publicity in the United States and abroad (Wright 1994, pp. 137–140).

The drafting of the Berg letter (as it later was called) was keenly watched by the international community of molecular biologists, and its publication was hardly a surprise for the close-knit scientific communities in Germany, France, and Britain. Because it demanded regulatory action, the Berg letter soon became an important element in the creation of the European politics of genetic engineering.

In Britain, scientists had already begun discussing the direction of regulatory policies before the Berg letter was published. One participant of the discussions remembers:

> The [Medical Research Council] knew they had to do something, and there were some discussions that in fact maybe the best thing would be for the [Medical Research Council] to set up some organization, because I had said . . . the best thing is to try and set down some kind of guidelines immediately—let people start, but under certain conditions, and to have it flexible.[5]

In reaction to the Berg letter, the Advisory Board to the Research Councils in Britain established, shortly after its publication in *Nature,* a working party chaired by Lord Ashby "to assess the potential benefits and potential hazards of techniques which allow the experimental manipulation of the genetic composition of microorganisms."[6]

In France, the letter was commented upon by *Le Monde* soon after its publication.[7] Translated into French by a young representative of a small group of French scientists engaged in recombinant DNA work, Phillipe Kourilsky, it was published in the journal *Biochimie* (Berg et al. 1974). In November 1974 a group of scientists wrote to the director of the Centre Nationale de la Recherche Scientifique to establish a system of controls for recombinant DNA work.[8] It was a research proposal by Phillip Kourilsky that had provoked François Jacob, in charge of the evaluation of this proposal at the Centre National de la Recherche Scientifique, to suggest regulatory action by the CNRS. Jacob remembers:

> The application had something to do with some work on bacteria and eventually work on putting the myosin gene on lambda. And I said it was completely all right and I had nothing against it, but I think it was better on principle that there were some laboratories able to handle it and other laboratories not able to handle it, and that a committee had to be established. So I refused to consider the application and we wrote a letter to the director of CNRS and to the DGRST saying that this was coming. That there was a moratorium which probably would be released within six months, or a year and that some attention ought to be given to this type of work, and that therefore some committee had to be established to deal with that.[9]

In a similar vein, discussions began within the German scientific-political community after the Gordon Conference on nucleic acids on the questions of hazards and on the potential actions that might have to be taken.[10] A scientist recalls the situation:

> The main discussions did . . . happen within the group of people who did related work or who were interested in doing such work. That was within the Deutsche Forschungsgemeinschaft which had a *Schwerpunkt* [specially sponsored program] in molecular biology. Then there were already at that time discussions with the people of the Ministry of Research and Technology, who were in charge mainly of financing applied research . . . those two agencies, the DFG and the Ministry expressed very much interest. But they wanted to await what will be discussed in Asilomar before defining the German position.[11]

A clear picture emerges from these events. First, the very scientists who were at the forefront of molecular biology research took the initiative in defining the meaning of the social, political, environmental, and epidemiological implications of genetic engineering technology. Initially some researchers had wanted a broader debate on the socio-economic, political, and safety aspects of genetic engineering (Wright 1994, p. 194). But a number of leading scientists in the field had a more limited conception of the problem. Mostly concerned about a debate that might interrupt further research in this field, they felt a need for some sort of action to preempt that possibility. At the same time, they defined the need for action as a need for safety regulations. Considerations of the political and ethical aspects of recombinant DNA work moved into the background, and a much simpler agenda of problems gained prominence: Which experiments should be conducted, and under what precautions? Which experiments should not be done at this time? How should a system of recombinant DNA regulation look? But the scientists who led the debate did not simply say that they wanted to continue with their research. Such a strategy would have failed to position the scientists in the center of regulatory action, where they would be able to define the genetic engineering problematic. The scientists' coding of genetic engineering's risk was linked to its contextualization as a scientific breakthrough with potentially enormous positive social and economic impacts. In other words, risk control was inscribed into the broader political imaginary—the socially available archive of the big stories that explain and justify a society's central goals and its identity. This hegemonic attempt to frame the genetic engineering problem culminated at the second Asilomar conference, in 1975.

The Narrative of Asilomar and the Politics of the NIH Guidelines[12]

Writing the Asilomar Narrative

The International Conference on Recombinant DNA Molecule Research, which convened at the Asilomar Conference Center on February 24–27, 1975, was a well-organized event. Broad and worldwide attendance was supported financially by the National Academy of Sciences, the National Cancer Institute of the National Institutes of Health, and the National Science Foundation. These funds not only paid for the administrative and travel costs, but also for the travel and living expenses of many of the American and foreign participants. Where this was not the case, the various national science foundations or the European Molecular Biology Organization covered the costs of attendance. It is important to note that many of the participants at Asilomar were later recruited by the various national recombinant DNA regulatory commissions.

Asilomar was not planned as a forum for philosophical-ethical reflection on genetic engineering. In a summary of the conference (Berg et al. 1975, p. 1981), the organizers outlined its purpose:

> The meeting was organized to review scientific progress in research on recombinant DNA molecules and to discuss appropriate ways to deal with the potential biohazards of this work. Impressive scientific achievements have already been made in this field and these techniques have a remarkable potential for furthering our understanding of fundamental biochemical processes in pro- and eukaryotic cells. The use of recombinant DNA methodology promises to revolutionize the practice of molecular biology. Although there has yet been no practical application of the new techniques, there is every reason to believe that they will have significant practical utility in the future.

This first paragraph summarizing the rationale of the meeting is followed by a statement on biohazards:

> Of particular concern to the participants at the meeting was the issue of whether the pause in certain aspects of research in this area. . . . should end; and, if so, how the scientific work could be undertaken with minimal risks to workers in laboratories, to the public at large, and to the animal and plant species sharing our ecosystem.

The third and last paragraph of the opening section offers a solution:

> The new techniques, which permit combination of genetic information from very different organisms, place us in an area of biology with many unknowns. Even in the present, more limited conduct of research in this field, the evaluation of potential biohazards has proved to be extremely difficult. It is this ignorance that has compelled us to conclude that it would be wise to exercise considerable caution in performing this research. Nevertheless, the participants at the Conference agreed that most of the work on construction of recombinant DNA molecules should proceed provided that appropriate safeguards, principally biological and physical barriers adequate to contain the newly created organisms, are employed. Moreover, the standards of protection should be greater at the beginning and modified as improvements in the methodology occur and assessments of the risks change. Furthermore, it was agreed that there are certain experiments in which the potential risks are of such a serious nature that they ought not to be done with presently available facilities. In the long term, serious problems may arise in the large scale application of this methodology in industry, medicine, and agriculture. But it was also recognized that future research and experience may show that many of the potential biohazards are less serious and/or less probable than we now suspect.

The text constructs a circular argument: the existence of risks involved in the practice of genetic engineering cannot be denied, but risk can be controlled by *containing* the newly created organisms. This containment of the potentially hazardous organisms can be accomplished by "erecting" *biological* and *physical* barriers to keep the newly created organisms from interacting with humans, animals, and the general ecosystem. "Biological barriers" is understood to mean the use of genetically engineered bacterial hosts that are unable to survive in natural environments and of nontransmissible vectors that are able to grow only in specific hosts. "Physical containment" refers to the creation of an environment that limits work with genetically modified organisms to controlled spaces. Such an environment could be established, for example, by the use of suitable hoods, negative-pressure laboratories, and other "good microbiological practices" (Berg et al. 1975, p. 1982).

The containment idea exemplifies the circularity of the risk philosophy of Asilomar. In the case of both biological and physical containment, the elimination of hazards, the non-occurrence of hazards, or the control of risk is attempted through technology; however, in the first case the very same technology that is used for genetic engineering is used to control the risks of genetic engineering. Not only are the very tools of gene technology (such as vectors and plasmids) of the nature and dimension of the processes with which they interfere; they are, at the same time, encoded as technologies that can control the very process of bio-construction. In this reading, the laboratory practice of genetic engineering becomes a self-monitoring system whose operations are defined by the discourse of molecular biology. In a further step, the scientific narrative of Asilomar states the principle that risk must be dealt with in a "reasonable" way. This requires, first, the application of techniques of containment and, second, the matching of effectiveness of the containment with the estimated risk (a difficult task that depended on acquiring new and additional knowledge; see Berg et al. 1975, p. 1982). In other words, the creation of a risk, its initial assessment, and its control depended on a knowledge of technologically constructed organisms whose composition required more technological intervention. The containment of risk thus necessitated the taking of more risks inherent in further intervention. In addition, controlling the hazards involved in genetic engineering required the pursuit of more research in molecular biology: risk could be fought only with risky technologies.

Furthermore, the containment policy called for a "reasonable" relationship between the degree of risk and the containment measures—a relationship that "balanced" risk and safety. Apparently, deviation from the Asilomar consensus—that is, divergence from the outlined hazard level of the containment strategy—was defined implicitly as "unreasonable." To be sure, we do not find any discussion in the Asilomar statement of what constitutes an "unreasonable" risk philosophy; however, by introducing the term "reasonable" to the discussion, a concept of risk to be avoided or excluded was sketched—namely, the "other" risk paradigm, which was in conflict in one way or another with the one outlined at Asilomar. Indeed, through this strategy, the technological definition of risk came to be equated with reason—a characterization that then fed into the circular mode of defining risk.

In addition, the Asilomar narrative offered an operationalization of genetic engineering's risk. In order to apply the given definition of risk to situations of risk, a system of types of risks, experiments, and containment was developed. There were four types of risk (minimal, low, moderate, and high) and four types of experiments (those involving prokaryotes, bacteriophages, and bacterial plasmids; those involving animal viruses; those concerning eukaryotes; and those that were to be deferred). Then the summary statement established a relationship among these types of risks, experiments, and containment.

Finally, the Asilomar statement went on to adopt, at least transitionally, a quasi-regulatory status: "In many countries steps are already being taken by national bodies to formulate codes of practice for the conduct of experiments with known or potential biohazards. Until these are established, we urge individual scientists to use the proposals in this document as a guide." (Berg et al. 1975, p. 1983)

It is important to see, for our reconstruction of the terrain for the government of molecules, that risk began to emerge as a topic for policymaking in a text assembled by a variety of geographically dispersed authors, most of them scientists, who for a historical moment had come together at the Asilomar Conference Center. They produced—in the name of scientific truth—a definition of the "nature of genetic engineering," in particular, a definition of the nature of the hazards connected with recombinant DNA technologies. Soon after the Asilomar gathering, state authorities led the effort to institute regulatory measures. This was a critical step in the politics of genetic engineering. Experts may exercise government, but they also are objects of government. The molecular biologists who had come together at

Asilomar could have had all sorts of wishes and ideas as to how they would like to proceed with the regulation or nonregulation of genetic engineering. But such wishes and ideas must be articulated within the established discursive economy of politics. The Asilomar narrative constituted a crucial rhetorical resource in the construction of the evolving European regulations. But equally important for the establishment of European laboratory codes was the translation or incorporation of the inscriptions of recombinant DNA's risks deployed at the Asilomar conference into a policy narrative, a regulatory statement. As I will show, in this process the scientists' reading of the genetic engineering problematic was connected to the state's project of protecting its citizens from harm, but also to other core elements of the given national political metanarrative, such as the importance of modernization for society. This process of creating a field for the deployment of expertise within the pre-established administrative-regulatory apparatus and the operating discursive economy happened simultaneously in Britain and the United States and later in France and Germany. Since the developments in the United States were most important for the adoption of a regulatory strategy in Germany and France, let us now look at the creation of a field for expertise on genetic engineering's risks within the US National Institutes of Health.

The Politics of the NIH Guidelines

On June 23, 1976, the National Institutes of Health, part of the US Department of Health, Education, and Welfare, issued its "Recombinant DNA Research Guidelines."[13] With this step, the United States became the first state to establish a regulatory framework for recombinant DNA research. In the following years, many other countries, including Britain, Germany, and France, issued guidelines strongly influenced by—and in many respects even identical with—the NIH guidelines.

The NIH guidelines started out by constructing a particular history and placing itself within that framework: "These guidelines replace the recommendations contained in the 1975 Summary Statement of the Asilomar conference."[14] With this rhetorical move, the NIH guidelines attributed quasi-legal authority to the Asilomar conference, while at the same time positioning themselves as the legacy of Asilomar.

But, in fact, the preparation of these new guidelines had already started before Asilomar: in October 1974, when the NIH Recombinant DNA Molecular Program Advisory Board (also known as the Recombinant DNA Advisory Committee, or RAC) was established to recommend guidelines for regulating the new genetic technologies. When the RAC held its first meeting after the Asilomar conference, it proposed that the NIH follow the Asilomar guidelines for research until "more specific guidelines" were available.[15] Eventually, the NIH guidelines formulated stricter conditions for regulating research.[16]

Again, the central challenge to regulation stemmed from the fact that the practice of genetic engineering involved so many unknowns. In the introductory statement to the NIH guidelines, NIH director Donald Fredickson developed a vivid picture of the debates surrounding the drafting of the guidelines, clearly indicating a division of opinion concerning the nature and the level of the risks involved. Eventually, he came to a puzzling conclusion:

> The experiments now permitted under the guidelines involve no known additional hazard to the workers or the environment beyond the relatively low risk known to be associated with the source material. The additional hazards are speculative and therefore not quantifiable. In a real sense they are considerably less certain than are the benefits now clearly derivable from the projected research.[17]

This statement expressed a genuine sense of uncertainty related to the question of hazards. However, setting this uncertainty in relation to the issue of potential benefits, Fredrickson went on to point out that the potential benefits, because they were more certain, outweighed the risks. In fact, the benefit issue emerged as a critical issue in the crafting of the regulations. The NIH guidelines argued that certain experiments should not be done, since they are considered to be too hazardous. However, the guidelines allowed,

> the remainder (of the experiments not considered to be too hazardous) can be undertaken at the present time provided that the experiment is justifiable on the basis that new knowledge or benefits to humanity will accrue cannot readily be obtained by the use of conventional methodology.[18]

This statement reveals other important elements of the regulatory philosophy: experiments can be undertaken if they create a "benefit for mankind" or help to acquire "new knowledge."

Along with "progress in science" and "benefit for mankind," three closely related principles were outlined in the guidelines: that appropriate safeguards should be incorporated in the design and execution of the experiments, as measures of biological and physical containment; that the level of containment should "match" the estimated potential hazard; and that the guidelines should be subject to at least annual periodic review, perhaps to be modified "to reflect improvements in our knowledge of the potential biohazards and of the available safeguards."[19] These principles were already contained in the Asilomar narrative. However, in contrast to the rather short Asilomar "rules of good conduct," the NIH guidelines offer a detailed "implementation" of these principles, a detailed translation into regulatory instruments. I will now further specify the "risk philosophy" underlying the NIH guidelines.

The guidelines distinguished between experiments that could be done and those that should not be conducted. The question concerning which experiments should not be conducted at the current stage of development was, of course, broadly debated at Asilomar and in the wider scientific community. The guidelines developed a list of experiments that should be prohibited:

- cloning of recombinant DNAs derived from the pathogenic organisms in Classes 3–5 of the "Classification of Etiologic Agents on the Basis of Hazard" issued by the Center of Disease Control, USPHS; or cloning of oncogenic viruses classified by the National Institute of Cancer as moderate risk, or cells known to be infected with such agents
- formation of recombinant DNAs containing genes for the biosynthesis of potent toxins
- deliberate creation from plant pathogens of recombinant DNAs that are likely to increase virulence and host range
- deliberate release into the environment of any organisms containing a recombinant DNA molecule
- transfer of a drug resistance trait to microorganisms that are not known to acquire it naturally, if such acquisition could compromise the use of a drug to control disease agents in human or veterinary medicine or agriculture.

In addition, the guidelines put restrictions on large-scale experiments (e. g., more than 10 liters of a culture) with recombinant DNAs known to produce harmful products. These restrictions, however, could be lifted in exceptional cases provided the experiments yielded direct social benefits and were approved by the RAC. For the time being, the guidelines had excluded from laboratory practices three broad types of experiments: experiments with certain "dangerous" organisms and certain categories of DNA, large-scale recombinant DNA work beyond the typical laboratory environment, and releases of genetically modified organisms into the environment. At this stage of development and knowledge, recombinant DNA work would be restricted within the walls of laboratories. Hence, the NIH guidelines focused on contained, non-industrial research work with genetically modified organisms.

At the core of the guidelines was a matrix that established a relationship between types of experiments and their matching measures of containment. The central purpose of these rules was to avoid an unwanted interaction between laboratory in vitro recombinant DNA and nonlaboratory in vivo DNA. The guidelines attempted to accomplish this goal with a strategy "based on both facts and assumptions," as the introductory statement of the director of the NIH states.[20] Again, it should be emphasized that the drafting of the guidelines was severely complicated by the fact that they addressed a new technology that was not well understood, either in itself or in its interaction with its environment.

What were the "facts" and the "assumptions" of the NIH guidelines? At the core of the "risk philosophy" was the following concern:

> In every instance of an artificial recombination, consideration must be given to the possibility that foreign DNA may be translated into protein (expressed), and also to the possibility that normally repressed genes of the host may be expressed and thus change . . . the characteristic of the cell.[21]

This inscription of the problem was based on the central assumption that DNA determines protein structure and on the additional assumption that protein structure can be artificially changed through a process of artificially recombining nucleic acids.

But the guidelines made a number of further assumptions. One of them was that the more similar the DNAs of the donor and the host, the greater the probability of expressing foreign DNA or de-repressing host genes. Implicit in this assumption was the idea that DNA experiments were more dangerous the closer the donor organism was to man. In a further differentiation of the risk model, a hierarchy was established on the basis of the principle of phylogenetic ordering, which defined experiments with eukaryotic recombinants as more hazardous than experiments with prokaryotic recombinants. Furthermore, this hierarchy created a classification of the experimental uses of embryonic tissue, which was considered to be less likely to be contaminated by pathogenic viruses and thus less hazardous than DNA isolated from normal tissue. In an additional distinction, experiments involving well-defined segments of DNA were considered less hazardous than so-called shotgun experiments, in which entire chromosomes were broken up with restriction enzymes and blindly "shot" into host organisms.

Focusing on the control of unwanted interaction between the laboratory and the outside world, the guidelines outlined two types of "containment": biological containment and physical containment. Part of the biological containment strategy was the selection of the *E. coli* K12 host as "the system of choice at this time," mainly because it was regarded as the best-known system.[22] Important in this context was *E. coli* K12's characteristic of including a nonconjugative plasmid system (meaning that this plasmid system cannot promote its own transfers, but requires the presence of a conjugative plasmid for mobilization and transfer to other bacteria—a constellation assumed to limit the spontaneous exchange of genes once artificial genetic recombination had been enacted) (Krimsky 1982, p. 186). Also important in the selection of *E. coli* K12 was the assumption that it was an enfeebled host organism not likely to survive outside the laboratory.

Three levels of biological containment were prescribed for the use of *E. coli* K12 host-vector systems. The first level of containment (EK1) was assigned to the standard laboratory strain of *E. coli* K12, which was considered much weaker than wild strains of *E. coli*. Important within the context of EK1 experiments was that some experiments involved prokaryotes that were defined as "like in nature"—an argument we will encounter again at later stages of the regulatory discussion. The second level of biological containment (EK2) was assigned to host-vector systems with strains of *E. coli* that were viable only under artificial conditions. EK3 containment implied the use of EK2 strains that had failed to survive in actual animal and human tests. However, when the regulations were drafted no EK3 system had been developed, and the regulations stipulated that any EK3 system was subject to prior certification by the RAC.

Physical containment denoted application of a set of standard practices generally used in microbiology laboratories, adherence to the so-called good microbiological practices, and use of special procedures, equipment, and laboratory installations to create physical barriers.[23] Accordingly, the guidelines distinguished the following levels:

P1 (minimal)—laboratories without any special engineering design features in which standard microbiological practices are used to control biohazards

P2 (low)—laboratories similar in construction and design to P1 labs, but with additional features, such as access to an autoclave[24] within the building

P3 (moderate)—laboratories with special engineering design features, such as separation from areas open to the general public, and with physical containment equipment

P4 (high)—laboratories located in separate buildings or in completely isolated areas within a building and with separate ventilation systems maintaining negative air pressure.

In the central classification section of the NIH guidelines, types of experiments (centered around specific combinations of DNA donors, vectors for implanting DNA, and host organisms of recombinant DNA molecules, which raised key risk-evaluation factors such as the closeness of donor DNA to humans) were related to types of physical containment. This relation of experiments to containment resulted in the elaboration of a hazard-evaluation scheme that established an equivalence between biological and physical containment.[25]

To be sure, these key elements of the NIH risk philosophy and their coding of the "risk realities" of genetic engineering were far from reflecting any consensus within the biomedical community. The various black boxes arranged throughout the argument of the NIH guidelines were constantly threatening to unravel. For example, there was a vigorous debate over the feasibility of physical containment. Critics argued that physical containment was inherently unsafe and could be easily undermined by carelessness or lack of training. An equally heated debate concerned the possibility of disarming *E. coli.* Further concerns were raised about how little was known about the ecology of *E. coli* and its plasmids and phages, particularly with respect the organism's ability to establish itself in the human intestinal tract. And questions were raised about whether, in view of the huge diversity of available organisms from which DNA might be derived, phylogenetic ordering could really be claimed to represent an accurate generalization about the hazards of cloning DNA (Wright 1994, pp. 160–180).

But what was at issue here was not only truth, but also power. The risks of genetic engineering were not "out there" to be discovered by truth seekers; they came into existence in an attempt to create a hegemonic interpretation of a phenomenon. The NIH guidelines attempted to bring closure to the genetic engineering debate by stabilizing the meaning of the hazards of recombinant DNA work through the creation of a new policy narrative. This policy narrative served to organize a chain of signification in which the risks of genetic engineering as seen by the discourse of molecular biology were linked to laboratory practices of genetic research, to the social and economic potentials of genetic engineering, and to state practices of protecting citizens from the potentially adverse impacts of genetic engineering. How was this stabilization of the meaning of the hazards of recombinant DNA work accomplished? First, by making two key assumptions: that physical containment was possible and that biological containment was possible. If either of these two assumptions turned out to be invalid, the risk philosophy would not credible. The director of the NIH had argued the guidelines were based on a mixture of facts and assumptions.

Essentially, the "facts" of the risk philosophy came from molecular biology, the "assumptions" from certain conclusions based on "facts" from other fields and on what was termed "experience." The "hard data" of the risk philosophy can be summed up as being informed by models based on central elements of thought in molecular biology, such as the "central dogma" as expressed in the models of DNA transcription and RNA translation.[26] The "phylogenetic closeness principle" is an example of this strategy of mixing "data" from molecular biology ("tested" models) with speculation ("untested" models). Of course, nobody could know what would happen in the case of a shotgun experiment, how well physical containment procedures would work, or what type of interaction would develop once a genetically modified organism had escaped the walls of the laboratory. These disputed issues revolved around the lack of epidemiological and ecological data and information; hence, speculation continued.

Since the risks of genetic engineering remained fuzzy, another linguistic device—which displaced the uncertainty surrounding risks—gained importance in the construction of the risk narrative: the cost-benefit metaphor. The cost-benefit metaphor equated costs with hazards and benefits with progress in recombinant DNA research. In this construction, recombinant DNA work also needed to be understood and justified in terms of its benefits to mankind. With this move, the evaluation of the beneficial aspects of genetic engineering became an integral element in the determination of the acceptability of risks connected with experimental practice. In practice, the implementation of the NIH guidelines was the responsibility of the principal investigator of a laboratory and the NIH. The NIH's initial review groups were required to evaluate the scientific merits of each grant application that involved recombinant DNA molecules, to assess the biohazards involved, to determine whether the proposed biological and physical safeguards were appropriate, and to refer those assessment cases it could not resolve to the NIH's Recombinant DNA Molecular Program Advisory Committee (RAC). With the exception of one political scientist, the membership of the RAC (chaired by the NIH's Deputy Director for Science) was drawn from the biomedical community. The RAC's tasks were to review and update the guidelines, to receive information on purported EK2 and EK3 systems, to evaluate and certify that host-vector systems met EK2 and EK3 criteria, to interact with the NIH's Initial Review Groups on unclear cases, and to review and approve large-scale experiments with recombinant DNA known to produce harmful effects (Wright 1994, pp. 163–190). By and large, the regulation of recombinant DNA work (involving the evaluation of its scientific merits and its potential benefits) was defined in such a way that it constituted a system of regulation of scientists by scientists.

This system, however, was connected to the state apparatus through the NIH, which had the task of regulating the scientific work it supported. Here we see the emergence of what I have called a multiple regime of governability. The NIH, or the state, was hardly the powerful institution in control of the everyday practices of recombinant DNA regulation. Rather, a flexible network of power had been shaped in which actors in the scientific community were linked to actors in the NIH. However, within this framework experts were given a central position in the definition of the risks of genetic engineering. A central force of the stabilization of this network was the narrative of the NIH guidelines (published on July 7, 1976), which linked a system of nodal points with the intention of fixing the meaning of "the nature of recombinant DNA hazards." This attempt to unify the political space of recombinant DNA regulation involved the telling of a rather complicated story that provided a reordering of events, actors, artifacts, and representations (nodal points) such as representations of nature, "assumptions," "facts," metaphors, the sites of the occurrence of recombinant DNA's hazards, the strategies to deal with the hazard problem, and the actors to be involved in the regulation process.

The Recombinant DNA Debate Continues: Creating Expert Enclosures

The above reconstruction of the early period of the recombinant DNA controversy in the United States tells much about the relationship between the emergence of risk and the deployment of expertise in the organization of the field of molecular biology policy. After the first successful genetic engineering experiments, differing interpretations of the risks of recombinant DNA technology proliferated, and the boundaries between experts and non-experts were relatively permeable. With the Asilomar conference, a regime of expertise came into existence—a relatively bounded locale of judgment where specialists gained exclusive authority in the determination of the meaning of genetic engineering's risks. In other words, discourse had played a critical role in constituting scientists as central actors in the regulation of genetic engineering. This development and the establishment of a regulatory system for recombinant DNA work did not lead to closure of the recombinant DNA debate. Rather, the genetic engineering controversy in the United States was about to change from a mode of negotiating to one of imposing truth claims. As I will show in the following sections and chapters, the constitution of the boundaries of genetic engineering's risks and the creation of an expert enclosure in the regime of governability of molecular biology were at the root of the genetic engineering controversy in the next few years. The attempt to stabilize the meaning of "genetic engineering's risk" in political discourse came at the price of damaging the communication between "the world of science" and other social worlds, such as "the public" and "the mass media." Links between these worlds can never be taken for granted; they are the results of careful negotiation, explanation, and translation of interests and objectives. Imposing truth is hardly a substitute for reconciling the proliferation of meanings and intermediating between conflicting policy narratives and discursive contexts.

In 1977–78 more than a dozen bills intended to legislate regulation of recombinant DNA work were submitted to Congress (Krimsky 1982, p. 198). These bills conceptualized the adopted regulations as insufficient and advocated stricter measures. However, at the same time, a number of experiments, scientific texts, and scientific meetings were mobilized to promulgate a new understanding of recombinant DNA's hazards and to push for deregulation on the understanding that the NIH guidelines went too far.

Three scientific meetings held in the period 1976–1978—one at Bethesda, Maryland, one at Falmouth, Massachusetts, and one at Ascot, England—dealt with the question of the hazards of recombinant DNA research.[27] In particular, these meetings dealt with *E. coli* K12 and with the results of an experiment, conducted by Stanley Cohen, that resulted in a shift in the definition of biohazards.

After the release of the NIH guidelines, the scientific community continued its assessment of the biohazard issue. Again, the "nature" of the risks of genetic engineering was about to change. Since *E. coli* had developed as "the organism of choice"—as the host to be used for recombinant DNA work—the biohazard deliberation centered around the question of the safety of its use for this purpose. There was considerable concern because of *E. coli*'s intimate association with all warm-blooded animals, including humans. Although the presence of *E. coli* in the intestines of warm-blooded animals is considered normal, it was known that the bacterium can cause disease if it invades sites outside its normal niche in the intestine. By the time of the Asilomar conference, some had already asked whether the addition of new genetic information into *E. coli* could turn the organism into a pathogen, whether that presented a threat to human health, whether that threat was confined to laboratory workers, and whether there was the possibility of an epidemic (Krimsky 1982, pp. 206–209). Hence, *E. coli* risk research became critical in the politics of determining recombinant DNA's risks.

The Bethesda meeting, in August 1976, began to narrow down the discussion of biohazards to *E. coli*. In June 1977, the Falmouth workshop addressed risk assessment of recombinant DNA experimentation with *E. coli*. Despite the lack of agreement among the participants, the NIH interpreted the outcomes of this meeting as confirming that *E. coli* K12 could not be converted into an epidemic pathogen by laboratory manipulations with DNA inserts and that it did not implant in the human intestinal tract.[28] The third important scientific meeting leading to a "reconsideration" of the risks involved in recombinant DNA work was the Ascot meeting, a major reason for which was the increasing discontent among virologists with the restrictions on the cloning of animal virus DNA in *E. coli*. Such research had been classified as high-risk activity in the NIH guidelines, and there was a move in the biomedical community to reconsider the controls that had been imposed. Again, despite considerable disagreement among the participants during the workshop on risks, the final statement of the workshop (published in the *Federal Register* in March 1978) gave a message of assurance: the cloning of viral DNA would "pose no more risk than work with the infectious virus or its nucleic acid and in most, if not all cases, clearly present less risk."

Besides the three conferences, other developments became part of the emerging new coding of the risks involved in recombinant DNA work. Important in this context was Stanley Cohen's and Shing Chang's claims that fragments of mouse DNA were naturally taken up by *E. coli* K12 bacteria and joined to pieces of a plasmid inside the bacteria. The key argument here was that recombinant DNA molecules constructed in vitro using the Eco RI enzyme simply represent selected instances of a process that occurs naturally (Wright 1994, p. 246).

This evolving "consensus" in the biomedical community, which reflected a stabilization of the dominance of Asilomar approach to assessing the risk of genetic engineering, prepared the ground for a revision of the initial NIH guidelines in December 1978.[29] The three conferences, the "scientific consensus," the years of accident-free experimentation, and the argument that other countries had adopted less stringent regulations became important elements in an emerging narrative that reconstituted the nature of the risks of genetic engineering.[30]

The revised guidelines offered a new reading of genetic engineering's risks and relaxed the guidelines in important ways. The regulatory model based on the assumption that special containment of risks was necessary because of unknown hazards involved in recombinant DNA work was increasingly displaced by an approach arguing that such activities contained no special hazards and hence did not require special containment. Another important meeting in the reframing of the debate on genetic engineering's risk took place in April 1979 at Wye College in Kent, England. With 150 scientists from 30 countries in attendance, this meeting on "Recombinant DNA" was co-sponsored by the Committee on Genetic Experimentation[31] and the Royal Society. The meeting's underlying intent was to dismiss the earlier concerns about genetic engineering by pointing out that the initial concerns were vastly overstated, a shift in position now supported by the European Molecular Biology Organization (Morgan and Whelan 1979).

However, with the next revision of the NIH guidelines, in January 1980, a decisive step in the relaxation of regulatory control of recombinant DNA work was taken. Again, broad evidence from scientific research and consensus among many researchers in the field was cited to support the claim that the measures already taken were adequate. Under the new regulations, no registration with the NIH was necessary. The only requirement left was that the investigators submit to the local Institutional Biosafety Committee a registration document with the information on the experiment to be conducted before initiating the experiment. The local IBC was to review such proposals, although such a review preceding initiation of an experiment was not required. With this regulatory move, some 80 percent of all genetic engineering activities taking place at the time could be carried out under minimal P1/EK1 conditions and were removed from NIH scrutiny.[32] However, the 1980s revisions retained other elements of the original NIH guidelines, including the restriction to approved host-vector systems, the prohibition of certain experiments, and the requirement of specific approval for others.

But in the following months the scientific community made a series of requests for new hosts, new vectors, previously prohibited sources of DNA, and previously prohibited large-scale procedures. Furthermore, the initial six prohibitions began to be weakened. In June 1980, the prohibition banning pathogens as sources of genetic material was reduced to cover only the two most lethal classes. Cloning of genes from pathogens deemed less harmful (including those causing anthrax, plague, tularemia, and yellow fever) was allowed, and the third prohibition (concerning plant pathogens) was deleted. From September 1979 on, the RAC reviewed requests for and granted exemptions to the sixth prohibition: that on the use of large volumes of genetically manipulated organisms (Wright 1994, pp. 392–395). By 1982, barriers for deliberate release of genetically modified organisms into the environment had been removed and the prohibition against intentional release of recombinant DNA had been replaced by a multi-tier review process. Deliberate releases required approval from the RAC, the IBC, the director of the NIH, and various subcommittees. By the mid 1980s there were no longer any prohibited experiments, although some experiments required approval by both an IBC and the RAC.

In 1981 the discursive shift in the framing of genetic engineering's risks was such that some very influential scientists were in a position to demand that the guidelines be converted into a recommended code of practice that would eliminate penalties for violations, institutional biosafety committees, and registration and would seek to reduce the containment requirements for many experiments and processes. Although this proposal was not adopted by the NIH, the guidelines were relaxed further. The IBC requirement and the obligation of NIH-funded institutions to comply with the guidelines were retained. Subsequently, the guidelines were revised several more times in response to very specific requests by individual researchers, companies, or institutions.[33]

To sum it up: At the center of the narrative construction of genetic engineering's risk was the assumption that the hypothetical danger to humans becomes greater the more closely the donor of the nucleic acid is phylogenetically related to man. The adopted regulatory measures essentially combined physical containment (safety measures in the laboratory) with biological containment (screened bacterial strains as carriers for the recombinant nucleic acid molecules). Thus, the representational strategies used to deal with the "risk question" were inseparably linked with the codings of life phenomena in the discourse of molecular biology.

Why was this narrative developed at Asilomar and in the NIH guidelines so important? Because it told a story about genetic engineering that contributed in an important way to defining the terrain for the governance of genetic engineering's risk. Policymaking is an attempt to manage a field of discursivity, to organize and fix a field of differences, to circumscribe the overflow and disparity of meanings. This is typically done through the creation of narratives. By defining problems, actors, and strategies, narratives create order and stability; thus, they make governing and hegemonic intervention possible. In the Asilomar narrative, very different actors or factors—such as vectors, human immune systems, the environment, and the public—were inscribed in such a way that they became numbers, categories, or beneficiaries of the very genetic engineering these actors or factors assessed. By means of this inscription, something essential had been accomplished: genetic engineering, first characterized by an overflow of meanings, was soon rendered more stable and workable and thereby transformed into an object for debate and diagnosis (Rose and Miller 1992, p. 185). The construction of the risks related to genetic engineering as "measurable" and "controllable" operated as a crucial technology in stabilizing the discursive field of molecular biology.

The dominance of this risk definition was not simply the result of policy strategies pursued by actors such as scientists or administrators. No doubt these scientists had certain goals and interests, such as to work with a minimum of inhibiting regulations and controls. But these interests were in need of justifications and explanations. The justifications and explanations came, to a considerable extent, from the discourses and practices of molecular biology. The discourse of molecular biology had established a logic of circularity by defining life in terms of macromolecules, which were, in turn, represented by technologies restricted to this realm. Now the discourse of molecular biology constructed the risks of genetic recombination along the very same conceptual-technological categories. As unexpected and unwanted interactions between laboratory work in vitro and human process in vivo, hazards came to be represented, as an epistemic thing, through an experimental system using as technologies things such as viruses and plasmids, thereby creating a representation in which hazards were controlled by the technology that produced their possibility.

The definition of risk as the assessment of a particular probability of a hazard was thus inextricable interwoven with its enabling technology. What had been established was crucial: a technological definition of risk. This technological framing of the hazard issue established host-vector systems and physical laboratory containment equipment as the obligatory point of passage for hazard control[34]: knowledge in this area was considered critical for safety. Hence, more knowledge about host-vector systems was necessary. This technological definition made any competent molecular biologist a spokesperson for recombinant DNA's safety.[35] In fact, what had been established was an expert enclosure. Such enclosures are relatively bounded locales or types of judgment, or elements of a regime of governability, within which the power and authority of a particular group of experts is concentrated. The arguments and assessments of these experts became obligatory modes for the operation of the policy network as a whole (Rose and Miller 1992, p. 188).

As molecular biology was inscribed into the regulatory philosophy, its continued support and application in research became the main precondition for eliminating risk. Consequently, the expulsion of hazard from the equation was by definition a matter of the "progress of science," meaning mainly the advancement of molecular biology as practiced and interpreted by the experts in the field. Thus, reduction of the risks of genetic engineering became inseparable from genetic intervention and manipulation on the molecular level. Subsequently, a number of studies, experiments, statements, and meetings were construed as confirmations of an ongoing "progress of science," which in turn allowed for a substantial relaxation of the initial guidelines. The discourse of molecular biology contained risk by progressively narrowing the definition of hazard. In this respect, controlling risk became a matter of controlling the play of signifiers, of promoting a specific representation of risk. Thus, the regulation of genetic engineering was in fact the introduction to a story about research support and deregulation. The experts on recombinant DNA research had successfully translated their own interests into those of the National Institutes of Health. On the other hand, the NIH—a government agency that, among other things, had the task of protecting the public from the potential dangers of genetic engineering—mobilized expertise in order to demonstrate its determination to protect the public from adverse impacts of science. This topography of the political space of biotechnology, which discursively positioned the experts on molecular biology in the center of regulatory actions, was to become the model for future regulatory policymaking in Europe during the second half of the 1970s.

Making European Genetic Engineering Governable

Governing requires the transformation of events and phenomena into information, data, and evidence. This is so because only such transformations make possible the core activities of governance: simplification, transference, displacement, comparison, concealment, management. As I showed in the last section, in the case of genetic engineering it was through inscription devices such as experimental systems, categories of risks, and statistical calculations of hazards that thousands of experiments in hundreds of dispersed laboratories could easily be moved to central administering institutions. There, authoritative decisions were made, permissions for research granted or withheld, and projects monitored. These activities, which constituted the field of molecular biology, would not have been possible without the various tropes, black boxes, agents, and epistemic objects that were inscribed and interwoven into the regulatory narrative's chain of signification.

Thus, narratives were an important device for the organization and structuring of the newly emerging sites of governing molecules. However, narratives are only temporarily stable; they are intertextual phenomena, always object to multiple interpretations. Interpretation takes place in specific discursive contexts. Accordingly, the Asilomar/NIH narrative was not simply adopted in Europe. Like all texts, it became the object of a process of reading and rewriting as soon as it was published. These modifications of regulatory narrative in Europe did much to determine what kinds of actors and institutions would be mobilized for regulatory purposes.

In the following sections, I will analyze how and with which impact the Asilomar narrative was rewritten in the various national debates about the risks involved in recombinant DNA work. Furthermore, I will show how these stories of risk were inscribed into the various national discursive economies. Since any successful stabilization of a policy narrative depends for its expression on the terms of the dominant political metanarrative, we also have to look at the hegemonic political discourses, the prevailing concepts of social order and legitimacy, the dominant state models, and the current practices of democracy in Britain, France, and Germany in the early 1970s to understand the construction of the different national regulatory strategies. These metanarratives served as "contextualizing tools" for the management of the discursive field of regulatory policymaking, which played a crucial role in the emerging regimes of governability in these countries.

European Initiatives in Recombinant DNA Regulation

In the European regulatory discourse, one important question concerned at what spatial-territorial level genetic engineering was to be regulated. By the early 1970s, the boundaries of the territory for which regulatory guidelines were to be developed were far from clear. The creation of specifically national recombinant DNA safety regulations in Europe was preceded by a brief period in which an attempt was made to articulate the risk presented by genetic engineering as something to be dealt with on the international or at least the European level. The idea of creating a supranational or an international framework for the regulation of genetic engineering was, to some extent, related to the heavy foreign and European participation in the Asilomar meeting. Indeed, one of the most significant post-Asilomar events was the "translation" of the conference's participants into "recombinant DNA safety experts," a status gained by mere participation in the conference. Many of the scientists who had been involved in the shaping and the practice of the politics of molecular biology during the 1960s became "national experts" on the regulation of recombinant DNA upon their return from Asilomar. Some of these experts advocated the "internationalization" of the framing of genetic engineering's risks developed at Asilomar via something like internationally shared regulatory guidelines. As a result of this perception, EMBO began to explore the possibility of "international" regulatory guidelines."

But EMBO also attempted to use the ongoing regulatory debate as part of its attempt to mobilize a European infrastructure for molecular biology. Strategies to control the risks of genetic engineering came to be closely related to more general strategies of structuring the political space of genetic engineering. A group of scientists active within EMBO took the initiative. One EMBO representative recalls a discussion on the safety of recombinant DNA work with tumor viruses that took place at the Cold Spring Harbor Laboratory in June of 1971:

> At the Cold Spring Harbor meeting I immediately realized that there would be no stopping this nonsense. And I knew there was going to be a problem coming up, a severe problem coming up. . . . So I proposed that—since we couldn't fight them—we decided to take the appropriate steps to insure that at least we could do some work when things did come to a head. And we wrote a letter, a couple of Europeans . . . wrote a letter to EMBO and told them what has happened, in terms of the technology. We indicated that there was going to be a call for regulations and we said that we needed to do this work in Europe and if we wanted to do that we would have to have a lab with a special set-up.[36]

In EMBO's understanding, regulating recombinant DNA was translated as the need for EMBO to build a European research facility. The argument for such a facility would eventually lead to plans for a European Molecular Biology Laboratory. The EMBO delegation reported upon returning from the Asilomar meeting:

> We believe that the coincidence of the development of recombinant DNA technology with the establishment of the EMBL provides a unique opportunity for the EMBL to promote European collaboration in this extremely important area of research. While facilities for conducting safely experiments judged to entail only low or medium hazards are likely to be made nationally available in Europe it is much less likely that facilities will be available in many countries for experiments of high scientific importance but also high potential hazard. We believe, therefore, that a containment laboratory meeting the most stringent containment specifications outlines in the report of the Asilomar conference, should be built within the EMBL.[37]

The recombinant DNA guidelines were successively loosened over the next few years. As the need for high-hazard facilities diminished, it became easier to construct such facilities on the national level. However, the European Molecular Biology Laboratory was eventually built (in Heidelberg), and risk arguments had been used to explain why the building of such a laboratory was important.

EMBO's strategy unfolded early. Even before the final version of the NIH guidelines became available, EMBO set up a Standing Advisory Committee on Recombinant DNA; it first met on February 14 and 15, 1976, in London. EMBO defined the purpose of this committee as "to advise, upon request, governments, organizations and individuals on the scientific and technical aspects of recombinant DNA research and to determine if the development of European recombinant DNA guidelines were desirable." The Standing Advisory Committee endorsed the philosophy behind the Ashby report, the provisional report of the Asilomar conference, and the La Jolla version of the NIH guidelines (December 4 and 5, 1975)—a philosophy that advocated combining support of ongoing work with recombinant DNA and the establishment of physical and biological containment measures. The committee came to the conclusion that, in face of the evolving, comprehensive NIH guidelines, the drafting of new European guidelines would not yield further or additional regulatory gains, and that, hence, the NIH guidelines should become the basis for an international code. EMBO made a strong point in establishing one common regulatory framework:

> We believe it is essential that those European national committees that are responsible for the control of recombinant DNA research should come into close contact, for instance through this EMBO Committee, through the EMBC and through the ESF to ensure that in Europe the conditions under which recombinant DNA research is allowed to proceed are as uniform as possible.[38]

To EMBO it did not matter which regulations were adopted, as long as they were not too strict and as long as they did not inhibit scientific cooperation:

> Any attempt to establish risks associated with recombinant DNA research is, today, based upon prejudice and conjectures rather than knowledge. Guesswork and intuition rather than objectivity have to be the order of the day, and, therefore, it seemed to many Europeans pointless to duplicate the task of assessing risks when it was pursued so energetically in the USA. . . . Moreover, the consensus represented by the draft NIH guidelines seemed to be a workable compromise which had the virtue of internal consistency. As the EMBO committee quickly discovered, this consistency is lost if attempts are made to change particular risk assessments that seem less reasonable than others. These considerations, as well as the recognition of the current hegemony of the USA in this field, made the acceptance in Europe of the NIH draft guidelines a matter of little trauma.[39]

EMBO's approach was to articulate the regulation of genetic engineering as an issue to be dealt with above the national level. Essentially, it recommended the national adoption of the NIH guidelines in Europe and hoped that an international common regulatory framework would eventually be established.

Initiatives to establish a specifically European framework of recombinant DNA regulation had little success, mainly because of the delayed reaction of the European Community to the problem of genetic engineering's risks. Although during the second half of the 1970s the idea that the European Community should draft a directive addressing the establishment of safety measures for recombinant DNA work was discussed, such discussions had little impact on legislation. Owing in particular to the continuous relaxation of the NIH guidelines, member states increasingly came to the conclusion that there was no need for EC legislation. The only tangible European piece of regulation produced during this period was the European Council of Ministers' 1982 recommendation concerning the registration of work involving recombinant DNA, which was, essentially, that a competent national or regional authority be notified before or (in cases of low-risk experiments) after recombinant DNA laboratory work was done.[40] Owing to the weak legal status of an EC Council recommendation and to the rather general nature of its request, the effect of the recommendation on national systems of regulation was rather small. With the gradual withdrawal of EMBO from the shaping of a European legal framework for genetic engineering, and with the establishment of a strategy to promote the NIH guidelines as "the guidelines of the world," the national level became critical for the deployment of political strategies for regulating recombinant DNA in Europe, as elsewhere.

Britain: Negotiating the Boundaries of Risk

In Britain the construction of a political space of recombinant DNA regulation took shape during the period 1974–1979, when the Labour Party was in power. It is against the background of the political discourse of this period that we have to understand the specification of Britain's negotiation-oriented regulatory strategy for genetic engineering. British modernization politics continued to focus on a strategy to reverse the decline of the British economy. But this core element of the British state discourse found particular expression during Labour's tenure.

The postwar economic expansion of Western capitalism was interrupted by the recession of 1966–67. The deterioration of the competitive position of British capital from 1967 on manifested itself in a continuous erosion of profitability and in a steady rise in unemployment (Overbeek 1990, p. 147). After the demise of the Conservative government of Edward Heath and a period of strikes and labor unrest, Labour focused on restoring of social peace by making generous concessions to the unions (ibid., pp. 166–169). Despite a severe recession that began in 1973, the Trade Union Congress (Britain's main labor union) continued to have a significant influence on the governing Labour party, and this influence resulted in a "new social contract" and a period of growing prosperity, a rising standard of living, and social reform (McKay 1993, pp. 20–21; Overbeek 1990, p. 167). Thus, in this period, the modernization discourse meant growth and industrialization, but with a "human face": workers' participation and benefits became an essential part of the arrangement. Several elements of this discourse were to become important contextualizing devices in the regulation of recombinant DNA research in Britain.

This process of contextualization becomes clear as we look at the framing of the recombinant DNA hazard issue and the related construction of a policy narrative. What exactly was the problem that recombinant DNA regulations should address? In Britain we see a translation of the recombinant DNA question into the problem of workplace health and safety. There was nothing obvious about this translation, as we will see when we look at Germany and France (or the United States), where issues of workplace health and safety played a much less important role in the elaboration of a hazard problem. Even in Britain, the initial developments in outlining why and how genetic engineering should be regulated pointed in a different direction.

For instance, the activities of the Godber Working Party on the Laboratory Use of Dangerous Pathogens indicated the selection of a different regulatory pathway. The Working Party (named after its chairman, Sir George Godber, a former chief medical officer for the Department of Health and Social Security) was formed in reaction to an escape of smallpox virus from the London School of Hygiene and Tropical Medicine in 1973. This event, which revealed underlying failures in the control of hazardous laboratory work, had stirred a broad public discussion in Britain. In late 1974, the Godber Working Party obtained evidence on the implications of recombinant DNA work from Hans Kornberg, a leading British molecular biologist. Published in 1975, the Godber Report recommended that a Dangerous Pathogens Advisory Group be established to control research in this area, and that this group be constituted to meet the need for advice on implementing control measures required in genetic engineering research (Bennet et al. 1986, p. 19). Thus, the Godber Working Party attempted to launch the British politics of recombinant DNA regulation. Critical in its argument was the establishment of a conceptual link between smallpox viruses and recombinant DNA work. This chain of reasoning was underwritten by the idea that any risks linked to genetic engineering would be analogous to known risks from other areas, such as smallpox laboratory work (ibid., p. 7). Furthermore, the Godber Working Party argued that existing legislation was not adequate to deal with genetic engineering and demanded that specific legislation be enacted.

Of course, the Godber Working Party was never given a mandate to start the British politics of regulating recombinant DNA. Nevertheless, it started the process of defining the genetic engineering problem. The Party had reasoned that the Health and Safety at Work Act (HASAWA)—then under discussion—would not provide a sufficiently quick and efficient system to deal with hazards stemming from genetic engineering. But its suggestion had no impact on the evolving system of regulations. Instead, regulation of recombinant DNA work was integrated into the HASAWA framework. With this move, regulating genetic engineering was defined as part of Labour's larger project of "human" modernization. This problem definition would implicitly lead to the identification of the key actors of the evolving system of regulation, and to a strategy of institutionalizing their interactions and practices. How can this development of the British regulatory policy narrative be explained?

We first need to develop a better understanding of the Health and Safety at Work Act in British political discourse. The HASAWA (which took effect in 1974) represented a consolidation of various existing pieces of health and safety legislation. It required employers to take all reasonable steps to protect the health and safety of their employers and the public from workplace hazards. The Secretary of State for Employment, in collaboration with the Health and Safety at Work Executive, could approve codes of practice for various occupations, industries, or processes. Such codes would not in themselves be legally binding, but could be used to bring before the courts employers who had failed to take reasonable safety and health measures (Bennet et al. 1986, p. 390).

The HASAWA dated back to the work of the Committee on Safety and Health at Work under the chairmanship of Lord Robens. The 1960s had seen an increase in the number of workplace accidents thought to be related to the emergence of new toxic substances and a heightened risk of catastrophic accidents resulting from the growing scale of industry, the storage of vast amounts of fuels, and so on. These developments provoked an investigation into the effectiveness of the existing legal provisions. In 1972, in the so-called Robens Report, the Committee on Safety and Health at Work reported that it had found two deficiencies in the existing system. First, the system tended to encourage people to think that health and safety was engendered by detailed regulation administered by external agencies. Second, the statutory system was felt to be unsatisfactory, because it depended on a myriad of statutory provisions while neglecting attitudes, capacities, and performance. Furthermore, the existing system was plagued by an excessively fragmented structure involving multiple ministries (Drake and Wright 1983). In response, the committee recommended reform measures aimed at developing a more efficient statutory system and creating a framework for a more self-regulating system.[41]

The following recommendations of the Robens Report created the basis for the HASAWA:

• The existing statutory provisions be replaced by a comprehensive set of revised provisions under a new act containing a clear statement of the basic principles of safety responsibility—principles supported by regulation and by nonstatutory codes of practice. The HASAWA then provided a tier of statutes specifying regulations and "guidance notes." The regulations may (a) prescribe minimum standards, (b) require the Health and Safety at Work Executive (HSE) to be notified of circumstances, activities, etc., and (c) make provisions for licensing or for the granting of prior approval of processes or activities. Unless the regulations provided otherwise, the HASAWA made admissible as evidence in civil proceedings any breach of regulations that caused any damage or harm (Drake and Wright 1983, p. 112). Furthermore, guidance notes could be issued by the Health and Safety Commission, by the HSE, by industry advisory committees, and by industry itself (ibid., pp. 118–119).

• The new legislation should address not only employers, but also employees and the public.

• There should be a new, unified administration. Through the HASAWA, a new National Authority for Safety and Health at Work with comprehensive responsibility for the promotion of safety and health at work was created. This authority, the Health and Safety Commission, was to represent the main interests concerned—employers, trade unions and local authorities. The Health and Safety Executive would serve as the main enforcement authority.

• Standards should not be enforced by extensive use of legal sanctions but through advice and persuasion. Legal enforcement should be reserved for serious infringements (ibid., pp. 35–36).

The HASAWA was seen to be different from both a "collective bargaining" position on labor law and a legislative approach. Historically, attitudes toward health and safety at work had moved from the stoical acceptance of bad working conditions, to collective bargaining whose intent was to improve these conditions, and finally to statutory intervention. However, the Robens Report (ibid.) highlighted the limits of the legislative approach: "There is no legitimate scope for 'bargaining' on safety and health issues, but much scope for constructive discussion, joint inspection and participation in working out solutions." Hence, the proposed system for safety representatives and safety committees was seen as a form of "consultation" rather than "negotiation" (ibid.). And of crucial importance was the act's institutionalization of the role of the labor unions in the process of health and safety regulation. One-third of the members of the HSC were trade unionists.

Framing the Recombinant DNA Debate

The discourse on workplace health and safety created an elaborate institutional system to administer the new regulations and gave labor a significant voice in its implementation. As we look at the evolving discourse on British recombinant DNA regulations, we can see that both trade unionists and civil servants from the Health and Safety Executive (which administers HASAWA) were important actors in the attempt to translate the genetic engineering problematic into a case for HASAWA regulations. Their narratives translated the concerns and interests of scientists and technical workers engaged in recombinant DNA laboratory work into interests and concerns of actors (such as unionists) concerned about health and safety at work in general. Unionists and HSE civil servants identified themselves as spokesmen for recombinant DNA's safety. In doing so, they could make references to the "humanized" modernization discourse; thus, they could use a strong semiotic resource in defining key elements of their agenda of recombinant DNA's safety within the prevailing political discourse.

Genetic engineering officially became an object of consideration in policymaking when the Advisory Board for the Research Councils set up a working party under the chairmanship of Lord Ashby in July of 1974. As its report stated, the Ashby Working Group inquired into the need for state intervention in the field of genetic engineering:

> The problems we have been asked to assess have already received publicity in the press and on radio and television. They are causing interest and some concern beyond the boundaries of the scientific community. Indeed they are an example of a wider question, namely: how can the social values of the community at large be incorporated into decisions on science-policy?[42]

The Ashby Working Group consisted mainly of representatives from the life sciences establishment, but it followed the decision of the Advisory Board for the Research Councils that the membership of the working group should "as far as possible" not include those who were using the genetic engineering techniques; it collected witnesses among representatives from most laboratories in Britain where the new techniques were being used. The central task of the Ashby Working Group was to assess the potential benefits and potential hazards of the new recombinant DNA techniques and to weigh the benefits against the hazards. The working group concluded, straightforwardly, that the new techniques would be greatly beneficial for science, medicine, agriculture, and industry. According to its report, physical containment procedures together with the "disarming" of the organisms used for experiments (biological containment) would enable scientists to control the hazards sufficiently. Hence, the working group concluded, the techniques should continued to be used. Obviously, this story was quite similar to the US approach, which developed simultaneously.

The next important step in problematizing genetic engineering can be found in the work of the Working Party on the Practice of Genetic Manipulation, reported in 1976.[43] The report of this working party contained recommendations for the establishment of the Genetic Manipulation Advisory Group (GMAG), a system of local safety committees and safety officers, a detailed code of practice for genetic manipulation experiments, and a scheme for their categorization.

Thus, the "Williams Report" represented the core of a distinct regulatory strategy that would be sharpened and developed in the following years. The regulatory situation thus consisted of two models developed at a relatively early stage of the international efforts for regulation, the UK and the US model, which at the same time "competed" for definitional "leadership" and which mutually reinforced one another in the general approach of framing the problematic of recombinant DNA's hazard.

On August 6, 1975, the Secretary of State for Education and Science formally announced the Williams Working Party in the House of Commons, defining its tasks as "(a) to draft a central code of practice and to make recommendations for the establishment of a central advisory service for laboratories using the techniques available for such genetic manipulations, and for the provision of necessary training facilities [and] (b) to consider the practical aspects of applying in appropriate cases the controls advocated by the Working Party on the Laboratory Use of Dangerous Pathogens."[44] The Working Party on the Practice of Genetic Manipulation was under the chairmanship of Robert Williams, director of the Public Health Laboratory Service in London. With the exception of Ronald Owen (deputy director of Medical Services, Health, and Safety Executive), the ten members of the Working Party came from the senior ranks of the biomedical establishment, covering such research areas as molecular biology, animal virus research, bacteriology, microbiology, and genetics. In addition, the Working Party included four representatives from "interested departments" (Health and Social Security, the Scottish Home and Health Department, Education and Science, Agriculture, and Fisheries and Food) and the secretaries of Education and Science. Four members had previously been members of the Godber Working Party, and many other members had previous experience in committees and conferences concerned with recombinant DNA.[45]

The Williams Working Party made a number of policy recommendations which led to the construction of Britain's recombinant DNA regulatory system. It rejected the idea of new legislation raised by the Godber Working group, demanded the creation of the Genetic Manipulation Advisory Group, and suggested that the government consider introducing statutory powers in addition to existing powers (such as those specified by the Health and Safety at Work Act).[46] The recommendations of the Williams Working Party were similar in many respects to those of the NIH guidelines; however, there were a number of significant differences:

• The standards of physical containment recommended by the Williams Working Party were generally more stringent than those recommended by the NIH. While the NIH guidelines emphasized biological containment more, the NIH's physical containment measures for the lowest-risk categories were substantially less stringent than those proposed by the Williams Rcport.

• The Williams Working Group, though recognizing the importance of biological containment, refrained from precise formulations in its report.

• In contrast with the NIH guidelines, the Williams Working Group did not draw a distinction between containment procedures for identical experiments with RNA from embryonic and adult animals.

• Whereas the NIH guidelines provided a detailed categorization of the experiments, the Williams Report focused on the control of each individual experiment by the Genetic Manipulation Advisory Group.

• According to the NIH guidelines, the principal responsibility for the implementation of the guidelines lay with the principal investigator, who was directly answerable to NIH study sections committees composed exclusively of scientific peers. Only in cases of doubt or dispute would the Recombinant DNA Molecular Program Advisory Committee (operating publicly but with only one nonscientist member) be consulted. In Britain, by contrast, a biological safety officer, working with a safety committee, was to make an initial assessment of every experiment and then refer the corresponding protocol to the Genetic Manipulation Advisory Group. The GMAG, composed of representatives of all laboratory employees and of the general public in addition to scientific peers, was to decide on the appropriate measures. Excepted from this process were experiments that were judged locally to fall in the two lowest categories; such experiments could be started before the GMAG's comments were required.

• The Williams Report required that all laboratory staff involved in experiments in all but the lowest-risk category be placed under the surveillance of a medical officer.

• In Britain, all in vitro recombinant DNA research in any laboratory was to be referred to the GMAG; in the United States, compliance with the guidelines was mandatory only for research funded by the NIH or the NSF—a distinction that, de facto, exempted industry from state control.

Probably the most important difference between the American and British regulatory systems consisted in the construction of the political rationale for decisionmaking on recombinant DNA's hazards. In the United States, scientific experts had constructed an expert enclosure which determined probabilities of recombinant DNA's risk based on the discursive practices of their disciplines. In Britain, the situation was more open: the representatives of molecular biology and related fields were not the only recognized spokespersons for recombinant DNA safety issues. In particular, labor unionists and "representatives of the public" were in a position to offer their readings of risk probabilities within the framework of regulatory institutions such as the GMAG.

The Genetic Manipulation Advisory Group represented a different political rationality than the Recombinant DNA Molecular Program Advisory Committee. Political rationalities address the proper distribution of tasks and actions among authorities of different types and consider the ideals and principles to which government should be directed (Rose and Miller 1992, pp. 178–179). According to the political rationality articulated in the Williams Report, membership in bodies regulating recombinant DNA work should not be limited to scientists; it should also be open to individuals able to take account of the interests of employees and the general public. The GMAG was defined as "a voluntary system of advice and control." However, located within the broader legal framework of HASAWA, the GMAG had the power to formulate additional regulations requiring the submission of experimental protocols and supporting information. The Williams Report concludes: "Given such a requirement, it seems very unlikely that the advice of the GMAG would be disregarded." But the Williams Report, pointing out that the HASWA did not cover hazards to plants and animals, demanded consolidating legislation in the form of statutory control, which would require compulsory consultation with the GMAG or a system of licensing for laboratories. In general the appropriate process of conducting recombinant DNA research under the new regulation was envisioned as follows:

• Within the laboratory, an experiment's scientific merits and potential hazards should be discussed with the laboratory' s Biological Safety Officer and with a local safety committee, both of whom should assume the key role in determining the desirability of a particular experiment and the deployment of appropriate containment procedures.

- If the experiments belong to category I or category II, the GMAG should be notified immediately, but work could continue.
- If the proposed work seem to fall in categories II and IV, the GMAG should be referred to for advice before the final decision to undertake the experiment was made.
- The GMAG then could either advise the laboratory to proceed as proposed, to adopt specific precautions, or to refrain from undertaking the experiment.[47]

Despite the fact that the GMAG in the mid 1970s did not represent an expert enclosure, its risk narrative reiterated a number of principles and assessments that were inscribed into the NIH guidelines. The Williams Working Party argued that, at the time, biohazards were difficult to assess but they were, provided the appropriate measures were adopted, "controllable":

> The underlying reason for the present review . . . was the concern of the scientific community that some experiments involving the techniques of genetic manipulation might lead, perhaps inadvertently or in an unpre-

Table 3.1
The British system of categorization according to the Williams Report (Report of Working Party on the Practice of Genetic Manipulation, August 1976).

Source of nucleic acid	Specification of nucleic acid	Host-vector specification	Category
Mammals	Random	Phage or plasmid/ bacteria, not disabled	IV
	Random	Phage or plasmid/ bacteria, disabled	III
	Purified	Phage or plasmid/ bacteria, not disabled	III
	Purified	Phage or plamid/ bacteria, disabled	II
Amphibians and reptiles	Random	Phage or plasmid/ bacteria, not disabled	III
	Random	Phage or plasmid/ bacteria, disabled	II
	Purified	Phage or plasmid/ bacteria, not disabled	II
	Purified	Phage or plasmid/ bacteria, not disabled	I

Table 3.1 (continued)

Source of nucleic acid	Specification of nucleic acid	Host-vector specification	Category
Plants and invertebrates and lower eukaryotes	Random	Phage or plasmid/ bacteria, not disabled	II
	Random	Phage or plasmid/ bacteria, disabled	I
	Purified	Phage or plasmid/ bacteria, not disabled	I
Mammals, amphibians and reptiles, birds	Random	Virus capable of infecting humans or growing in tissue culture cells	IV
	Purified	Virus capable of infecting humans or growing in tissue culture cells	III
Viruses pathogenic to vertrebrates	Random	Phage or plasmid/ bacteria, disabled	IV
	Purified	Phage or plasmid/ bacterial, disabled	III
Animal viruses nonpathogenic to humans	Random	Phage or plasmid/ bacteria, disabled	II
Bacteria specifying toxins virulent to humans	Random	Phage or plasmid/ bacteria, disabled	IV
Plant pathogenic bacteria	Random	Phage or plasmid/ bacteria, not disabled	II
Plant viruses	Random	Phage or plasmd/ bacteria, not disabled	II
Bacteria or fungi nonpathogenic to humans, animals, or plants	Random	Phage or plasmid/ bacteria, not disabled	I

dictable manner, to the release of harmful products into man, animals or plants. It should be stressed at the outset that most of the hazards that may be involved are conjectural. At present there is no experimental evidence that some of the most serious hazards that can be envisaged . . . are real; but equally there is no proof that they are not. Until further knowledge is gained about the use of the novel genetic techniques under discussion, it seems to us essential that rigorous precautions, based on the best estimate of possible hazard, should be observed by all laboratory workers using these techniques.[48]

The creation of the GMAG was justified in terms of flexibility: the Advisory Group would obviate guidelines (such as those in the United States) that were "too specific":

We see a need for a flexible approach. . . . We define certain cases in which work should not proceed until the GMAG has given approval. . . . As work proceeds, the experience gained should quite quickly build up into a body of "case law" on which further experimental protocols could draw. (ibid.)

In its categorization of experiments, the Williams Report constituted an intertext[49] with the NIH guidelines and closely followed their mode of reasoning. The potential hazards involved in recombinant DNA work were framed to be dependent on four factors:

1) the source of the nucleic acid from which the fragment to be linked is derived; 2) the degree of specification, or purity, of the nucleic acid; 3) the vector/host system involved; and 4) the manipulative procedures proposed.[50]

Here, again, we see at work the phylogenetic principle, which states that the closer an organism that supplies the nucleic acid is to the organism at risk, the greater the chances of hazard. In general, physical containment received more attention than biological containment, although both principles were considered to be essential for safe recombinant DNA work. Overall, the Williams Report tended to rely considerably on the GMAG to devise a clearer regulatory strategy for the future.

However, despite these interdiscursivities between the British and the American recombinant DNA regulation narratives, the British policy narrative had mobilized a different system of regulation. Since regulating recombinant DNA had been successfully translated into an issue of health and safety in the workplace, principles and norms of worker protection were incorporated in the regulatory narrative. Hence, the GMAG was composed of eight scientific and medical experts, four members representing the public interest, four members nominated by the Trade Union Congress, and two members representing the interest of management (one nominated by the Confederation of British Industry and one by the Committee of Vice-Chancellors and Principals).[51] The GMAG was formally constituted by the Secretary of State and Education as an independent body to assess risks and precautions related to genetic manipulation and to maintain appropriate contacts with the relevant government departments, which included the Health and Safety Executive and the Dangerous Pathogens Advisory Group. In addition there were assessors from pertinent ministries and a secretariat seconded by the Medical Research Council. Finally, in 1978 the Health and Safety at Work Executive published regulations that extended the meaning of work in part I of the Health and Safety at Work Act of 1974 to include any activity involving genetic manipulation. The regulations stated that "persons should not carry on genetic manipulation unless they had previously notified the Health and Safety Executive and the Genetic Manipulation Advisory Group."[52]

The Genetic Manipulation Advisory Group was formally set up as an independent body, with advising scientists and the Health and Safety at Work Executive having responsibility in matters relating to genetic engineering. The Department of Education and Science (DES) was responsible for providing the secretariat (through the Medical Research Council) and for appointing the members of the GMAG. This construction of the GMAG reflected a definitional compromise among employers, employees, and scientists: whereas employers and employees tended to favor the HSE's handling of issues related to genetic manipulation (an arrangement within an organizational structure they "were familiar with"), the scientists argued for "more flexibility" and for a solution that did not involve the HSE (Bennet et al. 1986, pp. 68, 71). In other words, the scientists favored the creation of political rationality for recombinant DNA regulation based on principles of an expert enclosure in which the determination of the probabilities of recombinant-DNA-related risks was confined to the discourse of the specialists in the field. Rather than being reconciled, these competing rationalities were inscribed into the emerging institutional structure, which became a site for these opposed rationalities to clash.

The Everyday Life of Regulation

For its first three years, the GMAG met at monthly intervals to discuss and decide on policy issues and to categorize proposed experiments. It was first set up for two years, to be reviewed after that period. The negotiations within the GMAG tended to be rather contentious. One union representative recalled:

> Getting GMAG to get together to begin with was difficult, because the various groups were not used to taking to each other . . . and we were dealing with a lot of theoretical situations. We did not have any evidence. So some scientist would say: I don't think there is going to be any risk in this. And I would say: What evidence do you have to support that? And he would say: Well, I don't think there is going to be any risk, because I have experience in science and I know what is what. And I would say: That is not good enough. I want some solid evidence. Which of course did not exist, and I knew it didn't. There were a lot of clashes like this one in GMAG we were dealing with.[53]

In other words, the abundance of black boxes in the established system of regulations made it likely that conflicts would arise as soon as any of the actors attempted to pose questions that might lead to the opening of any of these black boxes. And the way the British regulatory system was constructed, such questions could legitimately be asked.

The unions' attitude in the GMAG was certainly not to oppose genetic engineering in principle. The Association of Scientific, Technical, and Managerial Staffs (ASTMS), which played a central role in representing the unions' position, was mainly concerned with the safety of laboratories, especially since the smallpox outbreak in London in 1973. This association's positions can be summarized as follows (Bennet et al. 1986, p. 34):

- Genetic manipulation was a progressive technology that should be supported and advanced.
- There should be enforceable regulations under HASAWA.
- There should be a central advisory service maintaining a compulsory register of all genetic manipulations.
- This central advisory service should have strong labor union representation.
- Before submission to the advisory service, any experiment should have to be approved by the trade union representatives in the laboratory.

This political rationality hardly represented a rejection of recombinant DNA work per se; however, it implied an approach to regulation that was opposed to the idea of self-regulation by experts. In the construction of the GMAG, the boundary object of the "gene" as defined by molecular biology proved to be flexible enough to rally scientists, labor unionists, and representatives of the public in a way that they all could "get behind" the boundary object and work toward the common goal of combining facilitation and regulation of the expansion of molecular biology. Experts and non-experts were on the same committee, but they did different things. The committee structure of the GMAG was flexible enough to admit scientific modes of reasoning next more general arguments such as ethical arguments. At the same time, the very fact that the GMAG's process of risk evaluation was informed mainly by the knowledge of molecular biology sustained a boundary between the experts and the non-experts. Nevertheless, this boundary was permeable to an extent, and it could allow negotiation.

Up to February of 1978, the GMAG gave advice on 102 proposals from 27 centers. The GMAG focused on the key recommendations of the Williams Report. Soon the GMAG set up subgroups to examine particular issues in greater detail. Some of these subcommittees were concerned with the validation of safe vectors for genetic manipulation experiments and with commercial confidentiality (Bennet et al. 1986, p. 84).

In 1978, after extensive consultation, the Health and Safety Executive issued regulations and guidance notes for genetic manipulation.[54] "Genetic manipulation" was defined as "the formation of new combinations of heritable material by the insertion of nucleic acid molecules, produced by whatever means outside the cell, into any virus, bacterial plasmid, or other vector system to allow their incorporation into a host organism in which they do not naturally occur but in which they are capable of continued propagation" (ibid., p. 2). This definition covered the replication of recombinant DNA, but not other uses of genetically manipulated DNA or RNA. (The NIH guidelines, by contrast, had covered all research on recombinant DNA, encompassing the construction of recombinant DNA molecules, their replication in organisms such as *E. coli* K12, and their subsequent use; see Wright 1994, pp. 9–15.) To bring the regulatory system under the Health at Work Act, the guidelines redefined the HASAWA's notion of "work":

> For the purpose of Part I of the Health and Safety at Work Act 1974 the meaning of the word "work" shall be extended to include any activity involving genetic manipulation and the meaning of "at work" shall be extended accordingly.

The key provision of the regulations stated:

> No person shall carry on any activity involving genetic manipulation unless he has given to the Health and Safety Executive and to the Genetic Manipulation Advisory Group notice, in a form approved by the Executive.[55]

But soon the British approach to the regulation of genetic engineering came under pressure, particularly in the wake of the scientific meetings at Bethesda (1976), Falmouth (1977) and Ascot (1978) and the corresponding attempts to reframe the question of risks related to genetic engineering (see above). These meetings, naturally, constituted important semiotic support for a dismantling of the established regulatory framework. However, there was an important difference between the British and the American situations. In the United States, the recombinant DNA regulations were issued by the NIH and could be changed relatively easily. In Britain, the organization of recombinant DNA regulation within the HASAWA framework had led to a certain institutional entrenchment that would not be easily dismantled.

Undoing the Recombinant DNA Regulations

However, first changes came when the Williams Working Party's risk-assessment scheme was replaced by what was seen as a more flexible scheme. This scheme, initially proposed by Sydney Brenner, attempted to classify risks in terms of how they might develop; hence, it departed from the scheme of classifying hazards on a phylogenetic scale. The new risk-assessment scheme was based on a rough quantitative procedure for calculating the risk of any particular experiment. With the new scheme a quantitative expression of the relative risks of various experiments could be obtained. Typically, in the first step of the risk-assessment procedure a quantitative assessment of a particular experiment was made. In the second step, the containment category was identified. This identification was based on judgments of appropriateness and acceptability. By definition a political decision rather than a scientific process, it required the operation of the structures of the GMAG (Bennet et al. 1986, p. 106).

In 1979 a discussion was begun within the GMAG whose purpose was to develop a scheme with a new category of containment that would be equivalent to "good microbiological practices" (GMP). In due course it was decided that all experiments requiring GMP, all experiments in category I, and certain experiments in category II that did not involve the expression of gene products could begin as soon as the scientists had notified the GMAG. Previously it was necessary to await the GMAG's advice before commencing an experiment. Scheduled to take effect in 1980 (Bennet et al. 1986, p. 108), these changes were already reflected in the Second Report of the Genetic Manipulation Advisory Group (1979):

> During the past year GMAG, as a result of these discussions, has concluded that many experiments pose no hazard or only minimal hazards. . . . Certain experiments, however . . . still pose a hazard. The Group feels that, as a result of its activities, there is now greater awareness and acceptance of the need to observe safe physical and biological containment. Nevertheless, some uncertainties still remain and GMAG sees a need for a continuation of notification and for its authorization for experiments to proceed in the case of those likely to require the higher containment categories.[56]

After the new risk-assessment scheme was introduced, a decision to further relax the guidelines was made. Most experiments falling into category I would no longer have to be reviewed by the GMAG on a case-by-case base; only the local safety committee had to consider them and to submit an annual list of this work to the GMAG and the HSE.[57]

In the late 1970s, scientific discourse recoded the British recombinant DNA regulatory structures as "exaggerated." And there was also an important change in political discourse. With the departure of Labour as the ruling party and the entrance of the Conservatives under Margaret Thatcher in 1979, the future of the GMAG and the whole British system of recombinant DNA regulation came into question. The Thatcher administration embarked on a wide-ranging set of policies which were designed to create an "entrepreneurial society" and which marked a clear break with the bipartisan labor-capital model of the postwar settlement. The Thatcher government's discourse centered on deregulation as a strategy for giving economic agents greater freedom from state control—a strategy that radically questioned the necessity of organizations such as the GMAG.

But the institutionally entrenched GMAG survived for some years, although with a significantly reduced workload in the wake of the gradual relaxation of notification requirements. Large-scale work came under regulatory consideration in 1979, when a guidance note was published stipulating case-by-case review of large-scale work and site visits. Under pressure from industry, this system was refined in 1981,[58] when reconsideration of the initial proposal led to a policy of evaluating large-scale proposals in the same way as small-scale work (Wright 1994, pp. 33–37). Under the new provisions published in

1982, the GMAG became responsible for the biological aspects of proposals, while the HSE was put in charge of inspecting industrial plants.[59] The third and last GMAG report had a completely different tone than the first two:

> Since the Second Report was issued in 1979 . . . it has become apparent that the hazards specifically attributable to genetic manipulation of micro-organisms are, if they exist at all, far less than were conjectured when the Group was set up and can be contained by appropriate biological and physical containment."[60]

The recategorizing of experiments and the weakening of controls on industrial work led to a drastic reduction of the GMAG's workload. After June 1981, only three experiments were classified at levels higher than Category II. After December 1981, there were no further assignments to Categories III and IV and only two to Category II. Between 1981 and 1984, GMAG met only eight times (Wright 1994, pp. 11–44). During 1982 the change in attitude became further evident in the shift of oversight from the GMAG to the HSE. After the dissolution of the GMAG in 1984, a new body, the Advisory Council for Genetic Manipulation (ACGM), was set up. It consisted of five representatives of management, five representatives of the employees, and eight scientists. The new committee's responsibilities were widely regarded as consisting of any regulatory tasks other than the routine oversight of genetic engineering work. It seemed unlikely that the HSE, lacking government support, uncertain of its future, and under pressure from industry, would become a strict regulatory institution (ibid.).

What had happened was an interesting demonstration of how the objects, strategies, actors, and institutions of policymaking come into being and change over time. The position of the GMAG as an initially powerful institution in recombinant DNA regulation was the result of a complicated process of enrolling and mobilizing persons (e.g., scientists, industrialists), procedures (e.g., HASWA), ideologies (e.g., "modernization with a human face"), artifacts (e.g., certain host-vector systems), and representations (e.g., of hazards) by means of a regulatory narrative that emphasized the importance of tight regulations in the face of many unknowns with respect to recombinant DNA work. The GMAG constituted a stabilization of this discourse. The British system of recombinant DNA regulation never gave scientific experts the kind of definitional authority they had in the American system.

However, just as in the United States, a technological definition of risk was part of the established governmental rationality.

At the same time, expertise in the field of genetic engineering's risks was tied to broader social values and desires and "made up" trade union representatives and other "representatives of the public" as actors in the field of genetic engineering regulation. Thus, evaluating genetic engineering experiments and mobilizing expert knowledge for that purpose became parts of labor's project of "modernization with a human face." Expert knowledge in the field of molecular biology was not "imposed from above," but helped to define labor's modernization project. Regulatory policymaking was never simply a reaction to a (risk) problem; rather, it was an attempt to define, organize, and structure a problem and the actors and institutions who were supposed to deal with it. Also, in Britain the deployment of arguments of deregulation around the trope of the "progress of science" (meaning mainly the advancement of molecular biology as practiced and interpreted by the experts in the field) contributed to a shift in the site from which genetic engineering experiments were evaluated. This narrative increasingly undermined the British regulatory system from within, while leaving the HASAWA structures intact. "Representatives of the public" and of the labor unions continued to participate in regulatory decisionmaking, but there was less and less to participate in as—in the course of the "progress of science"—more and more of the initial hazard assessments "evaporated." The rewriting of the boundary object "gene" in the discourse of molecular biology shifted the boundary between experts and non-experts talking about the risks of genetic engineering from many to fewer and fewer cases. More and more experiments were conceptualized as not warranting the attention of the GMAG. Thus, neither these now-excluded experiments nor the boundary between experts and non-experts could become an object of negotiation. Increasingly, society turned into a black box in this redefined network of meaning of biotechnology policy and was constructed as the docile and safe recipient of the progress of biological research. As the initially constructed network of power disintegrated, the powers of the GMAG (and, later, of the ACGM) became increasingly symbolic. However, despite this process of deregulation, a well-defined institutional structure operating under the HASAWA remained in place and was officially in charge of recombinant DNA regulation in Britain. The continued presence of this governmental apparatus would turn out to be important as the recombinant DNA debate regained momentum in the mid 1980s.

Germany: In the Shadow of the NIH Guidelines

In Germany, the Bundesministerium für Forschung und Technologie (BMFT) and, attached to it, the Zentrale Kommission für Biologische Sicherheit (ZKBS) were designated as the regulatory agencies. These agencies were composed mostly of scientists who showed little inclination to negotiate the risks of genetic engineering with non-experts. Thus, the German organization of recombinant DNA regulation resembled the American model much more than the British one. How can we explain that the BMFT—one of the main funding sources for recombinant DNA research in Germany—came to assume such a critical role in recombinant DNA regulation? What was the impact of this arrangement of actors and institutions for the interpretation of genetic engineering's risk?

Differences in the discursive economies of Germany and Britain account for some of the differences in the framing of genetic engineering's risks in these two countries. In Britain the Labour government had improved living standards and maintained a conciliatory posture toward the labor movement. In Germany, the welfare state policies of Willy Brandt had been replaced by the conservative fiscal policies of new leader of the coalition between the Social Democrats and the Free Liberals (1969–1982), Helmut Schmidt. The highly internationalized economy of Germany continued to be a pivotal reference point in political discourse. In this narrative, international stagnation and crisis were construed as putting pressure on the state to expend all its means in fostering new, job-creating technologies. As was mentioned in chapter 2, molecular biology had relatively early on been framed as a "technology of the future," an interpretation which seemed to be further supported by the recent availability of new genetic technologies. The shaping of German biotechnology policy was part of a new discourse on planning and modernization that originated after the recession of 1967–68 and was further elaborated in the wake of the world economic crisis of 1974–75. In the context of Germany, then, modernization politics encoded an industrialization strategy that was driven by new technologies and international economic expansion. Fiscal austerity policies and attempts by the state to delegitimize the rising ecology movement (Roth 1985, pp. 50–51) created a discursive constellation in which policies that seemed to inhibit industrialization tended to be vilified. This discursive constellation affected the construction of risk in the emerging recombinant DNA regulatory narrative. From its inception, this policy narrative construed recombinant DNA regulation as a threat to the emerging German biotechnology strategy, a threat that necessitated the devising of a "careful" political strategy to deal with the "regulatory challenge."

The Deutsche Forschungsgemeinschaft, Max-Planck-Gesellschaft, and Bundesministerium für Forschung und Technologie operated in this discursive constellation, and their political strategies were informed by arguments, perspectives, and rationales that constituted German political discourse. They watched the discussions around the Berg letter closely, and the DFG made sure that representatives of the German research community would be in Asilomar. The committee responsible for the support of biotechnology in the BMFT suggested immediately after the publication of the Berg letter that its recommendation of a moratorium be followed (Binder 1980, p. 102). Despite extensive newspaper coverage of genetic engineering, in particular after Asilomar, the public debate did not reach the level of concern it reached in Britain and the United States. Nevertheless, just as in Britain, scientists played a crucial role in defining the agenda for regulatory action.[61] In fact, in the case of Germany the experts involved in recombinant DNA created, relatively early, an expert enclosure in the evolving regime of governability: a site from which these experts were in a position to define the political rationality underlying the practices of recombinant DNA regulation.

Regulation as Support: Mapping Genetic Engineering's Risk in Germany

The German government did not call upon a commission (such as the Ashby Working Group in Britain) to take an initial look at the "costs and benefits" of genetic engineering. Initially, the main institutional actors involved in the drafting of the recombinant DNA regulatory system were the DFG (the main supporting agency for basic research) and the BMFT (responsible for applied research and assuming considerable financial power over any construction-related activity that was by definition not on the agenda of the DFG). A participant in the early German deliberations after Asilomar described the initial decisionmaking process as follows:

> These are scientists who are sitting on the DFG as well as on the [BMFT], and we are on speaking terms with them on all sorts of matters. . . . I went to a meeting of a committee which is advising the [BMFT] on the support of research in biotechnology, fermentation etc. . . . This committee had set aside

two or so hours on reports about Asilomar and possible steps the [BMFT] may take. They had invited me as one having been there, together with somebody I think they had paid themselves. There were people from the DFG present. The man who was in charge of molecular biology in the DFG and the [BMFT] man discussed with us which agency in Germany should be in charge. And we agreed that for the time being it would be better if the DFG would be in charge, which is not a direct government agency.[62]

The DFG's Senatskommission suggested as a first step the "guidelines" sketched in the recommendations of the Asilomar conference:

First it was discussed in our Genetics Commission if we need similar rules as in the United States the NIH rules. We saw quickly that the DFG can only speak for the researchers which are financed by the DFG. We could argue that everybody receiving DFG money should follow DFG biohazard guidelines. But this would not have been valid for the Max Planck Institutes, for the industry, and even today the army, which says it is not involved in genetic engineering, and is not subject to the law. The DFG wrote to the chancellor saying it is glad to offer its expert knowledge, but it can only speak for those researchers it finances; hence, a larger system of rules has to develop making sure that genetic engineering in Germany is governed by these rules.[63]

After the NIH guidelines were published, the Senatskommission suggested that they be adopted in Germany. But soon it became clear that this strategy would not work. One of the key actors remembers:

There were those small things: for example, I started to build a P3 lab in our institute; the people in our local university building office wanted to know on which legal basis the lowered pressure, the 7 pounds per square inch, or whatever, I was demanding, on which basis I was demanding this. This was a small thing. But there were many small things. If you want to work in a country on the legal basis which is valid only in another country, then you run into difficulties. There was another point. In the DFG committee we had, as a guest, a representative of industry. It was felt that it would be easier for the chemical-pharmaceutical industry to acknowledge German guidelines than, let us say, specific British ones. . . . On a voluntary basis, which would be binding. . . . And . . . there were differences between the American and the British guidelines . . . everybody said "But you can't mix them and take the lowest denominator."[64]

In December 1976, the BMFT reacted to this situation by setting up a committee of experts to work on a draft for guidelines for recombinant DNA work in Germany. The ad hoc standing committee on Sicherheitsrichtlinien für Forschungsarbeiten über die in vitro Neukombination von Nucleinsäure (safety guidelines for research on the in vitro recombination of nucleic acids) was composed of eight representatives, seven from the biomedical community and one from the

BMFT. The guidelines were drafted and, after consultations with the states, scientific organizations, research support organizations, and key interest groups,[65] published as the Richtlinien zum Schutz vor Gefahren durch in vitro neukombinierte Nukleinsäuren (Safety Guidelines for Research on the In Vitro Recombination of Nucleic Acids) in February 1978. The final guidelines were, in several aspects, more stringent than the initial proposed guidelines.[66] But they constituted, nevertheless, something like an amalgam based on the lowest common denominator of the recommendations of the Williams Report and the first NIH Guidelines, in that the German guidelines set minimal standards of safety and oversight.

The German guidelines differed from both the NIH guidelines and the Williams Working Party's recommendations. They were much shorter, less deliberative, less skeptical, and much more assertive and optimistic about genetic engineering. The opening paragraph of the guidelines clearly set the tone for the rest of the publication:

> The technique of artificial production of new nuclide acid combinations . . . promises—the gain of important fundamental knowledge about living things—and hopefully allows in the long run for application, especially in medicine and agriculture. This is why the new technique is used and further developed in the Federal Republic of Germany—as in other countries. . . . But in the application of the new technique risks cannot be excluded. Since it is not yet possible with complete certainty to predict how the newly recombined nucleic acids will behave . . . careful security measures must be adopted.[67]

The regulatory narrative cast the guidelines not only as serving the regulation of recombinant DNA work but also as supporting it:

> The purpose of these guidelines is (1) to protect humans, animals and plants against recombined nucleic acids, (2) to support research, development and use of the scientific and technical possibilities offered by the recombination of nucleic acids, (3) to satisfy international responsibilities of the FRG in the field of DNA recombination.[68]

Thus, the guidelines implicitly defined a symmetrical relationship between the support of recombinant DNA research and its regulation. But, in a move familiar by now, this relationship was discursively constructed in such a way that the equality of the two principles guiding the legal text was necessarily subverted in favor of a hierarchical relationship of support over regulation through the discourse on which the principles relied.[69]

The representational strategies mobilized around the "risk question" were interwoven with the definition of life prevailing within molecular biology's discourse of life. Hence, the definition of risk as the assessment of a particular outcome of a hazard was inextricably intertwined with its enabling technology (technological definition of risk). As molecular biology was inscribed into the regulatory philosophy, its support and funding became the main conditions of eliminating risk. As a consequence, the removal of hazards from the equation was by definition a matter of the "progress of science," meaning mainly the advancement of molecular biology.

Indeed, in the following years policymakers gradually relaxed restrictions on genetic engineering, on the ground that the subsequent increase in scientific knowledge called for reevaluation of the initial hazards assessments. Risk (i.e., the assessment or definition of hazards) was controlled by the play of the signifiers of the discourse of molecular biology. Thus, the regulation of genetic engineering was in fact the introduction to a story on research support and deregulation.

Besides adopting the core element of the NIH and the British approach of regulation, the technological definition of hazard, there was a number of other strong similarities between the American, British, and German regulatory narratives. The German guidelines (Richtlinien in vitro neukombinierte Nukleinsäuren, 1978) resembled the NIH guidelines insofar as the former were understood to be binding only for institutions active in recombinant DNA work financially supported by the federal government. With respect to all other institutions, including universities and industry, the guidelines stated that it was "expected" that they would be followed. But despite this call for voluntary adherence, the guidelines were technically regulations released by the BMFT and based on administrative law (*Verwaltungsrecht*). However, in the legal literature even the status of the guidelines as administrative law has been challenged by the argument that the guidelines do not address themselves to government institutions and hence are not administrative law (Scholz 1986).

The German guidelines also followed the NIH guidelines in putting more emphasis on biological containment and less on physical containment (with the categories L1, L2, L3, and L4) than the British guidelines. This meant in particular that the four physical laboratory containment levels were less stringent in Germany than in Britain. At the same time, the provision was made that, pending approval by the

Table 3.2
German regulatory guidelines. Source: BMFT Richtlinien zum Schutz vor Gefahren durch in-vitro neukombinierte Nukleinsäuren, February 15, 1978.

Source of nucleic acid	Classification	Remarks
Tumors from primates	L4 B1/L4 B2	
Mammals, birds	L4 B1 or L3 B2	
Amphibians, reptiles, other vertebrates	L2 B2 or L3 B1	
Tumors and tumor cell lines from vertebrates	L4 B1 or L3 B2	
Invertebrates, lower eukaryotes	L2 B1	If the source contains pathogen or tumor tissue or if it produces known toxin or antibiotics the given classifications must be reconsidered and, if necessary, tightened
Plants	L2 B1	
Cell organelles	Classification according to cell type	If sufficiently purified source nucleid acids can be treated like cell nucleid acid after purification
Vertebrate viruses: oncogene retroviruses	L4 B1/L4 B2	
Viruses pathogenic for humans and animals	L3 B2	
Other vertebrate viruses	L3 B1	
Viruses of vertebrates and plants	L2 B2 L3 B1	
Prokaryotes and their plasmids and phages in exchange with *E. coli* K12	L1 B1	If the source is a pathogen to humans, animals, or plants or produces toxins, the classification must be changed to L2 B1.
Prokaryotes in no exchange with *E. coli*	L2 B1	If the source is pathogen to humans or produces toxins, the classification must be changed to L3/B1 or L2/B2

ZKBS, new host-vector systems could be deployed.[70] At the same time, the German guidelines in general adopted the NIH approach to biological containment; nevertheless, the German guidelines were much less detailed than the American ones and, thus, more similar to the British "case approach," which assigned the GMAG to further develop and specify the rather unspecific guidelines. The German guidelines, hence, followed the British approach by inaugurating a GMAG-like commission responsible for all experiments from the NIH level P2 B1 on. That commission was the ZKBS. Above these levels, the ZKBS was to consult, review, and register corresponding research.[71] For experiments from level L2 B1 on, the consent of two ZKBS members was necessary; from level L2 B2 on, the consent of the whole ZKBS was necessary.[72] At the same time, the composition of ZKBS was different from that of the GMAG; in particular the labor unions were much less represented on the ZKBS. Of its twelve members, eight were from science (four from the field of recombinant DNA work, four came from fields with experience with biological security measures, such as microbiology, cell biology, and hygienics); the remaining four represented the "public interest" (one representing the DFG, two representing industrial groups, and only one representing the trade unions). Hence, the ZKBS was much more a peer-review type of control system than the GMAG, and in this respect it was comparable to the NIH study group review system. In fact, with the creation of ZKBS an expert enclosure with the power to make its arguments and evaluations the obligatory mode for the operation of the network had been established in the emerging regulatory policy network.

In this partitioning of the political space of biotechnology, the regulation of recombinant DNA work became an integral part of the very same institutional configuration that had shaped the formation of the biotechnology discourse in Germany since the 1960s. Two of the scientists on the ZKBS, for example, were not only responsible for regulating the BMFT-supported research at the pharmaceutical companies Schering AG and Boehringer Mannheim Gmbh; at the same time, they were scientific consultants for the very same companies. Furthermore, labor unions lacked the kind of institutional framework that had given their British counterparts such an influential position in regulatory decisionmaking. While in Britain the "modernization with a human face" discourse was instrumental for the framing of genetic engineering as a problem of health and safety at work, in

Germany the prevailing discourse about the need to avoid excessive regulation worked against regulatory solutions with broader public participation. From the very beginning the labor unions adopted a much less critical attitude toward genetic engineering than their British counterparts. Furthermore, within the highly decentralized British system of labor unions, the Association of Scientific, Technical, and Managerial Staffs, which represented skilled laboratory research staff and scientists, had emerged as a highly engaged, informed, and critical force in the regulatory game, despite the fact that the Trade Union Congress had very little interest in genetic engineering.[73] Given Germany's centralized union system, policymakers were in a better position to mobilize broad union support for the evolving modernization policies. The union that would become most engaged in genetic engineering regulation, IG Chemie, kept a low profile from the beginning and defined its own position as resting within the government's modernization discourse. Together with the Deutscher Gewerkschaftsbund (the central umbrella organization of the German trade unions), IG Chemie, a traditionally "industry friendly" and organizationally relatively weak union representing chemical workers (Grant et al. 1988, pp. 164–175), became involved in the largely uncontentious system of recombinant DNA regulation that was evolving. A regulatory discourse that left the definition of genetic engineering's risk to a small circle of experts and linked regulation of genetic engineering closely with its support had become hegemonic. For the unions, scientific expertise was an instrument for ensuring the safe utilization of genetic engineering in research and industry and thereby bringing about modernization in Germany. The idea of modernizing Germany by means of biotechnology helped to relate the experts' interests in genetic experimentation to national interest.

Toward a Genetic Engineering Act?

There were several attempts to create a *Rahmengesetz* ("frame law") that would provide a general legal framework for public and private research, development, and production. Indeed, the BMFT issued two drafts for such a general law, one shortly after the issuing of the first German guidelines on June 22, 1978, and one on July 19, 1979.[74] But the fate of these two drafts reflected the operation of the same power constellation that had already had such a decisive influence on the shaping of the German genetic engineering regulations. The first law was discarded after sharp criticism, especially from the Deutsche

Forschungsgemeinschaft and the Max-Planck-Gesellschaft. In general, three arguments were mobilized against the planned law. First, the need for a law was questionable in the first place, since initial concerns about risks seemed to be exaggerated. Second, a law tied to legislative decisionmaking would turn out to be too inflexible, as progress in knowledge would require changes. Third, the sanctions against violations proposed by the draft law were considered to be too strict. In particular, criticism focused on Article 9, which specified for a violation of the law (even one that left no traceable consequences) a jail sentence of up to 5 years, which was considered excessively punitive (Deutsch 1986). The 1979 draft responded to the criticism by eliminating the stricter provisions, including jail sentences for certain violations. However, fierce resistance from industry and from the main research organizations led to the abandonment of the plans for drafting a law (Pohlmann 1990, p. 140). Undermining legislation of a regulatory code was a critical definitional victory in the "normalization" of genetic engineering: the absence of special legal provisions erased the distinction between recombinant DNA technologies and other biological and chemical technologies. At the same time, this strategy preserved the role of the representatives of the research community as mandatory spokespersons in issues concerning recombinant DNA.

Discussions about a regulatory code in Germany on the level of the BMFT and even in the Bundestag continued throughout the 1970s. Finally, in July 21, 1981, the BMFT declared for the first time that it no longer saw a need for a German law regulating genetic engineering. In its explanation, the BMFT claimed that initial concerns raised with respect to the dangers of recombinant DNA work were clearly exaggerated. At the same time, in view of the rapid developments in the field, it was premature to exclude the possibility of hazards. The BMFT also argued that, since there was still a chance that an international agreement on guidelines might be reached, plans to draft a German law should be postponed. Because the BMFT assumed that not only government-funded research but also all other research organizations and establishments and industry too were continuing to follow the guidelines, it drew the following conclusion: "Until a final assessment of the risk situation is possible any legal regulations should be deferred."[75]

Following the established policy narrative, the 1978 BMFT guidelines became the central regulatory mechanism for the regulation of genetic engineering in Germany. They went through four revisions

(1979, 1980, 1981, 1986), mostly following, with short delays, the changes in the NIH regulatory policy. As in Britain the rewriting of the boundary object "gene" led to a redefinition of the boundary between experts and non-experts from the committee level into society. This resulted in a generalization of the construction of the political space of biotechnology regulation as an expert enclosure and in a decentralization of regulatory decisionmaking from the committee to the laboratory level. One result of this process was a loss of sites for a legitimate debate involving the public on the risks of genetic engineering. In more and more cases of genetic engineering experiments, the boundary between experts and non-experts could no long be negotiated and thus lost permeability. What was going on in the labs was, increasingly, considered the business of no one other than the scientists working there.

Between 1981 and 1988 the ZKBS reviewed 1232 submissions: 178 of containment level L 1, 1030 of containment level L2, and 24 of containment level L3. In one case the ZKBS rejected a proposed experiment because it involved excessive risks and unpredictable factors. More than 95 percent of all submissions involved *E. coli* K12 host vectors.[76] Because of the ZKBS's secretive mode of operation, little is known about the style and content of its work. However, interviews I conducted suggest a rather hurried and superficial style of decision-making.[77] Indeed, under normal circumstances the examination of individual cases involved an oral presentation by two ZKBS members, who either accepted or rejected the proposal. A ZKBS member recalled:

> Usually we had discussions in the ZKBS. Usually individual members had looked at the details of the submissions in some detail and reported in the commission . . . when we had to deal with completely new situations . . . then operated slower and with subcommissions and external evaluations. . . . When it was clear in the first place that the lowest category would apply two ZKBS members received the details of the case. In other cases the submission went into the ZKBS plenum and was discussed there. . . . Usually there were oral presentations to be written down in a protocol to be accepted or altered by the ZKBS.[78]

The statistical data seem to support this assessment: between 1981 and 1988 all 1232 proposals submitted for ZKBS approval took no more than 26 meetings of the entire staff and five meetings of subgroups. Taking into account that the ZKBS had only very limited staff resources, it is very difficult to imagine how a commission of experts could evaluate more than 1200 projects in 26 meetings—an average of 50 per meeting.[79] What had evolved was a framework of recombinant DNA regulation that combined a normalization of genetic engineering with a system of state regulation that, among other things, created legitimization through procedure. As in Britain, the logic of the policy narrative's technological risk definition systematically reduced the scope and the intensity of regulatory activities. This development, however, did not lead to complete abolition of the regulatory institutions.

France: The Politics of Regulatory Minimalism

In comparison with Britain's and Germany's, France's regulatory approach to recombinant DNA research came closest to a system of self-regulation of scientists by scientists with an absolute minimum of regulatory requirements. While the risks of genetic engineering for humans and the environment "were" the same in France as in Germany or Britain, a different political-discursive constellation produced much less stringent regulations.

The initial pressure for regulation of recombinant DNA work in France came, as in the United States, Germany, and Britain, from scientists working in that field. In November 1974, alarmed by the Berg letter, a group of scientists wrote to the director of the Centre National de la Recherche Scientifique and to the Délégation Generale de la Recherche Scientifique et Technique asking for the establishment of a control system for recombinant DNA work. The letter was written after a committee that decided on the support of research within the CNRS turned down a research proposal involving recombinant DNA work in a decision motivated by the degree of uncertainty concerning biohazards related to such work.[80] One of the co-authors explained the motive behind the letter that was written immediately after the CNRS committee's refusal of funding for recombinant DNA as follows:

> One of the things we did was to contact all the people who were possibly interested in the very near future in this methodology. . . . And we wrote a letter to the DGRST asking that some sort of control be set up. We wanted to be controlled, okay?[81]

This desire for monitoring and control was due in part to a political controversy over recombinant DNA research. In the wake of Asilomar, the French public became deeply involved in discussions about the safety of recombinant DNA work, which received broad newspaper coverage. Concerns were especially raised from within the scientific community, especially by laboratory technicians, but also, for example, by people living in the neighborhood of research facilities working with recombinant DNA. From the perspective of the scientists involved in recombinant DNA work the situation looked somber:

> First, people got extremely neurotic about the whole business. For instance, in the Faculty of Science, people would refuse that their autoclave would be used to sterilize the products of experiments of this sort. . . . In this building people would refuse to take photographs of pictures of gels having been involved in these experiments. A campaign was organized against these experiments by a number of people who I suppose were deeply convinced that the issue was dangerous, but I am sure some of them were convinced that it was a beautiful matter for politics. . . . And let me say, by the way, that in all this business that got very unpleasant, the Stanford people were in a position to go on with the experiments while almost everybody else was stopped.[82]

France in the early 1970s was marked by a number of political controversies that indicated a high level of political mobilization and the emergence of new topics of political discourse. Arguments against productivism and heavy technologies were incorporated into the political ideologies of labor (especially of the Confédération Française Démocratique du Travail, the second biggest labor union), while environmental issues became part of the broader political discourse. In 1971, in a relatively early creation of a special ministry of the environment in the European context, France established a Ministry of Environmental Affairs. In 1974, 82 environmental associations supported a "green" candidate in the presidential election. Although the "green" candidate (René Dumont) received less than 2 percent of the vote, his candidacy had a significant impact on mobilizing popular support for environmental groups—the Friends of the Earth being the main beneficiary—and was crucial for expanding environmental ideas among both leftist and rightist parties (Duclos and Smadja 1985, pp. 136–137; Hainsworth 1990; Chafer 1984). These changes in political discourse indicated by the rise and expansion of "green" issues and parties happened in France slightly earlier than in Germany. In many ways the French social movements during the 1970s served as a model for activities in many other countries, including Germany (Leggewie 1985, p. 135). The alternative movement in France was definitely on the rise in the second half of the 1970s and in the early 1980s, with a significant growth of specifically aimed associations. By the end of the 1970s, there were several thousand associations, and "green" candidates were running in local, parliamentary, and presidential elections.

However, there was a gap between political mobilization and state responsiveness. The image of the independent and strong state in defense of modernization politics continued to dominate the French political imaginary and to inform policymaking. The broad anti-nuclear movement, for example, could not prevent the French state from turning France into the most nuclearized country in the world (Duclos and Smadja 1985, p. 138). In addition, the economic recession and a cautious and reform-averse policy course under the presidency of Valery Giscard d'Estaing and the prime ministership of Raymond Barre reduced the space for certain types of political reform.

The conceptualization of the recombinant DNA risk problematic must be seen in this discursive context. The state became active to "take control" of the recombinant DNA problematic. The particular form of this state activism was also determined by the fragmented and complex nature of the French political-administrative system (Mazey 1986). As much as the image of the "strong state" dominated the political discourse, this "strength" was always the result of complicated processes of network building. The "negotiating state" was the other side of the image of the "strong state." Institutionalized conflicts, poor vertical communication, and horizontally played out rivalry between the *grands corps* were dominant features of the French state (Mazey 1986, pp. 419–420; Crozier 1970) and made it crucial for the executive to form mutually beneficial informal alliances with constituencies and to establish patterns of cooperation rather than try to exert administrative control by formal means.

In response to the public controversy over genetic engineering, the Délégation Generale de la Recherche Scientifique et Technique set up an ethics commission in January 1975 which recommended the creation of a control commission for recombinant DNA work. Despite the double structure of the committee, the main responsibility lay with the commission de contrôle. The ethics committee, on the other hand, seemed to have primarily a symbolic value. As one of the key actors of that time saw it in retrospect:

> . . . who can decide on ethics, what does it mean? In practical terms, it had importance in that the people on this committee are rather famous, at least locally, and they are of the opinion that it was important to control recombi-

nant DNA experiments, that the conclusions of the Asilomar conference were okay, and that these should be followed up. . . . The major point was to show that the heads of the scientific community here were in agreement with what the technical committee was going to do. The technical committee is composed of fourteen guys who are doing the actual job[83]

In March 1975 the first meeting of the Commission Nationale de Classement took place. The task of this commission was to elaborate French guidelines for recombinant DNA research and to evaluate and classify research projects involving such work. The commission published its guidelines on December 13, 1977.[84] As early as May 1975 this commission had begun to evaluate recombinant DNA research projects, using criteria developed at Asilomar and in the NIH guidelines.[85] However, the French guidelines published in December 1977 were not regarded by the commission as the sole source for recombinant DNA regulation in France. Rather, they were seen as a sort of summary version of other existing guidelines, including the NIH guidelines and the British guidelines—a fact resulting from the shortness of the guidelines (not much longer than the introductory notes to the NIH guidelines) as well as from the commission's self-interpretation.[86] The guidelines followed the familiar definition of the levels of physical containment (L1, L2, L3, and L4, following the American P1, P2, P3, P4 model) and the definition of biological containment (B1 and B2, following the NIH EK1 and KK2 definitions). Apparently addressed only to scientists engaged in recombinant DNA work and not to a

Table 3.3
French regulations for rDNA work (Normes en matière de recombinaisons génétique in vitro adoptées par la Commission Nationale de Classement, December 13, 1977). Source: Ministère de la Recherche et de la Technologie, Normes de Sécurité pour les recombinaisons génétiques in vitro (Paris, 1985).

Source of DNA	Classification
Mammals, birds	L3 B1 or L2 B2
Amphibians, reptiles, fish	L2 B1 or L1 B2
Other eukaryotes	L2 B1 or L1 B2
Procaryotes	L1 B1
The cloning of pathogen DNA organisms and of cells creating viral genomes must be evaluated case by case.	

wider public, the French guidelines lacked any broader policy statements or explication of a rationale. Essentially, they addressed the question of risks involved in recombinant DNA work along NIH criteria. Following the now-familiar regulatory narrative of Asilomar and of the NIH guidelines, the French guidelines identified risk as a function of the source of DNA (reiterating the phylogenetic closeness principle), the number of independent recombinants involved in an experiment, the characteristics of the vector-host system, and the purity of DNA; they then went on to define measures of physical containment along the conceptual framework of the NIH guidelines.

The French Regulatory Model: Enclosing Experts

The main difference between the French system of regulation and the British and German systems lay less in the content of guidelines than in the organization of the enforcement and implementation of the guidelines, in particular with respect to the institutional setup, the scope of the regulations, and incorporated modes of public participation. In Britain the guidelines received their legal status from their embedding in the HASAWA framework; in Germany, they were binding for research supported by the federal government. In France, the state did not institute a legal solution based on public law, but adopted a private-law type of contractual solution, a rather unusual legal construction by French and international standards.

The idea behind this construction of the political space of biotechnology regulation was that the various public and private institutions involved in recombinant DNA work should sign a contract with DGRST which had at its core an agreement that the signing institution would follow French and international regulations in the field. According to agreed-upon procedure, scientists planning recombinant DNA work were to submit their project to the Commission de Classement in the form of a standard questionnaire providing the information necessary for an evaluation. Then two rapporteurs were nominated by the president of the commission to present the project to the commission. The second tier of the committee system, the Commission d'Ethique, was to be involved, as a sort of court of appeals, in the case of an ethical problem arising during the deliberations. In the next step, the Commission de Classement would inform the applicant, the director of the laboratory involved, and a local public institution of control, such as the committee for hygienics and security, or an equivalent committee. The local commission was to be made up of scientists and staff representatives. Whereas the Commission d'Ethique also comprised nonscientists, the Commission Nationale de Contrôle (Commission de Classement) was entirely composed of scientists, representing the disciplines of microbiology, microbiological genetics, molecular biology, etymology, and epidemiology, sent from the major research institutions, as well as representatives from the Centre National de la Recherche Scientifique, the Institut National de la Santé et de la Recherche Médicale, the Institut Pasteur, and the Institut National de la Recherche Scientifique.

Despite its double structure, it appears that the Commission d'Ethique did not really get involved in the process of regulation. As a matter of fact, in the time between the creation of the Commission Nationale de Contrôle and February 1978, some seventy projects were screened by the CNC, but the CDE was never involved.[87] As a result, the regulatory structure created a system in which scientists monitored other scientists. A 1978 confidential report of an interministry working group on the "control of recombinant DNA work" stated:

> The workings of the Commission de Classement of the DGRST is, in fact and in law, a system of self-control of the scientific community with respect to its work in the field of in vitro recombinant DNA work: on the one hand the commission is exclusively composed of scientists; Furthermore, its functioning is based on the cooperation of the major research institutions financing the project with the DGRST.[88]

Hence, the developing organization of recombinant DNA's risk was dominated by an expert enclosure whose problem definitions constituted the core of regulatory decisionmaking. But, unlike Britain and Germany, the French state refrained from any imposition of rules by adopting the "contractual instrument" to negotiate the regulatory system. This strategy to enroll the research institutions and private industry into the regulatory network turned out to be difficult and, at least initially, to achieve little success. The legal weakness of this system was twofold: first, its functioning required that all institutions involved in recombinant DNA work actually sign the contract subjecting them to the power of the Commission National de Contrôle. The contract was sent by DGRST to the large research institutions, such as the CNRS (the largest public research organization in France); INSERM (Institut National de la Santé et de la Recherche Médicale), the medical research institute, INRA (Institut National de la Recherche Agronomique), agricultural research; Institut Pasteur; CEA (Commissariat à l'Energie Atomique), and National Pharmacy Federation. However, in 1978, two years after its creation, a considerable number of research institutions, such as INSERM, CEA and, in addition, private industry, had not responded to DGRST's initiative. Second, even with respect to the organizations that had signed the convention, there were no legal means available to enforce the contract other than taking disciplinary measures against those researchers violating the rules. But even the disciplinary measures were, of course, limited to the state's laboratories and could not be applied against private industry.[89] Hence, owing to the failure of the state to mobilize a number of critical actors for the planned system of recombinant DNA regulation, this system turned out to be even weaker than the German one, taking into account that it gave state laboratories the freedom to decide if they were to cooperate or not on recombinant DNA regulation and that it refrained from raising liability questions with respect to industry. Apparently, an important element of French political discourse, the idea that the state should attempt to establish informal systems of cooperation with its constituency, had played an important role in the development of this situation. The constituency of the French state rejected its assigned role to cooperate, and the result was that for years to come the powers of the Commission de Classement were more formal than real. This demonstrates again that the power of any institution can never be assumed to be "given" but, rather, is the effect of a successful process of network building based on enrollment and mobilization by means of a dominant policy narrative which is able to mediate between competing codes by which reality is assigned meaning. In France the risks of genetic engineering were no less real than in the United States or in Germany; however, policymaking failed to stabilize a generally accepted relationship among problem definition, actors, institutions, and mutually agreed upon strategies to deal with genetic engineering's risks, even on the most minimal level.

Reforming the French Model?

The French system of recombinant DNA regulation continued to be debated. For once, the public seemed to be concerned about genetic engineering and to support stricter regulation. This was indicated by a public opinion poll on the attitudes of the public toward biotechnology taken by SOFRES in 1980. According to this poll, 33 percent had a favorable opinion of biotechnology, 36 percent rejected it, and 31 percent did not have an opinion.[90] In addition, there was a close relationship between concern over the new technology and the level of information: the better informed, the more concerned.[91] When asked if genetic engineering should become subject to strict regulations supervised by the state, 41 percent agreed; 38 percent responded that one should trust researchers. Asked about research money spent on genetic engineering, 13 percent supported an increase of the financial means for this kind of research, 32 percent were for leaving it at the same level as before, 19 percent were without an opinion, and 36 percent endorsed a decrease in research support.[92] Such polls are, of course, of limited value; however, they give some indication that the French public was not at all unconcerned and overenthusiastic about genetic engineering—a fact which was noticed by the state authorities.

In 1978 an interagency working group from concerned ministries (Research, Defense, Culture and Environment, Agriculture, Work, Industry, Health, and Universities) convened. The task of this group was to advise the government on the following problem: given the experiences of the years since the establishment of the Commission de Contrôle at the DGRST and the limited legal status of the current regime of regulation of genetic engineering, how can a more comprehensive legal and administrative regulation of genetic engineering be organized within the framework of the existing legislative possibilities?[93] After a thorough discussion of the legal limits of the current regulation, the Rapport Poignant discussed possible regulatory strategies. The report essentially concluded that genetic engineering could be regulated by a system with the Ministry of the Environment as the regulatory center or one with the Work Law and the DGRST at its center. As mentioned above, since the early 1970s there had been a push for environmental legislation with the newly founded Ministry of the Environment as its regulatory agency. But the report considered any regulatory system centered around the Ministry of the Environment at the center as "déformer l' esprit du système," especially with respect to difficulties in overseeing biological containment and workplace biohazards. In addition, the existing environmental legislation—especially the 1976 law on establishments specified to be regulated for the protection of the environment (Loi de 19 juillet et le decret du 21 septembre 1977 relatifs aux installations classées pour la protection de l'environment), with its system of public hearings—was considered "too cumbersome."[94] The report, however, was more favorable toward a scenario that formed around the existing Work Law (Code du Travail). Following this scenario, a blueprint was outlined to subject all recombinant DNA work to the commission de classement, develop the standardization of measures of physical confinement, implement the regulations on the local level, and provide a system of sanctions against violations. In this scenario, the report suggested the creation of a second, interagency commission not composed entirely of scientists (as the Commission de Contrôle was). The commission de contrôle, however, should essentially control the implementation of the regulations.[95]

The analysis of the (confidential) Rapport Poignant found itself partially reflected in another government report: the Rapport Royer, on the "Security of Industrial Biotechnology," published three years later. But, in contrast to the Rapport Poignant, the focus of the Rapport Royer was not on the outline of a new system of regulation. Rather, the emphasis was on arguing that the existing work, health, and environmental legislation, along with the DGRST's commission de classement, already offered a good deal of regulation for biotechnology. In the narrative of the Rapport Royer, there was no pressing need for quick, special regulation; the report argued that, for the time being, further legal studies, the observation of legislative efforts in other countries and on the EC level, and reflection on the possible structure of any new legislation should be in the foreground of regulatory reform activities.[96]

And this approach turned out to inform the French regulatory philosophy until the mid 1980s. Finally, between 1975 and 1980 the Commission de Classement signed contracts with the CNRS, the INRA, the Institut Pasteur, INSERM, and the CEA, and reviewed more than 250 experiments submitted by some 50 research laboratories. None of the submitted proposals were rejected.[97] With this move, gradually, a governmental network for the regulation of recombinant DNA had evolved which resembled its German counterpart. However, this success of enrolling crucial actors into the regulatory network went along with a process of deregulation. As in Germany, the original 1977 guidelines were revised and eased in 1980 and again in 1984 along the lines of changes in the NIH guidelines. The revised guidelines provided a basis for the operations of the core of the French regulatory system, the commission de classement in the DGRST (later the Ministère de la Recherche et de la Technologie).[98] This commission, composed exclusively of scientists, represented even more than its counterparts in Britain and Germany the evolved political rationality of recombinant DNA regulation: to create a regulatory network dominated by an expert enclosure, a site where the determination of the probabilities of recombinant DNA related risks was confined to the discourse of the specialists in the field.

Conclusions

In this chapter I focused on the shaping of policies regulating the risks of genetic engineering in the United States and in Britain, France, and Germany. My central argument was that, rather than inhibiting the advance of genetic engineering, the efforts to regulate risk must be seen as an important move in facilitating the project of genetic engineering. Risk-regulation policies were an important contribution to order and coherence in a new policy field, and they successfully defined a site where a debate about genetic engineering could be carried out.

The new molecular biology and its visions to transform society had already become highly controversial during the 1960s. In the late 1960s and the early 1970s, molecular biology was increasingly interpreted as a new, revolutionary technology with significant transformative potential. At the same time, concerns began to proliferate over the potentially adverse ecological, epidemiological, and social consequences of genetic engineering. This discursive constellation indicated the possibility of a variety of policy options, ranging from a moratorium on any future recombinant DNA work to the adoption of only minimal regulations in order to make rapid progress in research possible. By the early 1970s the regulation of genetic engineering had become the chief location for the social negotiation of the new molecular biology. Initially, the regulatory policies developed in the United States and in Europe opted neither for one regulatory extreme nor for the other—a situation which reflected an unresolved struggle over the meaning of genetic engineering's risks. But toward the late 1970s, regulatory oversight of recombinant DNA laboratory work was reduced to the scientific management of laboratory risk. This was so because a specific framing of the nature of the genetic engineering problematic had become hegemonic and had redefined the dynamics of policymaking which made recombinant DNA work not only possible but socially desirable.

This discursive construction of genetic engineering's risks shows a number of similarities between the United States, Britain, Germany, and France and helps us to understand how policy narratives come into existence and create order in a policy field. Looking at the United

States and Europe, I discussed how a particular regulatory policy narrative became hegemonic and mobilized specific sets of actors, institutions, interpretations, and artifacts for the purpose of regulatory policymaking. I showed that neither genes, risks, regulators, nor "the public" existed before specific sets of discursive practices articulated them into the political space where the social struggles over genetic engineering were carried out.

Thus, policymaking is never simply a reaction to a problem and is always co-productive of the problems to which it seems to react. Narratives and the discourses in which they are embedded are critical elements in such constructions of reality. Policy narratives tend to have a number of different authors dispersed across a regime of governability. Genetic engineering regulatory narratives were, to a considerable extent, drafted outside the political system by scientists who worked in the field of molecular biology. These scientists crafted their problem definitions by reference to the dogmas of molecular biology. By the mid 1970s the activity of regulating recombinant DNA work had assumed a very specific meaning. Based on the Asilomar narrative, the chosen representational strategies dealing with risk were inseparably linked with the concept of life within molecular biology's discourse on life. In this discourse, recombinant-DNA-related hazards were represented as controlled by the technology that produced their possibility. Thus, the definition of risk as the assessment of a particular probability of a hazard was inextricably interwoven with its enabling technology. The key result of this construction was a technological definition of risk. This narrative developed at Asilomar and in the NIH guidelines was crucial for the definition of the terrain for the governing of genetic engineering. The technological framing of the hazard established host-vector systems and physical laboratory containment equipment as the obligatory points of passage for hazard control[99]: If the right biological and physical safety measures were to be adopted, genetic engineering would be a safe technology.

But regulatory policymaking was not only based on the mobilization of scientific explanations. Molecular biologists qua scientific experts invested a great deal of their energy in translating practices of genetic engineering into expressions of modernization and social progress. Government officials, labor union officials, and other representatives of "the public" linked their participation in the regulation of genetic engineering to the social goals that were at the center of their value systems, such as environmental protection, workplace safety, or the modernization of society. As a result, a hegemonic constellation evolved in which a number of different discourses began to develop a mutually reinforcing relationship and defined the problematic of genetic engineering's risks and the institutions and actors in this new policy field.

This discursive production of risk had enormous implications for the distribution of power in the policy field. It positioned any competent molecular biologist or representative of related fields as a spokesperson[100] for recombinant DNA's safety, a development that was inseparable from the rise of molecular biology since the 1960s as the dominant provider of the truth about life processes. At the same time, by linking genetic engineering to fields such as workplace safety, the actors "representatives of the public" and "regulating state" were created, which played—with variations from country to country—the role of the "layman" and the "representative of the government" in the regulation process. Government officials oversaw the regulatory efforts and symbolized the state's interest in protecting its citizens from possible harm inflicted by genetic experimentation. Scientists (i.e., experts on molecular biology) provided framings of the problem that turned out to be crucial for the definition of the actors involved in recombinant DNA regulation and its modes of institutionalization.

Initially, neither in the United States nor in Britain, France, or Germany was this expert knowledge simply imposed on the scientific community and society in general. A system of communication was created within which the boundary object of the "gene" as rewritten by the discourse of molecular biology served as a link between networks of scientists, state officials, and representatives of interest groups and the public. First, it served to link experts and non-experts in a hierarchical relationship within the framework of regulatory commissions. Then it helped to redefine the boundary between science and politics in a process of generalization and decentralization. "Progress in science," meaning mainly the advancement of molecular biology as practiced and interpreted by the experts in the field, became a precondition for eliminating risks. Subsequently, a number of studies, statements, and meetings in the field of genetic engineering research in the United States and in Europe were construed as confirmations of an ongoing "progress of science" which in turn allowed for a substantial relaxation the initial guidelines. This process also reflects the mutual dependence between power and knowledge. The importance of scientists in the regulatory process was based on the mobilization of authoritative knowledge, which was broadly used to explain genetic engineering's risk. Without this knowledge, scientists hardly could have assumed such a critical role in the regulatory process. At the same time, the scientific arguments developed in the controversy were based on choices and interpretations that had much to do with the interest of many of the involved experts in continuing to do genetic engineering research with a minimum of inhibitions.

This textual construction of the genetic engineering question as mainly a risk problem allowed for translation and simplification of the multifaceted problematic of genetic engineering, including its ecological, social, and ethical implications, into a system of categories, risk groups, and enumerations related to types of experiments. Nonlinguistic material events, entities, or qualities such as the characteristics of a genetically modified virus or human immune system mechanisms were expressed in terms of numbers, classes, and risk groups and could be conveniently transferred to central government agencies such as the ZKBS or the GMAG. Able to draw on such data, experts appointed by the government made decisions on the acceptability of experiments. In this sense we can say that the particular modes of inscribing the risks connected with recombinant DNA technologies contributed to, rather than undermined, the hegemony of the framing of modern biotechnology as a socially acceptable "technology of the future." The loud, threatening, difficult, and confusing problem of the risks of genetic engineering had become one of the many objects for orderly routines of government.

Despite the many similarities between Britain, Germany, and France in the evolution of regimes of governability for recombinant DNA regulation, we also see some important differences. In this context, the American regulatory narrative served as a textual link and a central point of reference and orientation for European policymakers. In Britain, recombinant DNA regulation was initially more strict, allowing for greater public participation. This system was supported by the prevalent modernization discourse, which combined the idea of industrialization with elements of social compromise and participatory democracy. In France and Germany, owing to different discursive economies, regulatory measures tended to be construed as antidotes to modernization politics and thus were less strict and more guided by the idea of a system self-regulation by scientists. In both countries a number of attempts to establish a stricter system of control failed as a result of the difficulty of stabilizing new regulatory networks around broadly shared new definitions of the recombinant DNA regulation problematic.

At the same time, in all three countries the rise of neo-liberal political tendencies together with the hegemonic deployment of arguments of deregulation around the trope of the "progress of science" hastened the decomposition of the established systems of regulation. That decomposition was initiated by the discourse in molecular biology, which by the second half of the 1970s began to claim that it had achieved good control of potentially adverse impacts of genetic engineering experimentation. These discursive changes produced a new constellation of actors in the regulation system. Decisions concerning the safety of laboratory procedures were increasingly made outside the state apparatus by the scientists themselves. In this new regime of governability, the "freedom" of these decisions was defined by the parameters of molecular biology, which specified the proper procedures for laboratory practice. The boundary between experts and non-experts no longer could be legitimately debated; thus, it had lost its permeability. The practicing scientists had become the sole actors in genetic engineering regulation. Associations with the state apparatus continued; however, they took on an increasingly symbolic character. Nevertheless, these associations remained in place and helped to produce legitimacy by procedure. The sheer workings of the machinery of the state, the meetings of the different commissions, and their decisions and publications produced an image of a framework of scrutiny, which, in fact, had already largely mutated into a system of self-control of scientists by scientists at the laboratory level. In this new topography of the political space of genetic engineering, experts in regulatory commissions and on the laboratory level ceased to negotiate genetic engineering's risks with actors in other social worlds. This constellation would eventually lead to a serious destabilization of the politics of genetic engineering.

In 1982, when OECD experts drew up a summary of the recombinant DNA debates of the last decade, they came to the following assessment:

> In the early 1970s, when the technology of genetic manipulation was first acquired, predictions of scientists ranged from panacea to pandemic. A public storm followed and it was not at all surprising that national authorities in many countries, having been told that this technology was capable of creating new forms of life and that scientists themselves had requested a moratorium for this type of research, responded by setting up groups and committees to

consider the social and political acceptability of the risk. A public feeling of instinctive mistrust towards scientists promoting genetic engineering was widespread. However, finally after considerable public debate and advice from scientists, medical and epidemiological experts, the general conclusion has been reached that, provided suitable precautions are taken, the benefits of the technology far outweigh any conjectural risks. (OECD, Bull, et al. 1982, p. 60)

This assessment of the public's sensibilities concerning hazards connected to recombinant DNA work would soon turn out to be premature.

4

Myths, Industries, and Policies of Biotechnology: Between Basic Research and Bio-Society

Chapter 3 described the first phase of the recombinant DNA debate in the United States and in Europe during the 1970s. (The second phase of the controversy will be the subject of chapters 5 and 6.) What had started in the 1960s as a broadly conceived controversy about the relationship among the new biology, society, and ethics metamorphosed during the 1970s into a debate about appropriate regulatory measures to control the risks of genetic engineering for human health and environment. The hegemonic power of a well-crafted policy narrative was at the center of this controversy. Rather than inhibiting genetic engineering, the emergence of risk and its regulation turned out to be critical for the diffusion the new biotechnology into research and industry. In this construction of the political space of genetic engineering, the regulation of the risks of recombinant DNA technology by the state became the central site for the socio-political negotiation of genetic engineering. By the early 1980s, genetic engineering seemed to have lost its controversial character, and the institutions that were initially designed to allow for public participation in regulatory decisionmaking were abolished or diminished in importance. The genetic engineering controversy seemed to be closed.

Around the same time, policy discourse in the United States and in Europe began to depict genetic engineering as the core technology of a new industry, biotechnology, which was expected to have revolutionary impact on both society and economy. Inspired by what was perceived as the "new US biotechnology industry," European states in the early 1980s were increasingly assigned a central role in the support and the encouragement of biotechnology and its commercial usage. Whereas in the case of molecular biology research support in the 1950s and the 1960s (chapter 2) scientists were the main addressees of policymaking, now biotechnology policy operated in close relationship with industry. Genetic engineering was no longer confined to the laboratory. Biotechnology policymaking came into existence at the intersection between the private and the public, between developments in industry, in science, and in the political realm. The interpretations, framings, and scientific interpretations that were shaped in the various locations of this regime of governability played a critical role in the demarcation of the political space of biotechnology. They defined the traditions, actors, and institutions in the policy field. In this context, the already-established molecular biology policy field was an important point of reference and orientation. Supported by the state, research laboratories, field-test areas, hospitals, and farms became terrain for practices geared to genetic interventions at the subcellular level. By becoming instruments for producing (cloning) protein, microorganisms became laboratories and factories. The interventionist possibilities, which had been inscribed into molecular biology from the very beginning, seemed to turn into devices for transforming the structure of life on the molecular level.

Biotechnology policymaking attempted to create order and structure in this new and complicated field of power. The policy stories that were mobilized not only told about a project of transforming the economy; they also outlined strategies of redefining human beings, transforming nature, and creating a new form of political identity. Thus, the political space of biotechnology did not articulate a single logic and did not have a clearly defined set of goals; it was a terrain where a confluence of not necessarily causally related but nonetheless mutually reinforcing strategies and rationalities were deployed.

Accordingly, in this chapter I will focus on three interrelated themes: the discursive politics of framing biotechnology as a new industry and constructing it as an object for research and industrial policy; the difficult and controversial topic of the impact of biotechnology policies on the economy; and how strategies of creating political identity, redefining human beings, and transforming nature coexisted with the economic objectives of biotechnology policymaking.

My discussion of the relationship between language and biotechnology will underscore that discourse and "writing the text of politics" is not only about words but also about doing things with words. Discourse and writing are material practices that do not simply react to contexts. They are processes by which structure and organization are actively inscribed into semantically unstable environments. Neither "the economy" in general nor "the biotechnology industry" in particular "exists out there." For our understanding of the invention of biotechnology as a high-tech industry for Europe, it is important to see that new economic spaces, such as the economic space of biotechnology, do not come into being on their own. They do not have stable boundaries and identities; they are constituted performatively. Like other spheres, the economy is a terrain of political struggles, and it is governed not by a single logic but by a multiplicity of competing discourses. No economic space can be opened without being articulated in connection to other socio-political discourses in concrete historical conjunctures (Daly 1991). The US biotechnology industry, an improvement in a drug design, the pricing of a drug, and a drug's social acceptability are certainly not given facts; they are the outcomes of discursive struggles that can be understood only within the broader frameworks of discursive economies. Accordingly, biotechnology policymaking should not be conceptualized as some sort of reaction to "scientific developments," economic constellations, or group pressure. Though such factors play a role in policymaking, they constitute only part of a more complex picture which encompasses the multi-layered effort to negotiate realities of science, economy, and society. Policy narratives are crucial tools for negotiating these realities, for they always mediate between competing codes by which reality (such as economic contexts or scientific experiments) is assigned meaning. The realities of economic context or scientific experimentation are not simply "absorbed," denied, or reflected by discourse; they establish specific relationships with their codings or intermediations in discourse.[1]

The controversial question of the impacts of biotechnology policies also points to the need to pay attention to conflicting stories of policymaking. I will show that the discursive construction of biotechnology in Europe led to results that turned out to be quite different from what some policy narrations had initially envisioned. This gap between prediction and historical development stimulated the crafting of new policy stories. Hence, my analysis focuses as much on what the policies and intervention of various actors, groups, and institutions *intended* and *predicted* to happen through biotechnology as on the end results of their discursive practices, the actual distributions of funds, the institution-building strategies, and the cultural impacts of policymaking.

In the late 1970s, European biotechnology policy discourse had framed developments in US biotechnology that took place under unique socio-political and scientific circumstances as evidence for the emergence of a new high-tech industry: the new biotechnology industry. According to this story, Europe needed to "follow the US model," a project whose completion soon turned out to be a daunting task with unforeseen difficulties. Furthermore, in the United States as in Europe, biotechnology policy narratives contained a number of interpretations or assumptions, such about the structure of pharmaceutical and chemical industry and of international research and development, the workings of financial markets, and the "nature" of genetic engineering's conceptualization of things natural, such as humans, plants, diseases, and health. These interpretations constituted a chain of signification that gave meaning to the notion of the biotechnology industry as a "high-technology industry of the future." However, beginning in the mid 1980s a number of contextual and discursive developments began to undermine these interpretations given to biotechnology by policy discourse. These developments indicated that the various applications of genetic engineering in industry, agriculture, and medicine were hardly synonymous with the initial projections provided by policymakers, scientists, and industrialists in the late 1970s. By the early 1990s, biotechnology policy stories began to readjust and modify their explanations and interpretations of biotechnology. Increasingly, one of the central themes of biotechnology discourse, its characterization as a new "high-tech field," lost importance and was to an extent replaced and supplemented by a framing of genetic engineering as a new method used in the pharmaceutical industry or as "revolutionary, basic methodology to study nature."

Despite these partial reinterpretations in the early 1990s of biotechnology as at least not yet constituting a new "high-tech" sector, the policy narratives in the various countries continued to represent biotechnology as a project of transforming the economy, as a unique technology for the modification of nature, and as opening up ways to a completely novel understanding of human beings. The semantic integration of these various projects of biotechnology was made possible by what I call the "discourse of deficiency." As part of this discourse, biological narratives inscribed bacteria, animals, plants, and human beings as potentially in a state of deficiency and in need of a supplementary genetic technology. For example, in various human genome projects launched in the mid 1980s, the quest for health translated into a search for the genetic basis of unhealth, and, as a consequence, "normality" came to be defined as the absence of alleles[2] said to cause disease (Keller 1992, p. 298). Furthermore, policy narratives explained the transformative practices of genetic engineering, such as deliberate release projects, as essential to socio-economic development and to surviving the "international high-tech race." These simultaneous representations of and interventions into life constituted a "discourse of deficiency," because they implied a rewriting of life on a subcellular level in terms of "absences" and of "areas of improvement" in need of the intervention of genetic technologies (Rheinberger 1993, pp. 9–10). At the same time, policy narratives configured around the trope of "international competitiveness" and made various external and internal developments that otherwise would constitute signs of disruption within the collective identity "foreign."[3] Important developments of the 1970s, including the increasingly epidemic dimensions of cancerous diseases, environmental degradation, and the economic crisis, were removed from their complex sets of causes and transformed into "evils" or "others"—designations that virtually called for new strategies and approaches as offered by biotechnology. The rewriting of life in molecular biology's discourse of deficiency and the rewriting of political identity on the political level entered a mutually reinforcing relationship. The biotechnology policies emerging in the United States and in Europe established a semantic relationship between collective identity and the project of the rewriting of life through the new genetic technologies. According to this story of socio-economic development, society's identity was at once secured and modified by the practices of molecular biology. It was precisely these social symbolizations, metanarratives, constructions of identity, and underlying workings of power that provoked social resistance and were at the core of the new genetic engineering controversies of the 1980s.

Mythologies of US Biotechnology

The idea that the industrial applications of genetic engineering would lead to the establishment of a new industry, the biotech industry, was born in the United States. This "new" or "third-generation"[4] biotechnology was based two developments in molecular biology that occurred in the 1970s: the use of restriction enzymes to cut and splice genes ("recombinant DNA") and the use of cell fusion to develop hybrid cells with desired characteristics that would multiply themselves ("hybridomas").

A number of developments in US biotechnology seemed to underscore the huge economic potential attributed to the commercial exploitation of the new biology. Apparently, a new economic space had been opened up, primarily by a plethora of small companies funded by venture capital. For instance, in 1979, two years after its founding, Genentech announced its successful production of somatropin (human growth hormone) and thymosine alpha-1. By the early 1980s, genes had been cloned for producing proteins such as the already-mentioned human growth hormone, insulin, and Factor VIII (for the treatment of hemophilia). These developments paved the way for the emergence of new protein drugs based on proteins that occur naturally in the body's immune system, such as interferon, interleukin, tissue plasminogen activator (used to dissolve blood clots in arteries), and erythropoietin (used to treat kidney failure). Hybridomas led to the development of monoclonal antibodies, which, having antibody characteristics specific to a particular antigen, had wide application in separation processes and, of more immediate commercial value, in diagnostics and therapeutics. Outside pharmaceuticals, agriculture seemed to be opened up as a potential field of intervention by such techniques as the manipulation of plant genes to enable the rapid development of new hybrid plant species incorporating desired characteristics, such as resistance to drought or frost, and advances in animal biotechnology made possible the creation of transgenic animals (Sharp 1991a, p. 221). New experimental systems, new laboratory practices, and the scaling up technologies with appropriate automatization of manipulations, data storing, and processing appeared to create a powerful new system of intervention with actual or potential applications in primary industries (agriculture, forestry, mining), secondary industries (chemicals, drugs, foods), and tertiary industries (health care, education, research, advisory services) (Rheinberger 1993, p. 12).

Among investors in the United States the belief that biotechnology constituted a technological revolution comparable to that constituted by microelectronics and computers became considerably stronger. This development coincided with the creation of a strong venture-capital industry. In the early 1980s, Congress allowed pension, insurance, and endowment funds to invest up to 10 percent of their capital in venture deals. Enormous capital flows were directed into risky business projects. As a result, companies were funded earlier in their life cycles than public markets had ever done before. "Biomania" began as the first wave of biotech companies went public. Genentech's shares nearly tripled in price in the first 20 minutes of trading. Many other companies followed Genetech's path (Teitelman 1994, pp. 192–193). The US government joined in by giving substantial support to biotechnology research and development. In 1982–83, government funding of biotechnology was estimated at approximately $520 million per year, of which $511 million went to basic research. In 1989, roughly $5 billion was invested in biotechnology research, development, and manufacturing by US industry and the US government. Of this, an estimated $3.2 billion was provided by the government, with the National Institutes of Health providing the majority of the government support for basic biomedical research and training (Groet 1991, p. 201).

These developments in US biotechnology were soon to attain a mythical status in the European policy discourse. As Roland Barthes (1972, p. 143) has pointed out, myths do not deny things; on the contrary, they function by talking about things, simplifying them, and giving them a clarity, a factuality that eliminates contingency and the possibility of different interpretations. In Europe, the development of new protein drugs and Wall Street's favorable reception of biotechnology stocks were seen as indicating a revolution in the life sciences that would eventually lead to the establishment of a new high-tech industry comparable to that established by information technology.

The Discursive Economy of Europe in the 1970s

What was the European political discourse constellation within which biotechnology emerged as a new high-tech industry? After World War II, the political-economic discourses of the industrialized countries defined scientific development and economic growth as intrinsically linked to one another. The relationship was explained in terms of a strong correlation between the dramatic increase in R&D spending after the war and unprecedented economic growth (Coombs et al. 1987, pp. 205–207, 223–229).

After World War II, Western Europe and the United States began rebuilding their economies after a 20-year period of depressed consumer demand. Between 1945 and 1973, favorable economic conditions in the United States stimulated high growth throughout the world economy. The growth of the US economy helped the Western European countries and Japan to reconstruct their war-devastated economies. The reconstruction was based on American mass-assembly principles and was designed to achieve "economic miracles." Pent-up consumer demand for homes, household appliances, and other consumer goods stimulated production. Technological innovations in electrical equipment, machinery, scientific instruments, and transportation and in the chemical and aerospace industries contributed to rapid growth. All these conditions increased total demand, which in turn resulted in rapid economic expansion (Andrain 1985, p. 45). The world economy seemed to be in a never-ending boom, which seemed to be linked to scientific progress.

The political-economic discourse of the 1950s and the 1960s was dominated by the "science push theory" (Coombs et al. 1987, pp. 223–229), according to which the more R&D is "pushed" into the system the greater will be the flow of innovative products. And it was the state that had to "push" science. As a result, science and technology policies, especially during the 1960s and the 1970s, were being systematically developed in the industrialized countries. Political-economic discourse generally identified science and technology with economic growth, agricultural expansion, and medical advances within an overall framework of progress and modernization. Scientific and technological progress became increasingly represented as a central condition for a nation's viability and for its ability to compete in the international economy.[5]

With these changes in the economic and political discourse, a new inter-discursive economy was about to emerge—a new field of concomitance (Foucault 1972, pp. 57–58), in which an elevated status was attributed to the growth and advancement of science. Science and technology became increasingly defined as crucial elements in the creation of national pride, prestige, and viability. From the early 1970s on, it was increasingly believed that "simple" R&D policies were not enough for the eventual creation and commercialization of new products. Rather, at the level of governmental programs, science and technology were represented as elements in a complicated interactive set involving patterns of demand, market size, the organization of the university system, and the structure of industry—a representation that would eventually give way to the innovation and technology policies of the 1970s and the 1980s, which would respond to a new framing of R&D problematic (Rothwell and Dodgson 1992, pp. 227–228). In the first half of the 1970s, an oil crisis, surging inflation, and economic recession abruptly ended the dream of the never-ending postwar economic boom. European industry was hit especially hard by the economic crisis.

American semiconductor firms took advantage of the recession by cutting prices in order to improve their market shares. Many European companies pulled out of the "commodity chip" market and retreated into the relatively safer "custom chip" market. The semiconductor market effectively became divided into two noncompeting sectors: the "big league" (dominated by American companies) and the "little league" (the custom chip segment). In the early 1980s it became clear that microelectronics and its applications in manufacturing and information technology were going to be vital for all industries. Not surprisingly, the competitive advantage of American and (a bit later) Japanese firms in semiconductors, computers, telecommunications, and consumer electronics increasingly raised the painful question whether European industries were still viable in these crucial technological sectors (Sharp 1989, pp. 4–6; de Woot 1990). In the early 1980s, only true optimists believed that the European semiconductor, telecommunications, computer, consumer, and electronics industries could be effectively "saved." The declining competitiveness of the EC countries produced a negative trade balance from 1979 to 1985, which negatively affected not only high-tech products but industrial products as a whole. In sectors experiencing strong growth, such as electronic products, data processing, and office machines, the EC countries' share diminished by as much as 2.5 percent, while the United States and Japan saw their respective shares grow by 1.2 percent and 7 percent (Weiner 1990, p. 45).

It was around this time of "Europessimism" that several central characterizations were increasingly used to represent a situation in which Europe was seen to be behind in a technology race with its two main competitors: Japan and the United States. This political discourse was readily supported by economists and innovation experts who specialized in technological development, innovation, and trade policy. They expressed a consensus that Western Europe could survive and thrive only if it were to engage in the high-tech race and compete against Japan and the United States by vigorously supporting its national industries and research systems and encouraging them to innovate, cooperate, and compete. Success on the technological front became increasingly represented as a strategy that would help Europe save its identity. This could be accomplished only by fighting others, be it the United States, inert national industries, or critics of technology. Europe ought to overcome internal divisions, of markets or dissenting citizens, and confront as a united entity the common challenge of the international marketplace.[6]

The representation of a Europe engaged in a high-tech war with Japan and the United States was part of a textual strategy geared toward a neo-liberal reorientation that began in the United States and Europe around this time. The target of these strategies was the postwar capitalist model of growth. That model was represented as containing a set of barriers, including powerful trade unions, state monopolies, and a lack of investment incentives, which together inhibited market dynamics and kept the European national economies and Europe as a whole from adapting and facing up to new opportunities and challenges. Only if these barriers were removed could recovery be expected. Elements of this "strategy of recovery" enacted in various policy fields included increasing the profitability of private capital, creating a more favorable labor market by dismantling the trade unions, reducing corporate and middle class taxation, and privatizing of public industries and banks (Camiller 1989, p. 7). The neo-liberal strategy of recovery gave science and technology a strategic place. It was assumed that Europe could secure its survival only by facing up to the high-tech race, which, in turn, was possible only if the structural impediments to innovation were to be removed. Hence, technology policies, trade policies, industrial policies, and social policies all acted as mutually reinforcing discursive practices in a process that attempted to reshape the political order.

In the context of this industrial crisis, attention turned increasingly to what was widely perceived as the next important field of high-tech innovation after information technology: the new biotechnology. In particular, it was anticipated that the new biotechnology would have its strongest impacts on the chemical and pharmaceutical industries and on the agricultural sector—the strongholds of European industry. European dominance was especially evident in chemicals. Germany's BASF, Bayer, and Hoechst were the world's largest chemical companies, followed by the Britain's ICI (Orsenigo 1989).[7] However, in the early 1980s the European lead in these sectors was by no means taken for granted. By the early 1980s, as a result of the recession, the European chemical industry was in decline, and the European pharmaceutical industry, though outspending its US counterpart on R&D, was still behind.

The biotechnology policies drafted in France, in Germany, in Britain, and in the European Community in the late 1970s and the early

1980s were condensed into documents that later served as mythical reference points of orientation and direction in thinking about biotechnology. The programs that launched the biotechnology policies were not simply formulations of wishes or intentions. They were, more important, about the depiction of the field of biotechnology and about its re-presentation in a form in which it could enter the sphere of conscious political calculation. Programs consist of configurations of specific locales and relations in ways that enable the translation of preestablished political rationalities to new spheres of intervention (Rose and Miller 1992, pp. 181–182). Just as in the 1960s molecular biology was framed as a contribution to modernization politics, the new biotechnology was now inscribed in a way that made it the object of governmental intervention in the political context of the 1970s and the 1980s.

Genetic engineering continued to be framed as a contribution to modernization. But in contrast to the 1960s, when molecular biology was supported by classical measures of research support, in the 1980s, with a changed discursive constellation, modernization via genetic engineering was conceptualized as a problem for innovation policy, a policy field at the intersection of research, technology, and industrial policy. During the 1980s in Europe, political discourse defined the state as the central agency in a highly complicated project of coordinating a range of interrelated policy fields and institutions, such as financial markets, research systems and private companies. The practice of innovation policy displayed significant differences in Germany, France, and Britain and at the level of the European Community.

The European Community and the Making of Bio-Society

The European Community's support of biotechnology dates back to the Biomolecular Engineering Programme (BEP) of 1982–1986, which started out with a budget of 15 million European Currency Units and reached a total budget of 741 million ECUs for the period 1990–1994. As we have already seen, despite attempts between the 1950s and the 1970s to shape a European space of molecular biology and regulatory framework for genetic engineering, research and regulatory activities on the European level remained quite limited. The 1980s witnessed a "Europification" of the politics of genetic engineering as support policies and a European regulatory framework for genetic engineering began to emerge.

The European Community's substantial engagement with biotechnology, manifested in a multitude of programs and initiatives geared toward the stimulation of research, development, and production, raises the issue of the reasons behind the emergence of a supra-national level in the construction of the European politics of genetic engineering.

Representing European Science and Technology as a Governable Entity

In the 1950s, neo-functionalist theories argued that European integration would proceed exponentially, as the success of common policies in highly technical, noncontroversial sectors would put pressure on related sectors to formulate common policies (functional spillover). The newly integrated polices would then lead to the transfer of political authority from national to European institutions, largely as a result of political pressure from interest groups and political elites with their own interests in European institution building (Peterson 1991, p. 274). But, as my discussion of biotechnology policies in France, Germany, and in Britain will show, the course of biotechnology policy in Europe hardly corresponds to the neo-functionalist image of a gradual transfer of responsibilities from the national to the supra-national level. A more realistic approach would be to give a structural explanation of the emergence of biotechnology support policies, and to argue that biotechnology was seen as crucial to national interests and that supra-national collaboration was a form of collective defense.[8] But, as we will see, there is also little evidence for this interpretation.

The realist explanation, like the neo-functionalist explanation, leaves out several critical aspects that one must consider in order to understand EC biotechnology policies. Just as I see policymaking and power constellations as fragile processes of stabilizing differences, not assuming any "given" facts of governance on the national level, I do not assume any "existing" patterns of supra-national governance that could serve as "independent variables" to explain the outcomes of policymaking. By this I do not mean to dispute, for example, the "existence" of the Treaties of Rome or of the European Parliament. But I wish to emphasize that, for European biotechnology policy to emerge, Europe had to be inscribed as a governable entity. In this perspective, the government of the European Community, like the government of a nation, must be seen as a process, a specific institutional ensemble with multiple boundaries without a pre-given, institutional fixity. The articulation of European territory in the field of biotechnology must be seen as related, on one hand, to changes in the social forces, to the discursive economy, and to new social projects that re-imagined Europe as a governable entity, and, on the other hand, to the crystallization of past strategies for the creation of a European space of government. Likewise, we should not assume the "existence" of something like a European biotechnology industry or a European R&D system. As I will show, the determination of the boundaries of biotechnology research and industry in Europe was due as much to a contested, discursive process as to the specification of a "European level" of policymaking.

As I mentioned above, the representation of Europe as a competitor in an international "technology race" was important for the emergence of a European biotechnology policy. This representation culminated in the Maastricht European Union Treaty, which, among other things, gave the European Community explicit powers in industrial matters for the first time. Unlike the 1970s, when the EC's member states devised national-level industrial policies, which focused on such declining industries as steel and shipbuilding, the new EC-level industrial policy concentrated on "high-tech" sectors (Nicolaides 1993, p. 1).

Initially, the Treaty of Rome did not give the Commission of the European Community explicit powers to promote industry or R&D. But in the face of the economic recession of the 1970s, the European Council decided in 1974 that the Commission should pay systematic attention to research, and Ralf Dahrendorf became the first commissioner to adopt a research portfolio. Soon thereafter, Dahrendorf proposed the formation of a "think tank" that would shape the EC's evolving science and technology policy by examining future trends and needs in this field. There were also proposals to make this a fifth European institution; instead, it became the Commission's "Forecasting and Assessment in the Field of Science and Technology" (FAST) program. FAST, launched by the Council of Ministers in 1978, had as its main goal "to contribute to the definition of long-term Community research and development objectives and priorities and thus to the formulation of a coherent long-term science and technology policy."[9]

The articulation of biotechnology as a field of governmental intervention on the European level was not only related to a change in discourse concerning the need to establish a European R&D policy that would respond to the crisis of "old" industries and to the early 1970s recession. In the late 1970s the crisis of European microelectronics and associated technologies became highly visible and contributed to the emerging "Europessimism." The poor performance of the European electronics sector was a riddle. Since the 1960s the French, German, and British governments had subsidized this strategically important sector with considerable sums. Increasingly, this strategy of promoting "national champions" in each individual member state was questioned. Belgium's Commissioner for Industry, Vicomte Etienne Davignon, invited the heads of Europe's leading electronics and information technology companies to a series of round table discussions, out of which ensued the idea of a European strategic program based on collaboration among the major European companies, their smaller counterparts, and universities and research institutes, with a focus on the pre-competitive end of collaborative research (Sharp 1991b, p. 64). In 1984 the idea of this program (ESPRIT, 1984–1988) was grafted onto existing activities of previous EC research in order to form the First Framework Programme. It ran from 1984 to 1987, with a budget of 3.75 billion ECUs. With ESPRIT and the First Framework Programme, science and technology had been articulated into the activities of the European Community. Apparently, the new industrial policy discourse evolved around ideas of French-style interventionist collaboration and selective assistance of high-tech industries, rather than theories of promoting competition, and thus created a favorable environment in which focused policy programs could develop (Holmes 1993).

The Shaping of a New European Technology Policy Narrative and the Launching of a Biotechnology Strategy

Thus, in the mid 1980s a new narrative on European technology policy was being told: since European companies could not compete in an increasingly global market, they must be converted into competitive multinationals with a Europe-wide marketing base. One way to realize this story was to promote joint research ventures between European companies from different countries. It was hoped that this research collaboration would lead to joint ventures. This process would also have to involve research laboratories situated outside industry, and more industry-university cooperation would be necessary. But this inevitably meant that companies would have, at least at first, to collaborate with competitors. But this would be no problem, for the EC would only support pre-competitive research—that is, work on ideas that were far from becoming marketable products. Such research had the advantage of saving individual companies money and effort by having them pool R&D resources without giving away trade secrets. So the EC needed to encourage interfirm collaboration, to prevent the duplication of R&D, to coordinate R&D efforts, and to foster university-industry linkages.

Clearly, this technology policy narrative that crystallized in the mid 1980s established the Commission of the European Community as a crucial coordinating center in the shaping of biotechnology in Europe. European biotechnology companies, universities, and research laboratories active in "modern" biotechnology, protein engineering, bioreactor technologies and other models, instruments, and representations fundamental to the genetic engineering complex all became part of the network of European biotechnology being sketched by the Commission.

The problematizing activities that prepared the path for the construction of this network date back to 1974, when the Commission first developed a proposal for a program of genetic engineering, enzymology, and cellular and molecular pathology. A participant remembers:

> At that stage there was a small nucleus of lively scientists, in the Euratom and Ex-Euratom radiation biology program who put together around 1975–76 the first proposal for a Community research program in bimolecular engineering, a phrase used to cover genetic engineering and enzymology. In fact, they also wanted to do a medical element, molecular biology, that did not get accepted, . . . but the other proposals argued about for years were finally accepted in November 81.[10]

After several years of debate, an agreement was reached to launch a small program covering elements of genetic engineering and enzymology: the Biomolecular Engineering Programme, which supported research in the period 1982–1986. What made BEP possible was, to a considerable extent, its inscription into the larger EC research and technology offensive which was launched at the same time. In this context, a group within the FAST Programme, the Bio-Society Working Group, translated the devised biotechnology strategy into a part of the emerging R&D and industrial policies of the EC—a tactic that proved successful. One of the participants remembers:

> We got the backing of the cabinet of Davignon. The crucial thing really was that Davignon gave his backing. . . . Davignon had seen the need for a strategic response by the Community to the challenge to the European microelectronics and telecommunications industries. It was only after he had persuaded Maggie Thatcher and the other ministers to accept this huge 1.5 billion ECU program for ESPRIT that late '83 [and early] '84 what is happening in

biotechnology turned around. . . . There was a certain amount of inter-service tension around Feb.-March 83, but in effect it was DG XII[11] which drafted finally the COM 83–28, the first Commission communication (on biotechnology). I had also written around '82–'83 a paper called plan directive biotechnology which was one of twelve plans by objective underpinning the first overall framework program. Davignon had brought in the concept of top-down framework programming instead of just lurching from political priority to policy priority. . . . Davignon had a more systematic top-down system of priorities—what were our priorities? What needed to be done on the Community level?—and . . . for the underpinning of that looked for preparative papers, so the paper plan by objective biotechnology was one of these twelve papers . . . and basically went to the Athens Council in May '83 and said biotechnology is important and the Community must do something about it. And I suppose it had about thirty seconds of discussion in the Council of Ministers saying fine, if it is important, tell us what to do. So more a serious effort was written in the summer of that year and COM 83 672 was annexed and delivered in October and toward the end of that year it went to the Bonn Council saying we want six action priorities."[12]

According to the emerging EC technology policy narrative, the main objective of the BEP was to contribute, through research and training, to the removal of bottlenecks that inhibited the application of modern biochemistry and molecular genetics to certain sectors of agro-food industries and agriculture.[13] The Commission's major policy statement, "Biotechnology in the Community," argued: "While some international organizations exist which provide support for *basic* research (European Science Foundation, European Molecular Biology Organization), there is no equivalent support for basic technological development in the life sciences."[14] Clearly, the argument that there was a "need" for a biotechnology policy on the European level played an important role in the launching of the policy. However, to receive political support, the biotechnology program had to be translated into the idiom of the operating discursive economy.

Toward a Bio-Society?

This translation was done by FAST's "Sub-Programme C," titled "Bio-Society," which became the central framing device in developing, with the cooperation of scientists from various European countries, a political rationale for a European biotechnology policy. In the Bio-Society program, biotechnology policy was represented as a strategy to simultaneously create a new high-tech industry and a new type of society whose identity is modified and secured by the manipulation of genes. A participant remembers the launching of the "Bio-Society" program as follows:

We set up this forecasting group in science and technology (1978); a small team of graduate staff was brought together to find out what is happening in life sciences and biotechnology. . . . There were many biologists involved in the deliberations in the sense that we worked very much through the networks; we worked very much with the European Federation of Biotechnology and its working groups . . . so we had a very large network with people involved at the time we finished. In the December 1982 report, there was a substantial chapter on biotechnology . . . and we in effect implemented in the succeeding years almost exactly what was recommended in this report.[15]

The multifaceted FAST Programme put its narrative on biotechnology to work not only at the level of specific strategic recommendations but also in the development of a "political philosophy" of biotechnology. According to FAST, the current surge of interest in biotechnologies was related to the "biological revolution," which arose from "fundamental discoveries within the life sciences (molecular biology, genetics, biochemistry etc.) and certain applied sciences." Over 20 years, this "biological revolution" had led to "breakthroughs" in understanding and to "man's mastery" over living organisms, and had provided new solutions to strategic problems and requirements in health, agriculture, agro-food industries, chemical industry, energy supply, environmental management, and waste treatment. The FAST narrative proclaims: "There is little exaggeration in stating that *in biotechnology one can now 'invent to order,'* subject to some uncertainties on price and delivery time."[16] The specific meaning of "to order" becomes clearer in a FAST document that represents biotechnology not only as economically beneficial but also as form of social control. In the view of the FAST authors, Europe faced increasing uncertainty concerning future social, economic, and political developments, and biotechnology could play a stabilizing role:

Though labor troubles, unemployment, and energy crisis currently dominate the international economic scene, the coming twenty to thirty years will, it is thought, see two major changes: the computerization of society . . . and the biological revolution emanating from the boom of "life technologies." . . . Within the relatively near future, bio-technology could be used in a number of sectors such as: Human health and behavior: we could control the development of the human embryo, and perhaps within twenty years determine its sex. We could prevent certain malfunctions. We should be able to create new vaccines and inoffensive drugs to counter addiction to alcohol or tobacco—even to regulate moods and emotions. We could also improve the quality of life for the elderly, improve techniques for transplants. . . .[17]

Clearly, this part of FAST's policy story reflects a view of the present condition of contemporary society as troubled and "out of control." In addition to its economic benefits, genetic engineering was conceptualized as a potential contributor to a broader process of social regulation, mainly by virtue of its expected ability to control life processes and behavior on the subcellular level. It is here that the meaning of the term "Bio-Society" becomes clearer. "Bio-Society" should be understood as a shorthand for a future scenario "in which [an] increasing proportion of man's activities are based upon the sustainable processes and recyclable products of agriculture, biotechnology, and all other practical processes for manipulating and exploiting living materials."[18] Biotechnology, the report claimed, should not be seen as a novel development, "but as a field in which recent progress merely extends the quantitative and qualitative progress of man's historically increased mastery over nature. . . . We are thus confronted with a continuing need to improve our understanding and management of our system. Like any other natural ecosystem, we are a self-managing bio-society."[19] In short, from an "evolutionary" perspective, biotechnology is not an option, but a necessary "form of self-understanding" for securing the survival and progress of mankind. It is here that we see most clearly how, in the FAST narrative, biological knowledge serves to interconnect "technologies of the self," a new form of human self-recognition in the form of the molecular discourse on life, with references to such often-invoked totalities as "the economy" and "Europe." Thus, recognition of the truth of molecular biology presupposes a regime of power in which new knowledge of living matters is coordinated with the discursive construction of Europe and its potential for economic growth. In this construction of political identity, biotechnology policy becomes a strategy of revealing the truth of life processes and at the same time serves the self-management of society.

The emerging policy narrative related the notion of social control qua Bio-Society to a "practical" European strategy for biotechnology that addressed well-established elements in the political discourse on European politics. Europe, according to FAST's framing, has a strong tradition in medicine and pharmaceuticals. A biotechnology strategy ought to preserve Europe's strength in this field. In contrast to the health sector, agriculture has traditionally commanded a major place in EC politics and a dominant share of expenditures. Biotechnology has the potential to intervene in this context: "The definition and reconsideration of policy in agriculture has to include increasingly the strategic and rapidly changing technological dimensions: this is one of the central challenges to the Community."[20] In addition, biotechnology offers the chemical industry a chance to develop new technologies for the production of raw materials thereby potentially reducing the EC's dependence on petroleum. Belatedly, the development of energy-efficient biological routes to final products and the utilization of biomass fuels and wastes could play significant roles in satisfying local requirements. Likewise, potential environmental applications of biotechnology might help to confront the challenge of deteriorating environmental conditions.[21]

FAST's policy story also established important intertextual links to other European policy narratives by arguing that biotechnology promises to intervene in two important areas of European politics by allowing the EC to meet the strategic challenge of agricultural policy and to create a new strategy for the European chemical industry. One political strategy in this context, according to FAST, would be to provide chemical feedstocks from European agriculture to European producers at world prices. This practice would hasten the development of new chemical processes and, more important, provide an outlet for surplus agricultural production. Furthermore, new technologies might lead to the use microorganisms in mining. Europe could decrease its reliance on Third World resources by reducing imports from Third World countries, but also by installing new types of technology transfer: "Europe must consider in the context of long-term strategy the *parameters of acceptable dependence*."[22] Once the policy challenge was discursively constructed, FAST proposed a complex strategy of response that combined the following elements:

• manpower and training through exchanges between laboratories in various member states

• the launching of scientific and technological programs, particularly in the area of agricultural biotechnology

• the creation of a special unit responsible for the organization and coordination of biotechnology-related activities within the EC

• contextual measures, such as the regulation of recombinant DNA work, patent legislation, and the establishment of price regimes, that would make primary materials of crucial importance for the chemical industry available at world-market prices.[23]

Clearly, the Commission's strategy for biotechnology did not simply consist of focusing on R&D support, but expressed a grand design interfacing R&D policies, industrial policies, and regulatory politics.

One of the participants in the FAST team remembers the specific context for the development of the EC's biotechnology strategy:

> We took from the start a very deliberate decision to take a broad view of the life sciences and biotechnologies; it wasn't just on specific techniques of genetic engineering. It was certainly a concept of an integrated strategy in the sense that it is multidisciplinary in its roots, it is multisectoral in its applications, and it is consequently very difficult for vertical bureaucracies to handle, if it is the universities, the Commission, or the governments. But nevertheless, it has a unity and requires for some purposes an integrated strategy. The national approaches were more heavily constrained from the start by the need that there be a lead ministry. We in a sense being in the FAST program had from the start an interservice mandate, whereas, typically in the UK, for example, it was set up in the DTI. . . . You had everywhere the same kind of interdepartmental turf battles breaking out. It has happened in the Commission as well. In 1983 it was not happening because most people in a small bureaucracy like this are so busy with other things and basically had never heard about biotechnology or thought it is basically something for the scientists in DG XII to worry about. There had been the whole fuss about recombinant DNA after Asilomar, but the whole regulatory question was taken up in DG XII, which was prepared to draft a directive and placed it before the Council for a recommendation, so in the '78–'82 period the whole thing was handled within the research domain.[24]

With respect to the implementation of this strategy on the level of R&D policies, biotechnology was an expanding policy field of the Commission. A policy story about biotechnology in Europe had been constructed which began to suggest blueprints for socio-economic transformation and the positions and tasks of a variety of actors and institutions such as the Commission, science, and industry in the newly emerging policy field.

From Bio-Society to Bio-Industry

In the following years, the biotechnology policy of the Commission of the European Community outlined a large number of interrelated strategies to accomplish what was seen as Europe's transformation into a Bio-Society. Central to the Commission's policy stories was the mapping of a specifically European biotechnology research base and industry, which was defined as crucial to Europe's industrial future. In the document "Biotechnology in the Community," the Commission elaborated for the EC's Parliament and Council its views on a strategy for biotechnology as first sketched in the FAST documents. From the analysis, the program went on to elaborate a set of priorities for EC action and to construct the boundaries of a specifically European system of biotechnology R&D. These priorities fell into two major groups: "research and training"[25] and "creating a favorable context for biotechnology which together outlined a grand design of strategies and structures to shape European biotechnology."[26]

The "Research and Training" section distinguished *horizontal* actions (bearing on all branches of biotechnology and focusing on the creation of a supportive background for biotechnology research and thus eliminating barriers keeping industry and agriculture from exploiting fundamental advances in modern biology) from *specific actions* designed to stimulate specific in-depth developments in well-defined sectors of biotechnology related directly to sectoral policies of the EC. The horizontal measures included *infrastructure measures* such as bio-informatics (based on the assumption that the rate of development and exploitation of biotechnology will become increasingly dependent on advances in technologies of data capture, information storage, and retrieval), culture collection, gene banks, and basic biotechnology. This program represented the continuation of the BEP's focus on the removal of bottlenecks preventing the application of modern genetic and biochemical methods to industry and agriculture. Under *specific actions* a series of strategic issues were singled out, including interfacing between the agro-food and chemical industries; restructuring of the health industries through the development of new methods of genetic screening; and the development of new techniques, devices, or drugs, such as diagnostic tools or new drugs based on computer-aided molecular design, that could prevent diseases and substantially contribute to a reduction of costs.

The "Creating a Favorable Context for Biotechnology in Europe" section of "Biotechnology in the Community" highlighted a group of interrelated organizational and regulatory measures. One important element of a biotechnology strategy, the report claimed, was the creation within the Commission of a new organizational unit responsible for the concentration and monitoring of biotechnology.[27] A further contextual measure dealt with the supply of raw materials of agricultural origin to industry. In addition, the report argued, the EC needed to address the widely unsettled issue of intellectual property rights in biotechnology. The same was said to hold true for regulations affecting biological safety, laboratory research, industrial processes, and the production and circulation of goods.[28]

In the years that followed, the new policy narrative served as a central point of orientation for a strategy of biotechnology support at the level of the European Community. This strategy found itself, step by step and with certain modifications, specified, and to a considerable extent implemented, in a variety of specific programs. The biotechnology policy narrative was reinforced by a further consolidation of the EC's R&D policy on the institutional level. The Single European Act (SEA), which went into force in 1987, provided the EC with explicit, general powers in the field of R&D. According to SEA's policy discourse, the overall aim of EC R&D was "to strengthen the scientific and technological basis of European industry and to encourage it to become more competitive at international level."[29] Thus, community research activities became legally oriented away from fundamental research of the type done by EMBO and toward directed and applied industrial research. The changed policy discourse facilitated the ongoing attempt of the Commission to contribute substantially to the mapping of a specifically European biotechnology industry.

Several important discursive links were incorporated in the Commission's biotechnology narrative. The EC's biotechnology program was discursively coded in a number of ways. As industrial policy, it took a proactive approach toward expected restructuring in the chemical and pharmaceutical (core) sectors of European industry. As agricultural policy, it translated biotechnology as a possible solution for one of the central policy problems of the European Community: agriculture. As a new form of health politics, it promised new solutions for reducing health-care budgets. As a raw-materials strategy, it looked to decrease dependence on foreign (mainly Third World) countries. As a regulatory policy, it sought not only to fund biotechnology but also to manage its risks. As a supranational strategy, it aimed to build a strong and unified Europe. These representations of biotechnology as a "European project" cleared the way for its entrance into the sphere of political action. Biotechnology became a legitimate topic which the European Community had to deal with in a comprehensive manner.

The new biotechnology policy discourse also became instrumental in the shaping of new institutions and actors in the field of biotechnology policy. In 1984 a Biotechnology Steering Committee was set up to coordinate the policies and operational activities of the various units of the Commission of the European Community that were involved in biotechnology. At the same time, the new biotechnology policy narrative explained the ongoing creation of a system of biotechnology interest groups. FAST had already been given a mandate to establish an ad hoc system of collaboration between specialized research groups within the community, a system that would allow the groups to have active input into the program and would thus encourage coordination. This mandate was part of a strategy of enrolling European industry in the planned biotechnology network. Furthermore, as biotechnology was articulated as an agenda for EC action, industry began to elaborate a system of interest coordination and influence at the EC level (Grant 1993, pp. 27–46). In 1985, around the time of the launching of the EC's biotechnology policy, the European Biotechnology Coordination Group (ECGB) was created. The ECGB was a loosely organized association composed of the CEFIC (European Council of Chemical Manufacturer's Federations), the EFPIA (European Federation of Pharmaceutical Industries' Associations), the CIAA (Confederation of the Food and Drink Industries of the EEC), the AMFEP (Association of Microbial Food Enzyme Producers), and the GIFAP (International Group of National Associations of Agrochemical Manufacturers). Each of these groups, particularly the CEFIC and the EFPIA, was itself a powerful player in the lobbying game with respect to biotechnology. As the ECGB turned out to be rather ineffective in its lobbying, the CEFIC created a Senior Advisory Group in Biotechnology in June 1989. In contrast to the ECBG, the SAGB was based on direct and by-invitation-only membership. A forum representing seven of the most important chemical, pharmaceutical and agro-food companies operating in Europe (ICI, Unilever, Sandoz, Feruzzi, Rhône Poulenc, Hoechst, and Monsanto Europe), it soon became a highly influential actor in EC biotechnology politics.[30]

Thus, the mapping of a European biotechnology research infrastructure and industry provided important explanations and justifications for the creation of the new European biotechnology policy field with European actors and institutions. Policy discourse also oriented the establishment of CUBE (the Concentration Unit for Biotechnology in Europe) and the gradual formation of a system of interest intermediation by industry, a structure of policy decisionmaking in biotechnology with agendas for the support of biotechnology and its regulation which were integrated into one institutional framework. Thus were drawn the boundaries of a system of close collaboration between the Commission of the European Community, science, and industry that was intended to make European biotechnology a scientific and economic reality.

Conflicting Framings of European Biotechnology

While in democracies workplace-safety and environmental regulations are binding on companies and on research laboratories, and the space to negotiate their contents is limited, policy measures intended to support innovation in companies depend on their acceptance by those they address. But a critical component of the Commission's biotechnology narrative—the large chemical and pharmaceutical companies of Europe—seemed to have no intention of assuming the roles assigned to them in the EC policy discourse. Industry had been assigned the specific meaning of being a central part of the Commission's technological-industrial vision, but it rejected this role. The European chemical and pharmaceutical industry was in a process of restructuring, but it certainly was not in need of any substantial state support or intervention. From the early 1960s on, the pharmaceutical industry was characterized by three structural trends: a shortening of effective patent protection (due to the lengthening of time for the drug testing), a sharp rise in development costs, and a decline in the rate of introduction of new drugs. The pharmaceutical industry attributed these trends to a certain exhaustion of possible drug development strategies based on traditional biology and inorganic chemistry. Increasingly the new biotechnology raised hopes of setting new rules for drug development and reversing the downward trend that had characterized the industry's innovation process since the 1950s (Tarabusi 1993, pp. 131–132). Also, chemical companies began to invest in agriculture, and agro-food companies in biotechnology (Chataway 1991, pp. 1003–1005). The distinctions between the chemical, pharmaceutical, and agro-food industries became increasingly blurred.

But none of this implied that the pharmaceutical and chemical industry in Europe saw a need for a biotechnology policy on a European level. For instance, in a letter responding to a FAST biotechnology strategy paper summing up the Commission's planned biotechnology strategy, a representative of the German chemical company Bayer wrote:

> The commission should concentrate its efforts on the stimulation of scientific exchange, the improvement of information systems, and the harmonization of laws and regulations affecting research and introduction of new products. Furthermore I consider it of major importance to make starch and other agricultural feedstocks available to the European Biotechnology industry at world market prices. I see no necessity for the commission to fund research

projects additional to the national programs except in areas of applications to the food and energy problems of Third World countries.[31]

Such assessments failed to impress the Commission, and European biotechnology programs driven by the idea of creating a *European* biotechnology industry began to take shape. The Biomolecular Engineering Programme[32] was followed by the Biotechnology Action Programme.[33] Both of these programs were intended to promote and encourage the development of new technologies for the purpose of developing better biological and agricultural products.

At the same time, the large European chemical and pharmaceutical companies transformed themselves into global players. Industrial concentration had led to a situation in which multinational companies competed only with a handful of giants on the global level (no longer with numerous national companies) and the ability to exploit innovations globally in a rapid manner had become essential. This was especially important in cases where locally needed technology could not be provided by the corporate central research (de Meyer and Mizushima 1992, pp. 136–137). There is evidence that in the United States and in Europe large companies increasingly conducted biotechnology research and development "in house," and that, as a result, there was backward vertical integration of these companies into biotechnology R&D (Pisano 1991, p. 248).[34] However, in contrast with the United States, where a gradual integration of biotechnology companies into manufacturing was visible, there was little evidence of a forward integration of new biotechnology firms (NBFs) into manufacturing in Europe. This was a major reason for the dominance of larger firms in European biotechnology. The NBFs lacked access to "complementary assets," such as competitive manufacturing, marketing and distribution networks, and the ability to get a new product through the regulatory process and on the market. For exactly this reason, in the United States highly integrated patterns of interactions between small and large firms arose. The small firms had a comparative advantage in R&D (which had been transferred from the universities); the large firms had the complimentary assets the NBFs needed (Dodgson 1991, pp. 116–117). But in Europe the policy vision of a biotechnology industry characterized by the co-presence of a multitude of nationally operating small, medium-size, and large firms did not work. There was an unreconciled conflict of interpretation between the way the Commission and industrialists perceived the nature and the strategic requirements of the biotechnology industry in Europe.

Even successful smaller companies such as British Biotechnology and the French company Transgène never operated independently. Whereas in the 1980s Transgène was an integrated part of the Rhône-Poulenc empire, British Biotechnology collaborated with a number of large companies, among them Pfizer, SmithKline, Beecham, and Japan Tobacco. These collaborations had provided British Biotechnology with considerable revenue, assistance with scientific and technological development, and enhanced scientific and commercial credibility (Dodgson 1991, p. 121). In a similar way, American companies were responsive to collaborations with large European companies or were simply acquired by them. Between 1981 and 1989, American companies formed 51 strategic alliances with British companies, 48 with German companies, and 15 with French companies. During the same period, of the 291 US international biotech alliances (excluding those with Japan), small firms accounted for 164 alliances with large overseas corporations. Of the 107 alliances of large US firms, 51 were with small foreign firms. This trend, notably visible since 1987, reflects the rise of European biotechnology companies. However, the clear trend has been that small American firms have sought more and more alliances with smaller and smaller numbers of foreign corporations. Between 1986 and 1990, the percentage of West European partner companies rose from about 50 percent to 60 percent of all international alliances. Approximately 80 percent of all alliances involving partners that were not US biotechnology companies resulted in outward technology flows (Wagner 1992, pp. 529–531). Starting in the late 1980s, a new phenomenon emerged: the fusion of large companies, such as between SmithKline and Beecham, Rhône-Poulenc and Lafarge-Coppé, and Mérieux and Connaught Biosciences.[35]

This pattern confirmed that international arrangements were becoming the norm for companies engaged in biotechnology, regardless of type, age, ownership, or size. The firms that had participated in international arrangements tended to have greater resources and to have employed the most sophisticated recombinant DNA techniques (Wagner 1992, p. 530). This changing context for European innovation policy was not reflected in the biotechnology policies launched by the Commission of the European Community. While the European chemical and pharmaceutical multinationals saw the boundaries of the biotechnology industry as defined by the worldwide best opportunities for research and investment, European policymakers drew the boundaries of the bio-industry around Europe.

During the second half of the 1980s it became clear that the initial plan to encourage the collaboration between industry and universities did not work. The projects funded were mainly of basic research character and conducted in public laboratories and universities. Of the 262 research contracts awarded to laboratories, only 16 of the contractors were industrial firms.[36] Additional major obstacles to this type of collaboration were secrecy and property rights—areas that set clear limits on any inter-firm cooperation as practiced, for example, in ESPRIT.

However, the failure to intermediate between policy goals and policy context did not mean that the enacted policies did not have any impact. The Biotechnology Action Programme and the Biomolecular Engineering Programme created transnational cooperation between laboratories. In particular, a systematic effort was made through the concept of European Laboratories Without Walls (ELWWs) to gather research groups into common projects. During the period of BAP and BEP, 35 ELWWs were set up between laboratories that agreed to share methods, materials, and results and to allocate an important portion of their research to totally integrated joint efforts. When BEP was started, transnational cooperation in European R&D was virtually absent. Because of the emphasis put on cooperative research arrangements by the services of the Commission of the EC during the implementation of BEP, nearly every contractee had been involved in one or several cooperations by the end of the program in 1986 (Van der Meer 1986, p. 277).

What followed after BEP and BAP was a number of programs responding to the initial critiques made with respect to the initial biotechnology programs. Within the framework of Biotechnology Research and Innovation for Development and Growth in Europe (BRIDGE),[37] more than 600 research and training contracts were launched.

The focus of European biotechnology policy on agricultural problems is most evident in two biotechnology programs that dealt exclusively with agriculture: the European Collaborative Linkage of Agriculture and Industry through Research (ECLAIR) program[38] and the Food-Linked Agro-Industrial Research (FLAIR) program.[39] Clearly, the "main" biotechnology programs had already shifted their focus to agriculture. In the biomedical sphere the most important initiative was the Human Genome Project, approved in 1990 with funding of 3.3 million ECUs for 1990 and 1991. The task of the

Human Genome Project was to improve the human genetic map through the use of physical mapping, data processing, and data banks and to advance the methods and basis for the study of the human genome. Funding would also go toward the training of scientists and toward studies of the ethical, legal, and social implications of the research.[40] Originally presented as a "Predictive Medicine" program,[41] the Human Genome Project met with fierce criticism in the European Parliament and was finally withdrawn by the DG XII Commissioner Filipo Maria Pandolfi (Dickson 1989). The revised and finally adopted program included amendments, required by the parliament, that reflected the changed power structure of the European political system; it also reflected dramatic political changes during the 1980s that had resulted in a much more conflictual terrain for biotechnology research. (See the next chapter.) The new political climate was also visible in the 1990–1993 Biotechnology Programme, which, after a long delay by parliament, allocated resources for studying the social and ethical implications of biotechnology and also incorporated a special research section on ecology and population biology.[42]

With the continuous and exponential growth of the European Community's biotechnology programs, the Commission of the EC emerged as a major coordinating force in the political shaping of biotechnology in Europe. The discursive construction of biotechnology as something "European" had contributed significantly to the transformation of the Commission from an observer into a powerful actor in European R&D development. Furthermore, an attempt was made to enroll European scientists and companies in the pharmaceutical, chemical, and agro-food sectors and to enroll scientists as actors in the newly emerging European biotechnology policy field. These attempts had some institutional impacts, such as improved inter-European research cooperation and the development of biotechnology interest group system. But European biotechnology policy failed to manage genetic engineering's complex field of discursivity in a way that would trigger the envisioned cooperation by a number of dispersed actors in science and industry and eventually create the envisioned distinctly European biotechnology industry. The Commission's biotechnology policy had been successfully launched by inscribing itself into the discursive economy of policymaking at the EC level. Its policy narrative had represented biotechnology research and industry as part of the larger agenda of a European R&D policy. At the same time, in various biotechnology policy documents and statements, the boundaries of biotechnology were drawn around Europe. But these policies had failed to define the reality of European biotechnology for some of the key actors in the policy field. Significant differences in the codings of the realities of the nature of biotechnology industry between EC policymakers and actors in industry were an important reason why the economic effects of the EC biotechnology programs were limited. Nevertheless, these policies had considerable socio-political and scientific-infrastructural repercussions. While a European biotechnology industry failed to materialize, the programs developed in the Commission's policy narrative participated had a clear impact on the shape of the research infrastructure, on the diffusion of genetic engineering to a variety of new locations (from the human genome to the deliberate release of genetically modified organisms), and on the rise of the state (or a state-like actor such as the Commission) as a powerful actor in European biotechnology.

Germany: Modernization Discourse, Global Corporations, and Biotechnology

In the European Community's research and technology policies, biotechnology research and its industrial applications had been mapped as a distinctly European project. At the national level, there was a strong tendency among policymakers to draw the boundaries of biotechnology around the nation state. In Germany the shaping of the field of biotechnology policy can be tracked back to the late 1960s, when the Bundesministerium für Wissenschaft und Forschung (BMWF) launched its Neue Technologie program, which included biological and medical technology as a field of support. In loose coordination with the BMWF, the Deutsche Forschungsgemeinschaft and the Stiftung Volkswagenwerk initiated a discourse on the newly emerging biology and its potential industrial applications. (See also chapter 2.) This process culminated in a 1973 report by DECHEMA[43] that developed a powerful policy narrative about the status of modern biotechnology in Germany. This narrative and its implementation in a number of more specific biotechnology policy programs under the Bundesministerium für Forschung und Technologie established important discursive links between the new biotechnology and a set of more general arguments and strategies which constituted the repertoire of modernization discourse in the late 1960s and the early 1970s.

Like all discourses, the German modernization discourse was caught in a complex web of materiality and gained its meaning in a specific discursive constellation. The German "economic miracle" of the postwar period, manifested in the continuous economic growth in the 1950s and the 1960s, was often attributed to the success of the social market economy, a form of economic management characterized by the absence of excessive state intervention (Esser et al. 1983, p. 105). However, if we look at the practices of policymaking we see that since the mid 1960s economic planning and new modes of political intervention in the economy had been interpreted as crucial to the successful continuation of the German model of economic development. The "New Technologies" program of 1969 was part of this new state activism. The importance of science and technology for economic growth had also been stressed in the international discourse on economic development throughout the 1960s. But this international discourse of technology policy was not linked to any specific "policy recipes." Such recipes turned out to vary from country to country and, to a certain extent, from technology sector to technology sector (Ronayne 1984, pp. 228–230).

What was the meaning of the "new approach toward science and technology" within German political discourse? In this context it is important to see that Germany, unlike Britain and France, never developed a comprehensive concept of industrial policy (Horn 1987, p. 64). What Germany developed instead was called structural policy. Structural policy, which grew from the late 1950s onward, had three components: sectoral structural policy (which covered areas from rescue operations in shipbuilding to the launching of a nuclear power industry), regional structural policy (which was to adjust development differences among regions), and technology policy (which focused on support of science and on product development). The development of key technologies became a central part in the "structural politics of modernization" of the new social-liberal coalition formed in 1969 under Chancellor Willy Brandt (Horn 1987, pp. 49–51). The promotion of new technologies was conceptualized as a part of the larger framework of a policy aimed at creating the right structural conditions for economic growth. The program was future-oriented, yet at the same time it was represented as a reaction to similar programs in other countries. In particular, with respect to new technologies that often had uncertain prospects of commercialization, this policy of structural modernization argued that government should intervene to fill the gap left by the lack of industrial initiative (Hirsch 1970). According to this policy narrative, interventions were necessary to the R&D infrastructure and could be justified because they helped resolve problems of a "general public interest," such as those in the fields of environment and health.

The new industrial and technology policy narrative provided important reference points and links for the political mapping of biotechnology and the designation of the actors and institutions that should play roles in the future biotechnology policy. The key players of the new policy field emerged gradually in a complex web of interpretations, political strategies, scientific developments, and institutional change. By the late 1960s, biological research had been framed as an important task for state intervention. At the same time, the Deutsche Forschungsgemeinschaft and the Max-Planck-Gesellschaft increased their support for basic biological research, while the Stiftung Volkswagenwerk became a major player in the creation of the field of molecular biology. But there would eventually be a problem with this newly emerging network of actors and institutions in biotechnology. It had to do with history and political context. The Stiftung Volkswagenwerk, a central institution in the newly developing policy field, had to end its support of molecular biology. According to its constitution, it could be committed only to short- or medium-term support of basic interdisciplinary research. This feature of its self-understanding became particularly problematic for the Gesellschaft für molekularbiologische Forschung (GMBF), which had been set up by the Stiftung Volkswagenwerk. And the GMBF's steady growth and increasing orientation toward application created another problem. The BMWF, considered a possible alternative to the GMBF as a source of support, was unable to take the Stiftung's place for the very reason that its scope of action was constrained by the German constitution. The main responsibility for supporting research and technology rested with the Länder (the state governments), not with the Bund (the federal government). This provision—introduced to avoid a concentration of power in science policy, such as had been disastrously the case in the Third Reich—now turned out to be an obstacle to the federal government's efforts to expand into new technologies. In response to this legal situation (which, of course, also affected a number of other disciplines), and in the course of the ongoing process of redefining the role of the state in science and technology development, the constitution was amended in 1969 to provide for the possibility of federal research support for projects of "supra-regional importance," and to open up an avenue for federal government intervention in a variety of areas, such as large-scale research of the kind conducted by the GMBF (Fleck 1990, pp. 121–125).

These institutional adaptations and changes created a specific political context for the crafting of the new biotechnology policy narrative. By 1969, support for molecular biology had been translated into a topic of "supra-regional importance," and the GMBF received federal funds for the first time. Three years later, the Stiftung Volkswagenwerk and BMWF signed a contract stipulating that from 1975 on the federal government would assume financial responsibility for the GBMF. In 1971 the BMWF started a research program on "Biology, Medicine, and Technology," with funding of 130 million marks for 1972–1974. By then, biological research was already assumed to have potentially useful applications. The rationale behind this program was summed up by one of the administrators in the following way: "We felt that it was unavoidable that the theoretical study of physiological and morphological processes on the molecular level would eventually result in the discovery of new metabolisms and the identification of new products which would be of practical interest."[44] Another important institutional development that affected biotechnology policy was the establishment of the Bundesministerium für Forschung und Technologie as a central institution for the coordination and planning of research and development in 1972. This new ministry initiated a working group on biology, ecology, and medicine, with advisory groups from different ministries, industry, and research, to discuss potential new research fields and possibilities of application in biology.[45] With the creation of this group, an important institutional site had been defined from which a new biotechnology policy discourse could emanate.

But until the early 1970s the support of "biology and technology" by the state had lacked any systematicity or grand design and was mainly exploratory. One observer remembers:

> Our objective was not to tell people what biotechnics is and how to do it, but to ask: assuming the state wants to support biotechnics to stimulate long-range research topics, which otherwise would not be taken up, what kind of reasonable research problems are there, if the state wants to combine industrial innovation with the solution of public problems like health and nutrition?[46]

This situation changed in 1972 when the Deutsche Gesellschaft für chemisches Apparatenwesen (DECHEMA), a forum for cooperation between science and industry (Jasanoff 1985, p. 28), began to develop a more coherent framework for the support of biotechnology. At this time molecular biology and a number of other disciplines in the life science were undergoing dramatic changes that seemed to affect traditional biotechnology. But the scope and the potential impact of these changes were difficult to forecast. At the same time, there was a perception among policymakers in the BMFT that the German pharmaceutical industry, including some of the world's largest companies and food industries, maintained a conservative stance toward the new developments in biotechnology. Furthermore, it seemed that a number of countries were more advanced in the commercial exploitation of biotechnology than Germany. In fact, in 1971 a German delegation went to Japan to study that country's traditions of biotechnology and came to the following conclusions:

> We said biotechnology is everywhere. But the people who do it are either beer brewers or work in food technology and don't have any contact with each other. This is a typical German development. In Japan we saw that beer brewers and liquor companies also make antibiotics and amino acids, whereas in Germany we see isolation. Löwenbräu does not produce amino acids, Hoechst does not make beer. . . . But we felt that not all reasons why things went better in Japan were acceptable for us or potentially transferable. We came to the understanding that the Japanese are successful because of certain local and mental conditions. We cannot imitate that in Germany, we need to pursue different strategies to come to the same results.[47]

When the BMFT asked DECHEMA to develop a biotechnology program with a short-term component of two to three years and a middle-range component of some eight years, this was crucial step in the discursive construction of the new biotechnology as a field of research and industrial activity. The new program explored fields of innovation for industry, particularly in areas where financial risks for R&D investments were high and coordination among universities, other research institutions, and industry could be established (Buchholz 1979, p. 77). DECHEMA had became a strategic location of central importance for the crafting of a German biotechnology narrative.

The DECHEMA Report Attempts to Define the Reality of Biotechnology in Germany

DECHEMA selected a group composed of representatives from its own ranks, from the large chemical companies (Boehringer Mannheim, Bayer, Hoechst, Schering, Merck), from universities, and from the GMBF and other research institutes to draft a strategy for German biotechnology policy.[48]

The report, written between 1972 and 1973, published in 1974, and issued in a revised edition in 1976, was the first culmination of the efforts since the early 1960s to draw the boundaries of "modern biotechnology" and to link scientific disciplines, experts, and socio-political-economic prognosis semantically. Molecular biology received only scant attention, and genetic engineering technologies were generally not yet available when the first version of the report was written. However, genetic engineering was integrated in the second edition of the report, and, in general, the analysis and the spirit of the report became a central point of orientation for research and technology policies in modern biotechnology for the years to come. Internationally, it became a much-studied model for policymaking concerning the "new biotechnology" (Bud 1993, pp. 152–153). The invention of biotechnology as a force of social and economic transformation had begun to take shape.

The report's policy narrative broadly defined modern biotechnology as the nexus of microbiology (including microbial genetics); biochemistry, physical chemistry, and technical chemistry; and process and apparatus technology. It then postulated the need for a better-organized system of disciplinary linkages. Furthermore, the report evaluated and assessed Germany's strengths and weaknesses in the key areas of biotechnology and pointed at potential areas for development and progress. But the DECHEMA narrative went beyond analysis and prescription. It told a grand story of the origins, meaning, and future of the life sciences, and it delineated a path for modernization—a potential world model, established through the enframing of meanings, strategic silences, and evasions in the text.

The DECHEMA narrative left little doubt that modern biotechnology is a crucial new set or recombination of scientific approaches and technologies which are indispensable for a modern, industrialized nation such as Germany. According to this narrative, in the beginning there was biotechnology. It existed for centuries, as exemplified by beer and cheese production, and it continued to exist in a new and improved way. Biotechnology was crucial because it offered a whole new set of applications that could lead to economic development, solve environmental problems, and help fight world hunger. All this could be accomplished only through a new program of research and technological development, with the Bundesministerium für Forschung und Technologie in a central position to manage the new initiative in biotechnology:

In the Federal Republic biotechnology has been completely neglected in the past. . . . Because of its huge economic importance, this field absolutely needs directed support. Only in this way can the necessary level of economic and technical development be reached which is crucial for an industrial nation and this level of development is also the precondition for the further evolution and usage of modern biological methods.[49]

The DECHEMA study then elaborated a set of research fields to be supported. The report argued that the "biotechnology challenge" required the united efforts of the state, science, and industry, and that the Deutsche Forschungsgemeinschaft would provide support for many of the projects envisioned. Thus, in this construction of the political space of biotechnology, the key institutional players for biotechnology policymaking were located in their interrelationship to one another, and the state was defined as a coordinating agency and a clearinghouse.

In the following years the DECHEMA framework served as the main point of orientation for the BMFT's research and technology support policy. A number of biotechnology programs constituted further negotiations of the prevailing definitions of the reality of biotechnology research and development and of the roles attributed to the various actors in the biotechnology network that were outlined in the DECHEMA narrative. An important participant in these negotiations described the difficult task of the DECHEMA commission to discursively construct biotechnology:

The main problem in this early period was to develop a definition and a feeling for what is to be understood under biotechnics. Biotechnics is not an easily described field, but a synthesis between various disciplines and styles of laboratory practice. In the beginning it was pretty tough to listen to discussions between microbiologists and process technology people. . . . It took a long time until they had found a common language. This was one of the important impacts of the whole story, that all of a sudden we had a core of some 50, 60 scientists who talked about biotechnics and had found a common language tailored towards the needs of the BMFT.[50]

The year of the publication of the DECHEMA report was also marked by a number of events and changes in the political and economic context that affected the framing of biotechnology. The first oil crisis, rising inflation, the revaluation of the mark (which affected competitiveness in world markets), and the restrictive monetary policy of the Bundesbank all impacted on Germany. The economy fell into its deepest recession since World War II. When Brandt resigned as chancellor in May 1974, Helmut Schmidt took over (Horn 1987,

p. 51). Until the early 1970s Germany was a leader in the export of capital goods. The resulting positive trade balance and buildup of foreign currency reserves by the Bundesbank further raised the value of the mark, which then stimulated long-term investments by German companies abroad. By 1985 German direct investment abroad exceeded foreign investments in Germany by 50 billion marks. Between 1978 and 1979, because of the revaluation of the mark against the dollar, more than a third of these foreign investments were made in the United States (Altvater and Hübner 1988, p. 27).

Biotechnology and Modernization Discourse

The SPD-FDP[51] government under Helmut Schmidt responded to the crisis by developing a "directive, anticipatory structural policy" and a market-based program for economic growth. Technology policy became an important element in the new "politics of modernizing the economy." The emerging policy narrative located causes of the world economic crisis in the structural problems of the economy and argued that these problems could be solved by the systematic support of new technologies and the diffusion of these technologies into the economy.[52] During the SPD-FDP coalition, the objectives of this structural politics of innovation included the improvement of living, working, and environmental conditions, although these goals which were clearly subordinated to the objective of stimulating the international competitiveness of German industry. This focus changed in 1982 when, under Chancellor Kohl, the high-technology-oriented concept of a structural politics of innovation was significantly narrowed down to a technology policy oriented toward the needs of German industry in the international marketplace. But despite these changes, the German discourse on structural politics of economic modernization since the 1960s is marked by a great deal of continuity. The power of this policy narrative is also evident when one examines the policy instruments used for the implementation of the policies. Direct project support (such as for particular sectors of science and technology), institutional support (such as for the Max-Planck-Gesellschaft or the Grossforschungsanlagen), support for large research establishments (such as the GMBF), and indirect support for industry (for example, in the form of tax breaks) continued to be the classical policy instruments. In contrast to the conservative rhetoric favoring indirect support measures as being more akin to a market economy, the percentage of indirect support for industry between 1982 and 1990 remained similar to that provided by the Social Democratic–Liberal government.

It was against the background of this particular discursive context and constellation that in the mid 1970s the BMFT emerged as the "lead department" in coordinating and supporting of biotechnology, while other departments such as Economic Affairs, Education and Science, Defense, and Agriculture took on subordinate roles. Federal expenditures for biotechnology started off at 44 million marks, reached 92 million marks by 1980, rose to 152 million marks in 1985 and 256 million in 1990, and were projected to increase to 292 million marks by 1994.[53] In 1991, after DECHEMA organized a review process, a special biotechnology support program was established to support biotechnology in the former East Germany. With a budget of 5 million marks for 1991, this program was given a mandate to rebuild the East German R&D infrastructure and to integrate it with West Germany (Warmuth 1991, p. 6). And numerous biotechnology policy schemes at the level of the Länder must be added to the federal initiatives.

This exponential increase in government resources for biotechnology was consecutively administered at the federal level by three programs that were set up in the years that followed the DECHEMA report: the BMFT's "Leistungsplan 04—Biotechnologie" (1979–1983), the Applied Biology and Biotechnology program (1984–1989), and the Biotechnology 2000 program (1990–1994).[54] The BMFT programs were developed in close collaboration and mutual coordination with the Deutsche Forschungsgemeinschaft, the Max-Planck-Gesellschaft and the Verband Chemischer Industrie (Organization of the Chemical Industry).[55] From the perspective of the BMFT, German biotechnology strategy had four phases. Phase I began before the DECHEMA report and can be dated back to the early 1960s and the emergence of a politics of molecular biology. Phase II lasted from 1975 to 1985, when the methods of the new biotechnology became broadly available, and can be characterized by a focus on constructing a "research landscape." Phase III, from 1985 to 1990, emphasized the elimination of weaknesses in certain research areas and the distribution of funds to certain targeted areas. Phase IV (from 1990 to the present) concentrates on the diffusion of the genetic technologies into the economy and (increasingly) into medicine and, at the same time, on a deliberate effort on behalf of biotechnology policy to participate in debates on bioethics and risk issues.[56]

The three comprehensive programs formulated since 1979 share a rather broad framing of the biotechnology problematic. The Applied Biology and Biotechnology program stated its goals as follows: to make scientific-technological peak performances possible; to improve the conditions of innovation for industry; to support research and development projects in order to secure current and future needs of society; to evaluate the opportunities and risks of biotechnology, carry out risk assessment, and incorporate its results into safety regulations; and to train scientists and technicians in biotechnology. These core objectives can also be found in both the 04 program and the Biotechnology 2000 program. What stands out is the coding of the structural politics for economic modernization as *Vorsorgeforschung*—research anticipating future social needs. This concept defined biotechnology as an important discipline for such diverse fields as health, nutrition, natural resources and energy, mining, and the environment. Economic growth and the general well-being and quality of life of the Germans were all tied to the support and development of genetic technologies.[57] Quite typically, the Leistungsplan 04 states:

> In the last decades far reaching changes occurred in the field of biology. The simply descriptive comprehension of biological structures and processes has been pushed back by a comprehensive analysis of these phenomena by means of methods from molecular biology, chemistry, and physics. This breakthrough on the cell biological and molecular biological level has yielded results which now allow a deeper understanding of basic life processes. The combination of this new genetic, biochemical, and microbiological research with technological principles will open up new and promising applications in nutrition, environmental protection, medicine, and the acquisition of resources.[58]

This framing of molecular biology was then connected to specific sets of recommendations. Especially from 1985 on, the federal government began to single out research areas of strategic importance. The Applied Biology program put special emphasis on fields that were of crucial importance in a scientific context as well as for industrial innovation: genetic engineering, bioprocessing techniques, enzyme techniques, and new interdisciplinary research areas situated between biology and technology (e.g., bioelectronics and enzyme and gene design).[59] The Biotechnologie 2000 program continued this multidimensional support policy and added new fields for focused support, including molecular neurobiology, technology assessment, and the study of ethical issues related to biotechnology.[60] The policy instruments used to reach these goals were institutional support measures, indirect support measures for industry, and cooperative research and project support.

During the 1980s, 20 percent of the available research resources were spent on institutional support. Total R&D expenditures increased between the late 1960s and 1990 from 2.2 percent to 2.9 percent. At the same time, the percentage of industrial R&D in relation to the total expenditures increased from 55 percent to 64 percent (Süss 1991, pp. 99–100; Ronge 1986; von Alemann et al. 1988; Väth 1984).

The concept of *Grossforschung*—large-scale research facilities supported by the government—was another critical feature of the emerging biotechnology policy. This "Big Science" concept dates back to the nuclear research program of 1955–56 and the space research program of 1962 which established Big Science programs in order to push scientific and technological developments in particularly important directions quickly and vigorously. In the field of molecular biology, this idea was first pursued by the Stiftung Volkswagenwerk (which had created the GMBF), by the Gesellschaft für molekularbiologische Forschung [Society for Molecular Biology Research], which in 1976 became the Gesellschaft für Biotechnologische Forschung [Society for Biotechnology Research], an organization financed by the federal government and the state of Niedersachsen (Lower Saxony) in a 90:10 ratio.[61]

According to the biotechnology policy narrative first outlined by DECHEMA, the function of the Gesellschaft für Biotechnologische Forschung in the emerging biotechnology network was to build a bridge between basic research and industry. More specifically, the GBF was to mediate between science and industry and to push biotechnological laboratory processes to the level of potential industrial application, at which point these practices could be taken up by industry and further developed and specified for industrial purposes (Buchholz 1979, p. 98).

Indirect support measures were intended to assist industry in introducing new biotechnology processes, to stimulate the founding of new biotechnology-related companies, and to help industry in the training of R&D personnel. In the cases of both introduction of new biotechnology in existing companies and support for new companies, target areas (such as bioreactor development) were singled out for possible support.[62]

Finally, under the program of cooperative research (*Verbundsforschung*), certain research fields were singled out as candidates for collaborative research projects between public research institutions

and industry. As its politics evolved, the BMFT tried to make sure that most of its project support would go into Verbundforschung, a concept that became one of the core technology research instruments of the conservative-liberal coalition from 1983 on (Ronge 1986, pp. 334–337).

In addition to these policy measures administered by the BMFT and the policy initiatives at the level of the Länder, two other critical nodal points of the biotechnology network were closely related to support at the federal and Länder levels: the Deutsche Forschungsgemeinschaft and the Max-Planck-Gesellschaft. The role of these organizations was defined as to focus on basic research. Basic research at the universities was primarily financed by the DFG, whereas the MPG maintained its own institutes. Each of these institutions had played a crucial role in the initial shaping of molecular biology in Germany, and both continued to do so throughout the 1970s and the 1980s. Since neither institution had any special programs in genetic engineering or biotechnology, it is difficult to determine how much support either of them provided for biotechnology research. However, for instance, the DFG's annual report for 1988 states that one-fourth of all the research the DFG supported in the biosciences involved the use of genetic technologies; that implies expenditures in this field of some 90 million marks in 1988 and more than 100 million in 1990.[63] The combined biological and medical research budget of the MPG was comparable to that of the DFG, and if we make the conservative estimate that research supported by the MPG also implied use of methods of genetic engineering in roughly one-fourth of the cases we can assume expenditures for genetic-engineering-related research of about 100 million marks in the late 1980s.[64]

National Policy Narratives and Global Companies

Thus, the new biotechnology policy discourse had set in motion a multitude of R&D initiatives that were anticipated to lead to the emergence of a German biotechnology industry. The inscription of biotechnology as a "technology of the future" and the elaboration of an interconnected system of support programs were intended to build a shared perception of biotechnology and a system of informal cooperation among the state, research organizations, and industry. A "distribution of labor" among the state, industry, and science that might lead to the hoped-for emergence of a biotechnology industry in Germany was outlined. However, as happened at the EC level, the discursive

construction of biotechnology soon disintegrated into a number of competing language games giving different definitions of the realities and the potentials of a German biotechnology industry.

For the academic science community, genetic engineering seemed to be a "method of the future." The German model of biotechnology policy was effective in transforming the research infrastructure. The number of laboratories in which recombinant DNA work was done increased from 764 in 1987 to 1309 in 1990. Of these laboratories, 70 percent were at universities or Max Planck Institutes, 6 percent at large research institutions, 8 percent at other institutions, and 16 percent in companies (in 1990) (Gill 1991, pp. 92–96). But in industry, genetic engineering research and development was largely confined to a group of large, well-known companies. In the early 1990s the German biotechnology industry was dominated by 15 multinational chemical and pharmaceutical companies, such as Bayer, Hoechst, BASF, Merck, and Bohering-Ingelheim, with more than a billion marks of total turnover. These companies also had the largest number of recombinant DNA patents and laboratories. The number of startup companies involved remained very small, and they also tended to have few laboratories and patents.[65]

Apparently, while the universities, the Deutsche Forschungsgemeinschaft, and the Max-Planck-Gesellschaft happily participated in the policymakers' vision of a transformation of the life sciences through genetic engineering, industry pursued different strategies. Hoechst, potential entrepreneurs, and the many small venture-capital-driven biotech firms that were to come were important elements in the Bundesministerium für Forschung und Technologie's plans for the future of biotechnology. The BMFT's power was, to a considerable extent, built on the assumption that one day its policy narrative would become reality, and that biotechnology would actually help to build industries of the future in Germany. In reality, however, if new biotechnology companies were going to develop somewhere during the 1980s and the early 1990s, it was likely to be in the United States—with the support of German capital.

To a considerable degree, these policy failures can be attributed to competing codings of the political an economic context of biotechnology investments in Germany and to the failure of the German biotechnology policy narrative to intermediate these different mappings of genetic engineering. While the German policy narrative framed Germany as an excellent place for biotech investments, the

German financial system made financial innovation difficult and inhibited the emergence of a venture-capital industry. The conservatism of Germany's central bank and the close relationship between banks and industrial firms contributed to this anti-innovation syndrome. Another potential source of funding, the stock market, played only a limited role in the financing of German industry, partially because of the tendency of companies to obtain funds through bank credits (Canals 1993, pp. 89–90). Considerations of such structural problems for biotechnology policy had not sufficiently entered the policy schemes launched since the late 1970s. However, they seem to have affected the behavior of potential investors.

In absence of a sufficient number of smaller companies convinced that there was a future for biotechnology in Germany, state intervention in biotechnology went well beyond the limits of structural politics. The total 1990 R&D expenditures for new biotechnology (including state expenditures for universities) were around 1.5 billion marks, with industry spending some 250 million marks and federal and state governments around 1.3 billion.[66] This lopsided relationship of investments further underscores the importance of the state in the discursive invention of biotechnology industry. More recently, the relative absence of small companies in the biotechnology industry gave rise to initiatives by Länder such as Bayern (Bavaria) in order to help make risk capital available for biotechnology investments (Gottschling 1995).

Another important conflict over codings of the contexts of biotechnology development derived from the German state's goal of establishing a *German* biotechnology industry. While policy discourse drew the boundaries of biotechnology industry around the German nation state, the large German corporations, like most other large companies in Europe, defined the industrial applications of genetic engineering as a global project. In 1989 there were at least 40,000 persons engaged in R&D at German affiliates abroad (compared to about 300,000 in West Germany). This pattern was especially pronounced in the chemical and pharmaceutical industry. Hoechst acquired a majority share of the French pharmaceutical company Roussel Uclaf in 1984. Since 1982, programs of research cooperation have been established with Massachusetts General Hospital, Biogen, Chiron, Genentech, Immunex, Integrated Genetics, and the Salk Institute. When Bayer decided to move its biotechnology R&D operations to the United States, it acquired Cutter and Miles, established a research laboratory in West Haven, Connecticut, and agreed to cooperative relationships with

Genentech, Genetic Systems, and Calgene.[67] While the German medical and biological research infrastructure underwent significant transformation, a German biotechnology industry based on the US model remained a distant policy dream.

Britain: In Search of a Biotechnology Strategy

As the above discussion of biotechnology policymaking in the European Community and in Germany has shown, the development of these policies was not simply a reaction to certain challenges or contexts; it involved a complicated multi-layer process of reality production influenced by the selection and assembly of historical traditions, political and economic events and developments, forecasts, and strategic deliberations, which were interpreted and presented in the form of policy narratives. These narratives were typically constructed against the background of other important policy stories and political metanarratives that defined a country's major socio-economic and political realities. These complex conditions for the shaping of biotechnology policy could also be observed in Britain.

Whereas in Germany there is a clear pattern of continuity between the relatively late politics of molecular biology of the 1960s and the relatively early politics of biotechnology of the 1970s, in Britain biotechnology was not defined as a field for state action before the late 1970s. As was mentioned in the last chapter, the fragmented character of British government and British society led between 1974 and 1978 to efforts on the part of the Labour government to establish a "tripartite" form of cooperation among the leading industry-labor associations, the Confederation of British Industry, and the Trade Union Congress. Once Labour got into office, industrial policy became an instrument of managing the crisis of recession. From around 1976 on, it evolved into an instrument of both crisis management and structural change. In 1975 Labour had launched a new industrial strategy centered on the idea of a self-analytical form of planning that would be strategic and sectoral rather than aggregate and quantitative, would look for growth points and bottlenecks, and would focus on technology as a crucial element of an industrial strategy. The integration of industry and the unions in the elaboration of this new strategy strongly reflected the government's corporatist commitment. In 1976, the sterling crisis and intervention by the International Monetary Fund constrained public spending, and rising country-wide unemployment undermined the traditional rationale for regional industrial aid. From 1977 on, selective aid for industry switched from sector-specific schemes and firm rescues to discretionary schemes promoting innovation (Sheperd 1987, pp. 164–166).

Dreaming about America, Waking Up with the Spinks Report

It is against this newly emerging discourse of support for industry that we have to understand the construction of a biotechnology strategy in Britain. The emergence of the American biotechnology industry seemed to hold out the promise of a much-sought-after model of technology-driven innovation that Labour's new industrial policy intended to target. At the same time, a discourse on the absence of a US-type "biotech boom" in Europe began. Increasingly, scientists, industrialists, and government officials argued "that something had to happen" (Yoxen 1984, p. 223, n. 25). As one government official put it: "People felt that there was potential application of biotechnology which needed to be encouraged to a greater extent by using public funds. . . . It was a general feeling, we had microelectronics, the next is going to be biotechnology."[68] Another civil servant saw the developments in the United States as the driving force behind the building of a British biotechnology strategy: "I guess the impetus for it really came from all these investments in the small companies in the States and a feeling that there was a race we are losing in some way."[69]

In 1978, under a Labour government, British Secretary of State Shirley Williams announced the creation of an inquiry into biotechnology (Bud 1993, p. 158). A year later, the Advisory Council for Applied Research and Development, the Advisory Board for the Research Councils, and the Royal Society set up a joint working party (known as the Spinks Working Party after its chairman, Alfred Spinks, a former director of research at ICI) to review existing and prospective science and technology deemed relevant to industrial opportunities in biotechnology.[70] In contrast to the comparable molecular biology working groups of the 1960s, which had consisted entirely of scientists, the seven-member Spinks Working Party included Austin Bide, chairman of Glaxo and of the Confederation of British Industry; William Henderson, chairman of the Genetic Manipulation Advisory Group; and Arnold Burgen, director of the National Institute for Medical Research.[71] The composition of the group, as well as its stated mandate (to study the industrial applications of biological knowledge), signified a new orientation for the political coding of the new biology. The

Spinks Report became Britain's equivalent of the German DECHEMA report insofar as it played a pioneering role in problematizing and defining "the biotechnology challenge" and devising a policy narrative intended to create a British network that would to open up the economic space of biotechnology. In the Spinks Report, biotechnology policy was translated most prominently into a new type of industrial policy that would, in part, be an answer to the present economic crisis:

> For the United Kingdom, biotechnology is an area of high technology with large potential growth offering opportunities for the renewal of various existing industries and the creation of new ones. It can be used by both large and small business, and although initial capital and development costs are high for some processes, experience in the United States and in Europe suggests that growth can occur in small companies backed with only limited financial resources.[72]

The focus in such political representations of biotechnology as the Spinks Report was on what was perceived as the enormous potential of the new biotechnology to counter threats to health, to the environment, to agriculture, and to prosperity (Bud 1993, p. 206). Although much of the analysis revolved around initial trends and future possibilities, there were frequent allusions to an existing or at least threatening American and Japanese superiority in biotechnology—a scenario that was believed to undermine European industry and wealth. On this evaluation, British modernization, economic growth, industrialization of agriculture, and industrial innovation were unquestionably in need of further development by means of science and technology.

In these analysis of biotechnology, the constitution of biotechnology strategies proceeded within carefully crafted semantic fields that underscored the importance and the effectiveness of the new biotechnology. The strategies focused on the removal of "biotechnology obstacles" and the launching of "biotechnology offensives." Major obstacles to a successful biotechnology strategy were said to be located in the organization of research and industry as well as in the lack of "public understanding."

The Spinks Report came to the following devastating conclusion about the state of biotechnology in Britain:

> . . . the present structure of public and private support for R and D is not well suited to the development of a subject like biotechnology which, at the moment, straddles the divisions of responsibility both among Government Departments and Research Councils and the arbitrarily defined fields of

fundamental and applied research. Strategically applied research is in general ill-served by our research funding mechanisms, especially in areas where there are neither university departments to promote it, nor well-developed industries to provide market-pull. Biotechnology lacks university centers specifically for it and this results in a shortage of new ideas for industry and of suitably trained manpower. The absence of an established industry clearly identified with new biotechnologies allows Government policy in one industrial sector to have adverse implications for biotechnology in others and diminishes the attraction of the subject for the engineers. The result of such interactions is that biotechnology has not grown in the United Kingdom in the coherent fashion which we believe is merited by its potential. What is required at this state is a policy of "technology-push" reflected in a firm commitment to a strategic applied research. This will progressively produce potentially marketable products and processes and the policy should then be for a more "market pull" approach.[73]

This analysis suggested that Britain had neither a biotech industry nor a research system conducive to one, and that hence the national system of innovation should be rebuilt. What was required was an overall strategy. The key elements in this strategy should be a broadly defined set of fields where biotechnology would be supported (ranging from the chemical industry to agriculture and energy production), sufficient funds, strong engagement of the Research Councils through substantially increased support and coordination of activities in a Joint Committee for Biotechnology, coordination of government departments, and creation of an inter-departmental Steering Group. This strategy was depicted as crucial in the face of worldwide competition:

We are not convinced that in our mixed economy the task posed by these various problems can be met by the private sector alone. . . . We have some sympathy with the view, held for example in West Germany, that R and D support cannot be left to economic forces alone and that government intervention is appropriate for making industrial innovation feasible.[74]

Currently, the report argued, there were two models for the development of a biotechnology industry: the US model (with little government support, but with a strong base of industrial research and a sprawling venture-capital industry) and the European-Japanese model (with strong government involvement).[75] According to the narrative of the Spinks Report, Britain should not try to emulate the US model. The state should not limit its support for biotechnology to the creation of a suitable research infrastructure and R&D funding, but should expand to include entrepreneurial activities:

We have commented on the shortage of venture capital and other resources of financial support for innovation in the United Kingdom compared with some of our export competitors and on the absence of new biotechnology-based companies. There seems to be little prospect of similar companies being set up in the United Kingdom at the present time by wholly private finance. . . . We *therefore recommend* that the NEB (National Enterprise Board), with NRDC, should investigate the possibility of establishing in the United Kingdom, with some public funds, a research-oriented biotechnology company of the kind taking shape elsewhere.[76]

This strategy, which would also include a reconsideration of the regulatory regimes currently in place with respect to recombinant DNA work, would "continue as rapidly as possible" to reduce constraints on genetic manipulation experiments.

The Spinks Report was the first effort to enframe the meaning of a British biotechnology policy and to create a regime of signification that defined the interests and coordinated and stabilized the practices of a variety of subject positions configured in institutional complexes. Against the background of the structural weaknesses of British capital, the state emerged as the major force in the creation of a British biotechnology industry. But the Spinks Report's emphasis on tripartite industrial policy coordination and broad state intervention made it an anachronistic document soon after it was published. During the 1979 elections, the rhetoric of the "high-tech biotech race" gained a new meaning in the political discourse of Thatcherism.

Recasting Britain's Biotechnology Strategy under Thatcher

In Britain, the underlying economic conditions for a "social contract" between labor and capital, in particular as originally envisioned by the Labour government of Harold Wilson, never existed. The British economy was too weak and undercapitalized to sustain capital accumulation and profitability and, at the same time, continue subsidizing the social compromise that had been in effect since the 1940s. Owing to the effects of the 1970s global economic recession, British society had begun to polarize under conflicting pressures, and the social compromise eroded. This polarization opened the way for the emergence of the New Conservatism under Margaret Thatcher.

The strategic goal of Thatcherism was to embrace a distinctively neo-liberal, free-market system, to break the power of the working class, but also to curb industry's influence on the state. At the same time, this free-market approach was linked to traditional issues of Toryism: a return to the old values of Englishness, family, and nation (Hall 1989, p. 37). This reconstruction of an alternative ideological bloc focused on strong commitments to "the entrepreneurial society" and "popular capitalism." Its beneficiaries were the very rich and the new service class of skilled and semi-skilled manual workers in the growing private-sector industries—a social base concentrated mainly in the South of England (Jessop 1992, p. 34). This changing political discourse did not, however, imply a wholesale abandonment of policies of economic regulation. In general, this can be seen in the field of selective industrial policy during the consolidation of Thatcherism in the years 1979–1982 (Sheperd 1987, p. 169). Despite the official rhetoric of commitment to the virtues of market-led solutions to economic problems, the Conservative governments since 1979 have presided over what can be seen as a renaissance of intervention. What happened during that period was a shift from a "relatively extensive" to a "relatively intensive" regime of intervention. Whereas in the past the state's intervention in the economy revolved around mechanisms of subsidization, the mode of economic regulation and nationalization during the Thatcher period shifted to a new strategy.

The first element of the new strategy involved narrowing the field of operation of intervention by restricting its scope and its extent, but at the same time differentiating the field of operation. Second, economic intervention was deepened and made more selective, more discriminating, and more oriented toward "value for money." The heightened selectivity was combined with better modes of scrutiny, such as better forms of project assessment and follow-up studies on finished projects. Third, a new layered system of intervention composed of overlapping mechanisms of regulation was created. Fourth, granting an increase in discretion to the intervening mechanisms, the state gave the units within the regulatory system greater autonomy and more space in which to maneuver (Thompson 1990, pp. 135–142).

At the core of the economic strategy of Thatcherism was a "redifferentiation" of state intervention, which meant not simply "less state" and not state intervention geared primarily toward the interests of capital but rather a new mode of regulation—a new state strategy. The discourse of Thatcherism put a strong emphasis on "liberating" the state from the influence of both labor and capital as they were institutionalized in corporatist bargaining.

With respect to labor, this new mode of regulation centered on a systematic effort to remove, weaken, or radically reform those institutions which had been instrumental for unionism and which, in the view of the Thatcher government, had adversely affected labor costs, productivity, and jobs (Evans et al. 1992, p. 577). With respect to industry, the Thatcher government continued a tradition of government-industry relations dominated by assumptions inherited from the nineteenth century. At the core of this approach were the assumptions that preserving the commercial autonomy of the firms was of primary importance and that an "arm's-length relationship" between the government and industry was generally preferable (Grant 1989, p. 86). Furthermore, like labor, the Confederation of British Industry saw an erosion of its influence on governmental decisionmaking. In fact, government ministers systematically outmaneuvered and disarmed the Confederation. The new strategy of regulation did not involve a complete abandonment of all aspects of the previous mode of intervention. For example, subsidies were kept in the repertoire of regulatory mechanisms. But, like privatization and other aspects of the neo-liberal policymaking, the new "relatively intensive" strategy had to prove its value in everyday political and economic behavior. Reorganization on the level of macroeconomic theory and political representations of state-society interaction does not necessarily translate successfully into the microeconomics of managerial commitment and investment (Middlemas 1991, pp. 354–355). The assumption that liberating the state from the obligation of neo-corporatist bargaining would result in more autonomy for the state was, though important, only one tactic in the project of constructing new spheres of state autonomy.

It is against the background of this new discourse on state autonomy that we have to understand the Thatcher government's biotechnology strategy. Central to the evolving policy were practices that resignified the autonomy among units of the innovation system, such as the Department of Trade and Industry (DTI) and the Science and Engineering Research Council (SERC), and thus intended to make possible the implementation of elements of the new strategy of relatively intensive intervention. Objectives such as better scrutiny of supported projects or increased discretion granted to intervening mechanisms, such as granted to the Biotechnology Unit of the DTI, required the absence of antagonism within the innovation system and the successful enrollment of a variety of actors, such as of representatives of the R&D system, representatives of financial capital, and representatives of industry. This attempted coordination turned out to have only limited success in the case of British biotechnology policy, because the policy measures adopted were either contradictory or simply insufficient to reach the intended goals.

Soon after the ascent of the Conservatives, it became clear that the new neo-liberal political discourse had constructed a new meaning framework for interpreting biotechnology policy. The development of a British strategy for biotechnology followed the pattern of "relatively intensive" intervention closely. The government's first statement on biotechnology after the change in power followed many of the key recommendations of the Spinks Report but also indicated a change in strategy in pursuing the outlined goals. The Thatcher government declared its belief "that biotechnology will be of key importance in the world economy in the next century, and of rapidly growing importance before then."[77] In a so-called white paper (a government document for discussion), the following strategic recommendations were made:

Strengthen the scientific base of biotechnology within the existing framework of funding, mainly by concentrating (implicitly meaning shifting) resources.

Coordinate the biotechnology-related activities among various government departments, the Research Councils, the National Research Development Corporation, the National Economic Development Council, and the National Enterprise Board.

Coordinate and foster the collaboration among the universities, the Research Councils, and industry.

Encourage the increase of private investment in the biotechnology industry.

Remove regulatory constraints inhibiting biotechnological development, such as over burdensome health and safety regulations.

Foster international collaboration and competition.[78]

With this strategy the government had clearly moved away from the much more interventionist "Spinks philosophy," but at the same time it had outlined a project of inter-governmental coordination, closely linked to the coordination of the activities of the research councils, to the idea of pressing industry to increase research investments, and to the internationalization of research and production. This targeted strategy focused less on increasing government resources for biotechnology than on making more efficient use of existing resources and facilities, on the national as well as the international level. This strategy, however, required substantial intervention from the state; it also required the cooperation of industry.

The Thatcher government's new framing of biotechnology policy constituted an effort to redefine the interests of the "illusionary community" of those it claimed to represent. What was going on in this spacing of the state-economy interaction was a reconstruction of the nodal points interconnecting the national innovation system that redefined biotechnology policy from a mode of extensive intervention to one of intensive intervention. The generation of a "new climate for investments," the improvement of cooperation between the R&D system and industry, the establishment of new forms for subsidizing research in public and private institutions, and the establishment of a politics of "enforcing" higher R&D expenditures in industry were linked together in a chain of significations, which was to construct the new political space of biotechnology. In implementing these new strategies, the Thatcher government in effect rewrote biotechnology policy, reconfiguring its aims, its premises, its interpretative framework, and its subject positions.

But soon this attempt to stabilize the discursive field of biotechnology policy found itself challenged by a dissensus that reflected irreconcilable differences among a variety of key actors in interpretations of biotechnology. In a 1981 meeting of the British Coordinating Committee for Biotechnology (which represented various learned societies and professional institutions), representatives from government, science, and industry came to a rather bleak assessment of the future of biotechnology in Britain. In particular, the "failure" of the government to respond to the Spinks Report and to take measures such as those taken in Japan and Germany were highlighted. The lack of coordination between the Research Councils and between industry and research was frequently pointed out by the participants. Furthermore, the difficulties of financing innovative projects related to biotechnology received broad consideration. Despite apparent disagreements concerning the appropriate biotechnology strategy, the organization of biotechnology support policy on the governmental level began to take shape. After a change at the top of the Department of Industry from Keith Joseph (one of the key architects of Thatcherism, and a free-market advocate) to Patrick Jenkins, the government became more active in the field of biotechnology.

By the end of 1982 the shape of British biotechnology policy had become clearer. The Department of Industry had become the lead department, with overall responsibility for encouraging industrial developments and for those areas that did not fall clearly under the jurisdiction of the Department of Health and Social Security, the

Department of Energy, and the Ministry of Agriculture, Fisheries, and Food. The Department of Trade and Industry established its Biotechnology Unit, the objectives of the which were to raise awareness of opportunities in biotechnology, to fund innovation that encouraged more R&D in industry (with companies receiving up to 25 percent of the cost of R&D projects), to improve the infrastructure through support for industrially oriented culture collections, to identify and tackle strategic weaknesses in biotechnology, and to identify and promote sectors of biotechnology that represented particularly important opportunities for Britain (Senker and Sharp 1988, pp. 59–60).

These objectives intentionally overlapped with those of the newly founded Biotechnology Directorate of the Science and Engineering Research Council. Owing to its organizational setup, the SERC was singled out to become the crucial element in the government's biotechnology strategy. In the policy narrative of the DTI, the SERC provided the critical link between academia and industry (Salter and Tapper 1993, p. 50). The Biotechnology Directorate's task was to encourage academics to propose research topics with carefully defined programs in cooperation with industrial and state research establishments. Furthermore, it was part of the SERC's mission to monitor and coordinate all of its supported research through "clubs" and other mechanisms. The SERC was also required to keep extensive contacts with representatives of industry so that they were continuously aware of industry's needs (Senker and Sharp 1988, p. 16). Until 1985 the SERC allocated its funds for individual or cooperative research to a number of priority sectors (ibid., p. 24).

And there are further examples of the state's intent to promote more biotechnology R&D. The Department of Education made extra funds available to the University Grants Committee, which decided to allow special funding for biotechnology by launching a biotechnology initiative in 1982–83. The Agriculture and Food Research Council diverted members of its staff from lower-priority work to reinforce its biotechnology programs. The Medical Research Council increased its support for biotechnology, and the Department of Industry set up a new Biotechnology Unit at the laboratory of the Government Chemist to promote its long-term biotechnology program (launched in November 1982).[79]

During the 1980s and the 1990s there was an expansion of biotechnology-related R&D expenditures. Public-sector funding increased from 58.9 million pounds in 1985–86 to 110 million pounds in 1988 and to 146 million pounds in 1990; industrial spending increased from 48.6 million pounds in 1988 to 58.1 million pounds in 1989 and to 91 million pounds in 1991.[80] What is remarkable about these figures is that government outspent industry in biotechnology-related expenditures—a clear sign of government's determination to provide a "technology push" for industry. However, as I will show, the central problems of British biotechnology policy did not come from a lack of funding; they stemmed from the government's failure to successfully enroll the individuals and agencies that were essential to its biotechnology strategy.

Destabilizations of Biotechnology Policy

Soon after the shaping of a rationale and of a set of strategies for creating a biotechnology industry in Britain, the policy narrative's signification of biotechnology, its assignments of actor roles, and its codings of economic and political realities were challenged by a number of competing attempts to map biotechnology. One site of struggle was the British system of academic research and development.

Within the framework of British science and technology policy, basic research is funded through the University Grants Committee (UGC) and the Research Councils, both of which are under the Department of Education and Science. Whereas the UGC gives block grants to each university to cover both educational and research expenses, the Research Councils engage in a more targeted kind of research, focusing on particular areas such as agriculture and medicine. Industrial innovation falls under the responsibility of the Department of Trade and Industry through its Biotechnology Unit and the British Technology Group.

Following a recommendation of the Spinks Report, an Inter-Research Council Coordinating Committee for Biotechnology was established. Dissolved when it turned out to be largely inefficient, that committee was succeeded by the Biotechnology Advisory Group, which was succeeded in 1989 by the Biotechnology Joint Advisory Board (established by the SERC and the DTI and joined in 1990 by the Agriculture and Food Research Council, the Natural Environment Research Council, and the Medical Research Council). Hence, the attempts to find a suitable organizational structure for funding decisions in biotechnology lasted more than 10 years and, in the light of the current policy debates in the United Kingdom, still seem incomplete.

The MRC and the AFRC proved strong enough to resist persistent governmental efforts to bind them to a common decisionmaking structure headed by the Department of Industry. These Councils' moves in the direction of a more application-oriented approach, such as the MRC's cooperation with Celltech and the AFRC's with Agricultural Genetic Company, were driven more by the evaporation of government funds than by consent or a new corporate philosophy. Moreover, the lack of coordination that characterized the government's general biotechnology policy was also manifested in other actions of the Research Councils, such as the initially exclusive relationship of MRC with Celltech (1980–83) and the relationship of the AFRC with the Agricultural Genetics Company (established in 1983). Celltech was established by the government in 1980 after it was realized that the process for creating monoclonal antibodies—discovered in a MRC laboratory—had never been patented. The original agreement gave Celltech the right of first refusal on any MRC discoveries. This was canceled in 1983, although Celltech maintained close ties with the MRC laboratories. But this structure hindered other companies from working with these two councils and blocked the Research Councils from transferring technology from the R&D system to industry. As a result of this situation, there was a significant overlap between the research activities of MRC, the SERC, and the AFRC in such areas as basic molecular biology and genetics, as well as in certain aspects of biochemistry, biophysics, and cell biology. After the establishment of the SERC's Biotechnology Directorate, the SERC's relationships with the MRC and the AFRC began to deteriorate, since these councils felt that the SERC was targeting research areas that fell within their domains.

There were also discursive struggles over the coding of the economic context of the British biotechnology industry. The financial markets interpreted the future of a British biotechnology industry much more cautiously than the government. Influenced by the US model of small companies' opening up the economic space of biotechnology, the British biotechnology policy narrative had defined small companies as crucial to the government' s biotechnology strategy. But there were difficulties with this construction. Like the United States, Britain historically relied on private capital for the financing of industry. As Gerschenkron (1962) has pointed out, Britain did not develop a system of long-term finance to promote industrial growth, because none was needed in this first case of industrialization. (See also Zysman 1983, pp. 171–231.) This anti-industrial trend had been reinforced by the development of the city of London as a large offshore center catering to the expansion of the Eurodollar market. The private capital markets, left to their own devices, pursued policies that favored greater capital investments in property, overseas industry, and government debt over investments in domestic industry. The Thatcher government's economic policies fitted perfectly into the structure of the financial system (Cox 1986, pp. 42–47). The specialized role of Britain as the principal site for international foreign institutions was consolidated under the Thatcher government through abolition of exchange controls, deregulation of financial institutions, and favorable tax treatment for speculative investments. This emerging constellation resulted in a financial system geared more to overseas investment and to a further attenuation of industrial banking, in an underfunded and uncompetitive industrial base, and in an adversarial political system that was incapable of arriving at a national consensus concerning economic policy (ibid., p. 35).

Furthermore, the emerging venture-capital industry, construed as a fruit of deregulation, failed to fulfill its promises as a source of capital for new risk investments. In dramatic contrast to the United States, despite strong growth of the British venture-capital industry since the 1980s, relatively little investment had been allocated to the financing of technology-related startup or early-stage deals by UK venture-capital firms. In contrast to the early days of the business, in the 1990s funds have been more likely to put money into management buyouts and company development than into startup projects. But even the general growth of the UK venture-capital industry must be put in perspective. In 1991, for example, all of Europe's venture-capital funds (88 million ECUs) did not match the $114 million that Genentech spent on R&D in the same year (Irvine 1990; Kenward 1992; Murray 1992). The reluctance of British financial capital to fund industrial production turned out to be a major obstacle for the government. After all, according to the biotechnology policy narrative venture-capital-driven investments were to play a critical role in the creation of the British biotechnology industry. While the British biotechnology policy narrative had coded venture capital as a critical source of funds for biotechnology investments, the financial markets resisted this definition of their role.

Furthermore, small and medium-size companies, which played a prominent role in the government's policy narrative, were not only faced with a virtual absence of venture capital. Even the available instruments of financing were not flexible enough to match the needs of an emerging industry. This was exemplified by the requirements a company had to meet in order to be listed on the London Stock Exchange, a crucial source of money. Whereas in the United States the National Association of Securities Dealers (an exchange for small stocks) has few rules, the London Stock Exchange requires companies to show revenue, profits, and minimum trading experience before they can be listed. As a result, the only medium-size biotechnology company on the London Stock Exchange was British Bio-technology Group, which secured a listing in 1992. In the same year, the British biotechnology company Cantab opted instead for a NASDAQ listing. Concerned that other companies would follow the lead of Cantab, the London Stock Exchange began to reconsider its policies. Under the new regulations, however, only companies operating in the human-pharmaceutical sector could apply. A London Stock Exchange official put it this way: " . . . we believe that only biotechnology companies developing human drugs in clinical trials can provide the level of comfort we desire."[81]As a result, British Biotechnology was followed at the London Stock Exchange by other biopharmaceutical and diagnostics companies, including Celltech and the slightly renamed Canatab.[82] However, companies operating in agriculture, diagnostics, and manufacturing remained excluded. Furthermore, pharmaceutical companies were required to demonstrate their R&D competence with patent applications or awards and to be conducting clinical research on at least two drugs developed in house.[83]

Not surprisingly, a recent survey found fewer than 50 new biotechnology companies in operation in the United Kingdom. The vast majority of these companies had only a few employees, and most preicted flat growth. Only half a dozen of these companies appeared to have the ambition and potential to become significant, multi-product companies.[84] Without the instrument of an initial public offering (in the US industry the typical and profitable route of exit for venture capitalists), venture capitalists in the UK refrained from biotechnology investments and concentrated on other areas, such as computers. With minimal governmental support for small biotechnology companies and with a lack of interest or a "hands-off" attitude prevailing among those venture capital companies involved in the funding of companies, virtually the only major external source of funding for newly founded biotechnology companies was other, larger manufacturing enterprises.

Using this source of funding usually meant becoming an object for acquisition (Oakey et al. 1990, p. 159). As a consequence, within the general context of the comparatively rigid British financial system, various state schemes to promote collaboration between science and industry could have only limited results. Whereas large companies turned out to be attracted to government programs, the small and medium-size companies that had been considered especially valuable by the government were unable to engage in such collaborations, mostly because of a shortage of funds (Senker 1991, pp. 108–113). This particular organization of the financial markets and their further deregulation generated practices that led to considerable destabilization of the state's biotechnology strategy. The government's biotechnology policy had constructed the financial system as a nodal point in its writing of British biotechnology industry, but the financial markets simply did not generate the capital necessary for biotechnology investments. As a result, only a few small and medium-size new biotechnology companies developed in Britain, and "the biotechnology industry" was constituted mainly of large, well-established, internationally operating corporations. Not only had multinational companies such as Glaxo, SmithKline Beechham, ICI, Welcome, Boots, and Fisons taken a strong interest in biotechnology; they also profited from governmental support of genetic-engineering-related R&D.

France's Struggle to Become a Leader in Biotechnology

France's biotechnology policy, like Britain's and Germany's, reflects the attempt to map biotechnology as a national project. The French political metanarrative provided a suitable context for this strategy. One of the few things on which the fifty or so French governments since 1944 have agreed has been the need for an industrial policy and, more generally, the need for a growth-oriented politics of modernization (Messerlin 1987, p. 76; von Alemann 1988, p. 161). Furthermore, the representation of a modernization process driven by science and technology has been a characteristic feature of French political and economic discourse since the mid 1950s, particularly since the Fifth Republic and Charles De Gaulle's presidency (Gilpin 1968, pp. 188–217). Central to this political imaginary has been a powerful state at the helm of the politics of economic development.

Even the early French efforts to shape a policy field dealing with molecular biology were inseparable from industrial policy discourse.

The emergence of a molecular biology policy in the 1960s was closely related to the government's efforts to develop *industries de pointe* (high-tech industries) to make France internationally competitive. Later, under the presidency of Georges Pompidou, the narrative of industrial policy forged the concept of "national champions," according to which the state, to confront rising international competition, sought to promote the emergence of one or two enterprises in each major industry. These national champions would protect the small and medium-size producers that supplied and bought from them. The creation of national champions took place within the framework of *plans sectoriels*—government blueprints for the structures of particular industries. In the chemistry industry, for example, Rhône-Poulenc was "picked" as a national champion (Aujac 1986, pp. 14–15).

As the national-champion policy gained ground, the selected companies seemed to be less profitable than other companies in the same line of business. Declining efficiency offset economies of scale, and there were even some serious setbacks (for example, in the computer industry). During the presidency of Valery Giscard d'Estaing, the oil crisis of 1973, the disintegration of the international monetary system, and the accelerated internationalization of the economy led to a new approach to industrial management, with a heavy reliance on market mechanisms and macroeconomic policy. From 1978 on, a new industrial policy discourse provided a new theory of industrial development in France: the *politique des créneaux*. According to the new narrative, national champions were no longer able to dominate all parts of a markets; rather, they were supposed to develop expertise in particular niches of the market. Such expertise would allow them to become truly competitive in those segments of the world market (Aujac 1986, p. 16). In research policy, a similar effort was made to define niches that would allow companies to focus on particular areas of research (Rouban 1988, p. 160).

The Protein Crisis: How to Shape a Biotechnology Policy Narrative

It was in the situation just described that biotechnology was articulated as a field for comprehensive state intervention. Several key documents served to outline the grand design for a French approach toward biotechnology. In 1978 the Nobel Prize-winning biologists François Gros and François Jacob and the medical doctor Pierre Royer developed, in a report for the President of the Republic entitled *Sciences de la Vie et Societés,* a broad and philosophically charged outline of the future of life sciences and their importance for France (Gros et al. 1979). In a detailed supporting document by Joël de Rosnay entitled *Biotechnologies et Bio-Industrie,* the grand policy ideas developed in *Sciences de la Vie et Societés* were further elaborated in terms of a possible industrial strategy (de Rosnay 1979, pp. 7–11). These two reports adopted a comparative perspective by looking at scientific and industrial developments in France and in other countries. In political significance, they constitute a French equivalent to the German DECHEMA Report and the British Spinks Report. In the narrative of *Sciences de la Vie,* the new biology is in the process of transforming man's understanding of living things and of himself and, thus, represents a new step in man's hitherto successful control of the world (Gros 1979, pp. 14–15). The implications of the new biology for society will be far-reaching:

> What is being created today in several developed countries is a new industrial field. . . . Modern biology is disposed with large potentialities, is very diversified, and grows out of a number of what probably constitutes the most advanced disciplines in the sciences. . . . The great breakthroughs in biology can be compared with those in physics and chemistry in the 40s and 50s. . . . As a consequence, the development in research has created a veritable revolution in biology. . . . The fields of application of biotechnology are extremely vast, probably comparable to those of chemistry at its beginnings. (de Rosnay 1979, pp. 7–11)

Biotechnology, the narrative continued, will help to respond to the three current major crises of today: the energy crisis (due in particular to the exhaustion of nonrenewable resources), the crisis of nutrition (due to the bad weather of past years), the energy costs involved in the production of grains, and the increasing costs of protein production. The report argued: "Our country depends up to 85 percent for its proteins on imports. . . . A cow of 500 kilograms is capable of producing half a kilo of proteins within 24 hours, whereas 500 kilograms of microorganisms in a fermenter can produce 5 to 50 tons of proteins within the same time. Hence, the production of proteins with the help of unicellular organisms constitutes a strong answer to the protein crisis. . . . " Finally, in regard to the crisis of the environment, the argument was made that naturally occurring microorganisms could be used to fight environmental degradation (de Rosnay 1979, pp. 20–21).

But the beneficial impact of biotechnology could not be taken for granted. The chances for the diffusion of the new biology into the economy, and hence the chances of resolving the major crises, were hindered by a series of obstacles. According to de Rosnay, the development of biotechnology in France was inhibited by a pathology of interrelated causes: the traditional focus on mathematics and physics at the universities and, related to it, the neglect of applied biology; the conservatism of the agro-food sector, which had changed little since Pasteur; the lack of an infrastructure for technology transfer; the universities' tradition of focusing on biomedical and biochemistry—fields that turned out to be indifferent to transferring their knowledge to industry; and the poor quality of biological research done in industry, from the agro-food sector to antibiotics research. Furthermore, there was a gap between science and industry: "We lack 'scientific entrepreneurs,' as much as commercial entrepreneurs who are prepared to take scientific risks." (de Rosnay 1979, p. 95) The de Rosnay report concludes: "What is missing . . . is a global strategy for biotechnology that integrates technological progress with an evolution beneficial to society." (ibid., p. 25)

The two reports went on to outline a strategy for France that linked the reality constructions of *Sciences de la Vie* to political strategies and to the designation of actors and structures in the new policy field. The idea of *créneaux* (niches), a concept from the dominant industrial policy discourse, served as an important point of reference. It was argued that some of the niches selected should reflect current strengths of French research and industry and that some of them should be fields for necessary development (Gros 1979, pp. 252–261). To implement this "Plan Biologie," interrelated strategies ranging from public and private financing to industrial plans for developing sets of priority and diversification in industry were proposed (de Rosnay 1979, p. 120).

Mitterand, Nationalization Discourse, and the Golden Age of French Biotechnology Policymaking

The overwhelming victory of the Socialists in the elections of 1981 ushered in a new period in French industrial policy. A new Ministry of Research and Technology was set up,[85] and Jean-Pierre Chevènement, the new Ministère de la Recherche et de l'Industrie, dreamed of making it a French version of the Japanese MITI. In 1982 the ministry was renamed "Ministry of Industry and Research."

The Socialists did not believe in a *politique de créneaux;* instead they created a new industrial policy narrative: the *politique de filières,* a strategy based on providing aid to entire vertical streams of production. According to this model, no industrial niche is viable on its own.

Each depends on the vitality of "upstream" and "downstream" industries. Hence, the *politique de filières* recommended, the government should intervene at every necessary point in the vertical stream of production (Aujac 1986, pp. 17–18). According to this framing, the political task was to reconstruct the productive apparatus in such a way that it would be fully integrated from the stage of raw materials to that of finished products (Lipietz 1991, p. 33). A new industrial policy narrative had been shaped, and it soon became a central point of orientation for the emerging biotechnology policy.

After the 1981 elections, there was a round of nationalization that included five major private firms (among them Rhône-Poulenc and Saint-Gobain) and all the major banks. The idea was that the nationalized industries should act as the "locomotive" of the economy, while the private sector would be under strict control of state-owned banks. Thus, nationalization revived and formalized the old concept of national champions (Messerlin 1987, pp. 101–102). And the Policy and Planning Act for Research and Technological Development in France (passed on July 15, 1982) created a framework that would ensure the decentralized diffusion of technology into the various regions of the country.[86]

The old idea of a biotechnology strategy for France fitted well into this design of political intervention. François Gros, director of the Institut Pasteur and co-author of *Sciences de la Vie,* became President François Mitterand's science adviser (Rouban 1988, p. 162). Gros remembers:

> The report had a great impact in foreign countries, such as in Israel, Switzerland, Belgium. . . . In France there was no impact, but a year or two afterwards the industrialists said that this is an important report, we have to do something. . . . Then François Mitterand came into power. Some months later I was invited to lunch by the future culture minister Jacques Lang and Mitterand. The reason for this invitation was that Mitterand had read the report commissioned by d'Estaing and found it very interesting, and he asked me to work with him, he had understood that this is something important for the future of the society. . . . Six months after formation of the government in 1981, Jean-Pierre Chevenment consulted with me. We talked about the future of biology and Chevenment decided to launch several mobilization programs, and this happened very fast, Chevenment had a lot of energy.[87]

In July 1982 the Ministry of Research and Industry launched the Programme Mobilisateur: l'Essor des Biotechnologies, which has been described as "something like a master plan for biotechnology research and industry development in France."[88] The Programme Mobilisateur, scheduled to last 10 years, had the not-exactly-modest goal of raising France's share of the world biotechnology market to 10 percent by 1990.[89] It outlined paths of development, strategic goals, actors, and institutions that were at the core of the new policy domain. In a press conference held to present the French government's new program, Chevenment announced that the government was prepared to spend 600 million francs over the next 3 years for biotechnology and that it expected industry to do its part by spending a billion francs per year on biotechnology.[90] Unlike the draft reports issued during the Giscard presidency, which were devoted to niche strategies in industrial policy, the new program re-inscribed itself into the new dominant industrial policy discourse of a *politique de filières*. In its opening statement, the Programme Mobilisateur defined its mission as taking charge of "the ensemble of the aspects of a *filière* that goes from the biosciences to the bioindustries through the intermediary of the biotechnologies."[91] Whereas the earlier reports sought to identify niches in which to concentrate resources, the Programme Mobilisateur proposed sweeping coverage of the field of biotechnology, from research in genetic engineering, microbiology, enzymology, enzyme engineering, immunology, fermentation research, and seed and plant improvement to industrial applications in such fields as pharmaceutical products, agriculture, chemistry, and environmental applications.[92] The scope of the support program became especially obvious in the analytical parts of the document, which pointed out serious weaknesses in a number of research fields where only a handful of companies were engaged in biotechnology R&D. However, the *filière* program sketched various axes of the proposed biotechnology infrastructure with the intention of integrating various public research laboratories, including the Centre Nationale de la Recherche Scientifique (CNRS; the largest public research organization in France), the Institut National de la Santé et de la Recherche Médicale (INSERM), the Institut National de la Recherche Agronomique (INRA), and the Institut Pasteur. The Ministry of Research emerged as the "center of calculation" in this effort to construct a biotechnology network and to open up its economic space. Thus, the main actors in the creation of a French biotechnology industry and the roles they were to play in the new industrial sector had been identified. Between 1982 and 1985 the new program's available resources increased steadily (1982: 43 million francs; 1983: 70 million; 1984: 80 million; 1985: 110 million).[93] The declared strategic goal of the program was to create an interface between science, technology, and production.[94]

Besides direct support from Fonds de la Recherche de la Technologie, indirect aid for biotechnology came from the Agence Nationale de Valorisation de la Recherche, which had been set up in 1968. ANVAR's task, under the supervision of the Ministère de la Recherche et Technologie, was to develop research and to promote innovation. ANVAR had at its disposal a variety of innovation instruments, of which the most powerful was a subsidy for companies that was repayable if a project was successful. The purpose of this scheme was to support activities prior to industrial production. Although mainly directed to firms, this aid was also given to laboratories and to small businesses.[95] Exact statistics on ANVAR aid and other indirect measures are not available, but it is estimated that the total amount of funding from these sources was at least equivalent to the direct Fonds de la Recherche de la Technologie subsidies.[96] In addition, in October 1985 the Ministry for Research and the Ministry for Agriculture launched the multi-year Aliment 2000 program, the purpose of which was to promote development in the agro-foodstuffs industry. Between 1986 and 1989, 104 million francs were allocated to biotechnology.[97] The expenses of the main public research bodies (CNRS, INRA, INSERM, Commissariat à l'Energie Atomique, and Institut Pasteur) increased between 1982 and 1987 from 5.6 percent to 10 percent of the total budget allocated to public-sector agencies (1982: 570 million francs; 1983: 780 million; 1984: 861 million; 1985: 993 million, 1987: 1507 million).

These various forms of state support for biotechnology and its industrialization can hardly be dissociated from the state's massive efforts to restructure the chemical and pharmaceutical industry. The new biotechnology policies were closely linked to the broader nationalization discourse. During the 1970s, official policy focused attention on telecommunications and aerospace while leaving developments in the chemical industry to the forces of the market. After Mitterand's victory, the chemical and pharmaceutical industry became part of the state's industrial policy, which entailed the nationalization of Rhône-Poulenc and the creation of a new division of labor among French chemical producers. The ninth five-year plan (1984–1988) identified chemicals and pharmaceuticals as two keys to increasing the international competitiveness of French industry. Massive state support and restructuring enabled Rhône-Poulenc, nearly bankrupt at the time of its nationalization, to a break even in 1983 and to register profits of $348 million in 1985 (Adams 1991, pp. 94–95). In other words, the companies that were singled out to serve as the industrial end of the *filière biotechnology*—which, according to various evaluations, were not exactly in a position to implement the government's biotechnology strategy in the marketplace—became objects of the larger industrial policy and received state support in amounts that dwarfed the subsidies explicitly intended for biotechnology.

Toward "Nationalization" of the French Biotechnology Industry

In the second half of the 1980s, the French biotechnology policy narrative underwent a number of important changes. In the early 1980s biotechnology had been coded as a new field of research and industrial activity with a potential to transform the whole economy. Now a much more modest interpretation seemed to develop. At the same time, there seemed to be less reliance on individual, private entrepreneurs to create a biotechnology industry. And the international context had changed in a way that forced strategies of adjustment on biotechnology policymakers.

An important assumption of French industrial strategy was that the international economy would provide a suitable environment for France's ambitious national policy goals. While the French state was tightening its grip on the economy, external and internal constraints led to a U-turn in Socialist policymaking in 1983. Because the government's Keynesian policies had created a huge trade deficit, the government was forced into a series of devaluations. The working and middle classes turned to oppositional forces and to demonstrations. The state set off a new course of *rigueur* politics intended to bolster free trade and thus to bring down inflation and to increase firms' declining profits by means of a decline in household purchasing power. At the same time, the overambitious *filière* philosophy of industrial policy gave way to a new discourse of restructuring. "Modernization" became the official watchword.[98] In a way, these changes in political discourse indirectly benefited the biotechnology strategy, particularly by facilitating the restructuring of the chemical and pharmaceutical industries.

But first, after the parliamentary victory of the right in 1986, the U-turn toward a neo-liberal orientation continued, as evidenced by significant budget cuts, particularly in areas of research. Half of the 15 billion francs to be cut from the budget for 1986 involved research.

The Fonds de la Recherche de la Technologie biotechnology program, slashed to 50 million francs in 1987, never recovered even after the Socialists returned to power in 1988 (1988: 57 million francs; 1989: 50 million; 1990: 50 million; 1991: 45 million).[99] Beginning in 1986 the state launched other national programs with biotechnology components, such as EUREKA (1988: 5.5 million francs for biotechnology; 1989: 10 million) and the "Technology Leaps Forward" program (1988: 17 million francs; 1989: 15 million), but contributions in the area of biotechnology were relatively modest.[100] The other arm of the state's biotechnology program, ANVAR, was seriously reduced in its budget and operations while the right was in power.[101]

Eventually, other challenges to the government's biotechnology strategy arose. For instance, some began to question the scope and efficiency of the Programme Mobilisateur with respect to the transfer of knowledge from the research system to industry. One of the organizers of the Programme Mobilisateur summarized this as follows: "The success was partial. It was very good to mobilize the people, to mobilize their interest and attention. It was less good in terms of yielding specific results at the level of companies. . . . "[102] As had happened in the cases of Germany, Britain, and the European Community, it seemed that the resources pumped into the French research system had led to a fundamental reorientation of the R&D system toward the new biology. However, it appeared that the various mechanisms and programs of technology transfer from the large research establishments to industry were, with the exception of the Institut Pasteur, limited in their efficiency.[103] Furthermore, the enrollment of industrial actors failed to go beyond nationalized companies. Industrial research expenses continued to be lower than the international average; this held especially true for the agro-food industry, which was not dominated by any large or nationalized companies.[104] Apparently, there was little determination to invest in expensive new research directions. In addition, small startup companies—with the exception of Transgène, started in 1981 by two scientists—had not attained any substantial importance. But even Transgène was controlled by the government: 36 percent of its shares were held by the government-controlled Paribas, AGF, and Elf-Acitaine corporations, and another 15 percent by government organizations. Moreover, Transgène's main clients included Roussel-Uclaf, Institut Merieux, and the Institut Pasteur, all of which were government controlled.[105] In other words, the invention of a "French biotechnology industry" was "successful" where the government was in a position of impose its vision on the actors in industry without depending on a negotiation of the economic realities of genetic engineering. It seemed to work much less well where the cooperation of industry was voluntary.

Reinforced by the dominance and the increasing strength of Rhône Poulenc, Sanofi, Roussel-Uclaf, and Transgène in the therapeutic field, biotechnology policy began to focus state support on diagnostics and vaccines. Besides this development, the agro-food industry, despite its disappointing R&D performance, continued to be a target of a rather broad range of support measures aimed at modernizing a variety of production processes.[106]

Just as in Germany and in Britain, the dream that many small venture-capital-based companies would emerge to constitute the backbone of a French biotechnology industry did not materialize. The companies that did emerge (some 40 by the early 1990s; see Raugel 1990), including Transgène, tended to be backed by government and industrial groups (often government controlled). In addition, a number of contextual factors were responsible for the absence of small companies in biotechnology. The specific coding of these factors in biotechnology policy narrative contributed to the lack of success of policymaking. First, the state's biotechnology policy initiatives were undermined by the financial sector. The financial system was characterized by strong control over the total amount of credit and the interest rate and by rigid regulation of capital funds. French companies, in general, needed external financing to cover the costs of investment. Dividends were subject to double taxation, first as profit and second as income, and this made the purchase of shares less attractive. However, more neutral fiscal clauses and stock-market reforms improved the situation for private investors. Furthermore, before the deregulation of the French banking system around 1984 there was no large financial market. A good part of consumers' savings was directed toward to the acquisition of homes, and household financial investments were steered toward traditional bank accounts or government bonds. After deregulation, a more differentiated financial market developed, and gradually a venture-capital industry took shape (Canals 1993, pp. 126–128). However, as in Britain, small companies had restricted access to the stock market. This made capitalization on investments with a stock market listing (as on NASDAQ in the United States) impossible; hence, it created a need for a longer and closer associations between investors and startup companies—often with venture-capital companies working in close collaboration with state innovation-aid organizations such as ANVAR. This is exactly the kind of relationship that developed between risk-capital companies such as Sofinnova and Agrinova and their clients in the biotech industry.[107] Despite the development of a venture-capital industry, however, the reliance on those companies as well as on banks (which in France have not traditionally had a close relationship to industry) remained limited.[108] More important were strategic alliances between small French startup companies and large French companies—alliances in which the large companies would provide part of the funding. The emerging structure of the French biotechnology industry was coupled with a trend in which large companies acquired interest in dynamic foreign biotechnology companies. This pattern was pioneered by Transgène and originated in the large French agro-food and pharmaceutical companies' philosophy of not engaging in biotechnology and of leaving this field to specialized companies.[109] For example, until 1991 Elf-Sanofi (the biotechnology subsidiary of the state-owned oil company Elf-Aquitaine, with a turnover of 3.548 million ECUs) was one of the shareholders of Transgène. In 1990 Elf-Sanofi acquired Genetic Systems and combined its purchase with its Pasteur Diagnostics subsidiary, which made it the world's second-largest producer of AIDS tests. Roussel-Uclaf (turnover: 1.886 million ECUs) is a 54 percent subsidiary of the German company Hoechst and has partners the United States, including Cetus, Calgene, and Immunogen. In 1992 Roussel Uclaf signed an R&D agreement covering biopesticides with the Institut Pasteur and the American firm Ecogen. Rhône-Poulenc holds 51 percent of the Institut Mérieux (turnover: 708 million ECUs), which, after acquiring Pasteur Vaccins (in 1985) and the Canadian company Conaught (in 1989), became the world's leading producer of human and animal vaccines. The Lyon group had also taken control of two biotechnology companies, American Virogenetics (1989) and Transgène (1991). This acquisition made Transgène part of the Rhône-Poulenc empire. Last but not least, there was Rhône-Poulenc itself (state-owned, with a turnover of 11.400 million ECUs), which had increased its contributions to the life sciences (human and animal health, agricultural products) from 38 percent in 1988 to 43 percent in 1990. Its aggressive expansion policy made Rhône-Poulenc a leading chemicals and pharmaceuticals group. Rhône-Poulenc opened a 30-million-ECU research center near Paris in 1988 and is planning another center in Japan. It also acquired the agrochemistry business of Union Carbide (in 1986) and the pharmaceutical company Rorer (in 1990), and in 1989 it took control of the vegetable-and-flower-seed company Clause in a joint venture with Orsan.[110] In 1993 Rhône-Poulenc Rorer acquired a 37 percent stake in Applied Immune Sciences, with an option to acquire up to 60 percent before June 1997.

Obviously, after the mid 1980s French biotechnology policy strategy and the closely connected biotechnology strategy gave up the *filière* strategy and narrowed its attention to the health and vaccines industry, leaving the crown jewel of the French R&D system, the Institut Pasteur, in a key position and the most dynamic "startup company" Transgène integrated into a vaccine group. The industry maintained a strategic interest in plant genetics, reflecting France's position as the world's second-largest producer of seeds and its emphasis on the agricultural sector. The food industry, devoting only 0.2 percent of its revenues to research, remained the weak link in the biotechnology strategy.

This reconstruction of the political space of biotechnology culminated in the Bioavenir program, launched by the French state in 1991. Bioavenir was an unprecedented collaboration between Rhône-Poulenc and France's major publicly funded research institutes. It was a five-year research project with a budget of 1.6 billion francs, 660 million of which came from the French government. The formation of Bioavenir consolidated Rhône-Poulenc's collaborative relationship with the Institut Pasteur, the CNRS, INSERM, and CEA, and brought various institutes in different fields of interest (human health, products for agriculture and biological catalysts) together under its leadership. The establishment of Bioavenir had as its primary goals transforming basic science into industrial applications and "catapulting Rhône-Poulenc into the 21st century."[111] With this strategy of "exporting French nationalization" Rhône-Poulenc became the world's seventh-largest chemical company, but also, at the same time, the most heavily indebted of France's state companies, with debt at 80 percent of equity in 1992 (Celarier 1993, pp. 52–59). Hence, the invention of a French biotechnology industry climaxed in the creation of a "European champion"—a heavily indebted multinational corporation that, under the tutelage of the French state and in close collaboration with some other large companies, provided the "industrial end" of the government's biotechnology policy. The French strategy of mapping biotechnology as a national project fitted well into the structure of the dominant industrial policy discourse, which endorsed the idea of heavy state intervention in the industrial field.

Stories about Biotechnology

In this chapter I have examined the interactions among the myths, industries, and policies of biotechnology. By focusing on a number of interrelated policy stories and on how they ordered the economic and social realities of genetic manipulation, I have explored the relationship among genetic engineering's discourse of deficiency, strategies of redefining human beings, political and economic discourse, and contexts such as economic developments, scientific innovations, and financial markets. My central argument has been that the political space of biotechnology was set up by a number of different and partially conflicting discursive practices. These practices did not reflect some sort of central project; they constituted a contingent constellation of mutually reinforcing textual trends and developments. For example, predictions about the industrial future of biotechnology and about molecular biology's theory of genes were not causally related developments, but they nevertheless had a mutually supportive influence in gaining political support for the manipulation of genes. This discursive constellation was the surface where the objects, institutions, actors, and strategies of biotechnology policymaking emerged.

In European policy discourse, biotechnology was encoded as a central field of high technology that would determine the futures of all nations. Mythical interpretations of "US biotechnology" gave rise to comprehensive biotechnology policies in Britain, in France, and in Germany and at the level of the European Community. In the biotechnology policy narratives, comprehensive models of biotechnology were outlined: How will biotechnology change society and economy? Which actors will collaborate in the realization of this model? Which strategies need to be pursued? How will all this lead to a happy political future? The stories that emerged from these questions not only discursively constructed the boundaries of biotechnology industry as a European or national project; they also assigned roles to actors, established their identities as separate actors and as members of a larger policy ensemble, defined social, political, and economic realities, and sketched ways of dealing with these realities. By assessing the quality of research, counting laboratories, evaluating industries, compiling tables, and delineating possibilities, the anarchic reality of biotechnology was broken down into specific dimensions, numbers, plans, industrial sectors, and agencies that could be supported, revised, eliminated, or broadened. In short, biotechnology was to become governable.

However, as I have shown, after almost 20 years of "the biotechnology revolution," biotechnology policymaking is characterized by the co-presence of a number of partially conflicting policy narratives and framings of genetic engineering, which have failed to produce a broadly shared perception of "biotechnology" among the relevant policy actors. These stories and interpretations were not simply deployed by utility-maximizing actors; they were crafted against the background of specific discursive economies and socio-economic and political contexts. Following my definition of policymaking as a process of intermediation between policy narrative, discursive constellations, and discursive contexts, I have shown that by the late 1980s the discursive construction of the European biotechnology industry had been seriously undermined by a number of developments and trends in biotechnology. These destabilizing factors became objects of the public policy debate.

One area of contestation among policy stories was that the first 15 years of the various European governmental programs for biotechnology had yielded far less in terms of products, technological advances, and new companies than was initially envisioned. The new genetic technologies had wrought a major transformation in biology and medicine. Less successful seemed to be the translation of the new research into industrial products or, as initially anticipated in most countries, the creation of a new kind of industry—a biotechnology industry—composed of many new companies spawning new jobs and contributing to economic growth. Instead, European biotechnology seems to be dominated by a small group of well-known multinational companies located in the chemical and pharmaceutical industry. By the early 1990s—unlike the United States—Britain, France, and Germany had only a few dozen small startup companies, of which only one or two (such as Transgène and British Bio-Technology) had acquired international renown and standing.

In the late 1980s, when European policymakers began to assess the apparent failure of biotechnology policies to stimulate a new industry, the lack of venture capital and the rigidities of the capital markets figured prominently in their explanations. It was argued that the European financial institutions had been ill-prepared to support the creation a "new biotechnology industry." Furthermore, particularly in

Britain and France, the organization of research and the relationship between research and industry seemed to have created obstacles to the development of a "US type" biotechnology industry.

But this story of institutional inadequacies was only one among several stories that could explain the difficulties of European biotechnology policies of the 1980s. A number of other contextual trends and discursive developments conflicted with the interpretations and assumptions that were central to the European biotechnology policy narratives. Among those trends and developments were certain practices in international biotechnology research and production, the gradual reframing of genetic engineering's status as a "high technology," and the emergence of new readings of the nature of molecular biology, its claims, and its theories of natural processes.

First, the mapping of biotechnology in the dominant policy discourse as something distinctly European, British, French, or German conflicted with strong tendencies of globalization in the international chemical and pharmaceutical industry. With the United States widely regarded as a highly attractive location for biotechnology-related research and production, European biotechnology policies were not very successful in stimulating the creation of a biotechnology industry at home.

But European biotechnology policy had not simply become a victim of rigid financial markets or of processes of globalization and their perception by investors. Today, biotechnology seems to have undergone a fascinating semantic transformation from a "high-tech industry" to a "high-tech method" to be used in "traditional industrial sectors," such as the pharmaceutical industry. Furthermore, a number of voices have started to point to the possible limitations of molecular biology as a research paradigm. Finally, these new assessments of biotechnology are reinforced by a long list of product failures.

One core assumption in the biotechnology narratives concerned the quality of biotechnology as a "high technology." In the case of the semiconductor industry, for example, the basic technology was developed at the birth of the industry and could then be acquired by other firms for a reasonable price. A number of common theoretical principles underlie biotechnology, but these common principles have not proven readily adaptable to different industries and do not easily translate into a core technology suitable for widespread modification into commercial niche products. Furthermore, the transition from the first initial conception of an idea to the successful sale of a product is often a protracted process. Hence, most of the new biotechnology firms do not have products on the market and thus cannot rely on profits as a main investment source.

The typical situation for a new biotechnology firm is as follows: The absence of profits during development means that large sums of external investment are required to cover expenses, while the absence of products renders the raising of capital difficult. Taking these peculiarities of biotechnology in the late 1980s and the early 1990s into account, some have asked why this field is subsumed under the same "high technology" label as, for example, the semiconductor industry. One problem with the definition of biotechnology as "high technology" lies in the fact that most current definitions of that term depend on the measurement of innovation inputs, not innovation outputs. A high R&D input does not necessarily generate industrial-growth-producing outputs (Oakey et al. 1990, pp. 151–156).

McArthur (1990, pp. 817–881) proposes to distinguish between widely diffusing and newly emerging technologies. Widely diffusing technologies, such as information technology and telecommunications equipment, and microprocessors, are adopted across a wide range of manufacturing and service activities and have economy-wide impact. Newly emerging technologies are in the initial state of their development, their diffusion is limited, and they are incorporated into a limited range of products. Though they may eventually diffuse widely, many technologies will be used only over a limited area of the economy. When one takes into account the current diffusion of genetic engineering, with its particular emphasis on human therapeutics and diagnostics and its rather sluggish growth in the agricultural sector, the food industry, and the field of environmental biotechnology (Glaser 1992), the characterization of biotechnology as a widely diffused "new technological paradigm" seems, to say the least, premature.[112] While the major biotechnology policy narratives in Europe continue to emphasize genetic engineering's character as a technology giving rise to a new "high-technology industry," trade journals and specialists in the field have gradually and cautiously begun a reassessment of this story. According to them, the initial predictions concerning biotechnology were too optimistic. This reading constitutes an important element in a newly emerging discourse that constructs biotechnology not as "a new industry" but as a "methodology to study nature" and as a new industrial method, with benefits to be expected either in specialized areas or in the distant future.[113]

Another new and important story was told by a group of scientists in the medical field. During the early 1990s, scientific discourse in biology and medicine had begun to raise a number of important questions concerning the scientific and technological principles that form the theoretical basis of genetic engineering. The reductionist approach of molecular biology became a focus of criticism. Reductionism in molecular biology represented biological systems in terms of the physical interactions of their parts. In this view, the fundamental understanding of biology comes only from the level of DNA, the alleged blueprint for living systems (Tauber and Sarkar 1992, p. 228).

At the core of this scientific narrative is the following construction:

DNA → RNA → Protein → Everything else, including disease.

According to this scheme (also known as the "central dogma" of molecular biology), a sequence of DNA functions either by directly coding for a particular protein or by being necessary for the use of an adjacent segment that actually codes for the protein.[114] In the United States, by the late 1980s, the preponderance of the National Institutes of Health's research budget was going to projects that reflected this dogma (Strohman 1993, p. 117). Following the logic of a reductionist life narrative, many researchers in molecular biology truly believe that studies on gene cloning and regulation will lead to important medical advances. This genetic research paradigm, however, is questioned by a number of theoretical counter-arguments that have developed over the last years, and also by findings of epidemiological studies. The goal of the genetic paradigm in medicine—to detect single gene associations with human disease—finds itself challenged by the current patterns of disease distribution. For the major diseases, such as cancer, cardiovascular conditions, and most psychological illnesses, there is no evidence for either single-gene mutations or chromosomal aberrations as a unitary cause. Rather, the major diseases are polygenetic and complex, have environmental determinants, and are not approachable by genetic analysis alone (as the reductionist narrative of molecular biology suggests they are) (Strohman 1993, pp. 119–120). The Human Genome project argues that genetic effects can be separated from effects emerging from gene-gene, gene-gene product, and gene-environment interaction. Furthermore, it is assumed that genetic programs are operative, and that polygenic diseases can be mapped onto Mendelian components or onto grouped components. But most of these assumptions have been rejected by population geneticists and are gradually being rejected by biologists working on fundamental problems of molecular and cell biology (ibid., p. 130). One of the main criticisms leveled against the Human Genome Project is that its framing of disease ignores the complexity of the issues and the limitations that can be imposed on the possibility of such theoretical computation. A crucial example for such difficulties concerns protein folding. In the representation of molecular biology, the assumption is made that the primary structure or amino acid sequence of a protein determines its tertiary structure or three-dimensional conformation. Therefore, it should be possible to calculate the conformation from the sequence information. However, despite three decades of research efforts, the problem of protein folding has yet to be solved (Tauber and Sarkar 1992, p. 223). But protein conformation is the key to receptors, which are essential for successful drug design.

Such "research bottlenecks," biological counter-narratives, and patterns of epidemiology have become parts of a new story that not only questions biotechnology as a high technology but also, and more fundamentally, raises doubts about its theoretical foundations and about related predictions on the transformative potentials of genetic engineering.

At the level of product development, this constellation has given rise to a story about biotechnology's "long and winding road to the marketplace."[115] Vivid accounts of how winding this road can be found in the annual reports on "product failures." For example, in 1995 in the United States, Synergen and Immunex abandoned the development of sepsis treatment; Immunex terminated clinical trails of its tumor necrosis factor receptor approach; Chiron abandoned work on sepsis treatment; Regeron stopped development of its ciliary neutrophic factor treatment for Lou Gehrig's Disease, which showed no appreciable difference over the placebo in clinical trials; and Alpha 1 Biomedicals announced that its thymosin alpha 1 for hepatitis B did no better than the placebo during Phase III clinical trials, and the same held true for Magainin's MSI-78 imetigo therapeutic and for Greenwich Pharmaceutical's rheumatoid arthritis drug, Therafectin. The same year witnessed only a few product successes.[116] Such failures and the limited successes in product development in the United States constituted important destabilizations of worldwide biotechnology policy. Apparently, even the widely praised and mythologized US biotechnology industry seemed to be based to a considerable extent on the predictions of scientists and the hopes of investors rather than on the success of genetic engineering in creating new drugs and products. In the 1990s, after the period of gross exaggerations by scientists, venture capitalists, and entrepreneurs in the late 1970s, a more sober mood set in, partly because of disappointment with the achievements of the biotechnology industry.[117] The reconfiguration of genetic engineering from a "core method of a new industry" to a "revolutionary method to study nature" is the semantic reflection of these experiences of product failure. Such changes in the interpretation of biotechnology certainly affected policymaking, as can be seen in the example of France's evolving strategy of focusing on health and on vaccines.

So, can we assess the European biotechnology policies of the 1980s as a policy failure, or as having only a small impact? Again, this depends on the kind of story we choose to tell. We certainly can argue that in France, German, Britain, and the European Community the dominant biotechnology policy narratives had not anticipated or taken into account a number of important contextual factors, such as the structures of domestic financial markets. At the same time, important addressees of policymaking, such as potential scientist/entrepreneurs, had considered these contextual factors so important that they had refrained from responding positively to government initiatives. In addition, product failures began to point at the possible obstacles to turning genetic engineering into a broadly applicable industrial technology. These failed products became objects in a new story about the limits of the research paradigm of molecular biology. Finally, although the biotechnology policies did not have the anticipated *economic* impact, their *socio-political* effects were considerable.[118] This consideration leads back to the interpretation of biotechnology policy as not only a form of technology and industrial policy but also a broader cultural strategy.

The socio-political impacts of genetic engineering are difficult to measure, and they have less to do with tangible economic success or breakthroughs than with changes in the culture and practices of science and society. For example, the new genetics has designated medicine, and with it the human body, as the area with the most immediate hopes for application and product development. This has led to large-scale scientific projects, such as the Human Genome project. Also, we see today a trend toward the geneticization of diseases in the professional realm, and also on the level of popular representations of science (Duster 1990; Nelkin and Lindee 1996). In other fields, genetic engineering has moved very close to application. The mid 1980s witnessed a worldwide surge of deliberate releases of genetically modified organisms into the environment.[119] These trials produced only a few crops with a chance to be successful on the market; however, they reflected a thorough transformation of research practices in plant biology in many countries, and they contributed to the public perception of genetic engineering as a project of "engineering nature."

Thus, the story could be told that between the late 1970s and the early 1990s in Europe and in the United States the mapping of the boundaries of biotechnology resulted in the construction of a differentiated infrastructure of a biotechnology complex, involving research and production facilities, universities, factories, field sites, animals, plants, and human bodies. Genetic engineering's "discourse of deficiency" had suggested a rewriting of life on a subcellular level in terms of "absences," of "areas of improvement" in need of the intervention of genetic technologies. In tandem with political-economic discourses such the "international high-tech race" rhetoric or the invention of the myth of the biotechnology industry, this discourse of deficiency had contributed to the creation of a society whose identity and self-image began to be influenced by the culture and the practices of molecular biology. The Human Genome Project and its underlying constructions of health and disease were major steps in this direction. The emerging Bio-Society did not necessarily have a Bio-Industry at its core, but a regime of governability that systematized and coordinated genetic manipulation as a technological practice in a wide variety of interrelated sites under the tutelage of the state. In this context, from today's perspective, seemingly unrealistic stories about biotechnology as a "new high-tech field" may have been as important as "successful" deliberate release trials in agriculture. In this mix of reality and fantasy, bio-engineering's deeds, its dreams, its myths, and its often conflicting predictions were soon met by the fierce social resistance of political activists. The second phase of the recombinant DNA controversy had begun.

5

Deconstructing Genetic Engineering

The policies for regulating and supporting biotechnology that began to develop in the United States and in Europe in the second half of the 1970s derived a considerable part of their legitimacy from references to "the public," to "public interest," and, more generally, to the health, the nutritional needs, and the economic situation of the "population." Furthermore, the various representations of the economic and social impacts of genetic engineering had constructed a society that would not only profit from biotechnology but would also undergo a deep transformation—an expectation depicted in the image of "Bio-Society."

This semiotic appropriation of collective identity did not remain unchallenged. From the early 1980s, throughout Europe, a disintegration of this coherent notion of "the public and biotechnology" was evident when biotechnology once again became the object of a broad public controversy. This second phase of the recombinant DNA controversy was characterized by broad political mobilization and by a fundamental challenge to some of the basic elements of the biotechnology policies intended to regulate and promote biotechnology.

The unification of a political space always depends on the institution of chains of signification in the form of narratives that give broadly accepted, hegemonic definitions of the political realities, rationalities, and types of actors and institutions involved in a policy field. As has been discussed in the previous chapters, the construction of the political space of biotechnology was a result of the deployment of a number of discursive strategies, which did not necessarily articulate the same social goals or problem definitions but which nevertheless reinforced one another and produced the specific political topography of biotechnology as a policy field. Entrepreneurs, medical doctors, and research administrators may have had rather different perceptions and expectations concerning biotechnology; however, their arguments, interpretations, and dreams supplemented one another and began to inform policy behavior. Policy narratives were central in linking the different discourses into a coherent whole that provided a dominant definition of the reality of biotechnology. However, policy narratives tend to be contested, and conflicts over meanings and necessary actions in a policy field may destabilize policymaking. In chapter 4 I argued that this was the case with European biotechnology research and industry support policies during the 1980s. But the difficulties and failures of biotechnology support policies did not lead to an open crisis, and the narratives of biotechnology policy adjusted to the new situation by providing new interpretations of the goals and potentials of biotechnology.

The situation changed when a new group of actors began to challenge past biotechnology policies in a more fundamental way. In this chapter and the next I will show how a counter-narrative resulted in a hegemonic crisis of biotechnology policymaking that nearly led to a collapse of the political shaping of biotechnology in Europe. The policy counter-narratives I have in mind go beyond episodic resignifications of meanings attached to elements in a policy field by attempting to establish an alternative, counter-hegemonic model for the organization of a political space. They do not originate in one place, but are assembled in a number of different sites. They testify to the fact that regimes of governability and their hegemonic constructions are temporary stabilizations of power relations that often encounter challenges and resistance. The biotechnology counter-narratives constituted a critique and a deconstruction of the practices, dreams, mythologies, and futurology of genetic engineering that had been outlined so forcefully in the various discursive constructions of biotechnology.

These developments indicated important changes in the topography of the political space of genetic engineering—changes that had a particular impact on the definition of the risks of genetic engineering. In the early 1980s, in a process of boundary drawing, practitioners of molecular biology had established the regulation of recombinant DNA as an expert enclosure. Experts create enclosures when they are capable of establishing relatively bounded locales or fields of judgment that concentrate and intensify their power and authority (Rose and Miller 1992, p. 188). In the late 1970s the rewriting of the boundary object "gene" in the discourse of molecular biology had redefined the boundary between experts and non-experts in a way that made the social negotiation of genetic engineering increasingly difficult. As I will show in this chapter, such enclosures are always provisional and can easily become objects for contestation. The overflow of meanings that typically result from such contestations lead to destabilizations in policymaking, as happened in Europe when a number of new "spokespersons for genetic engineering" attempted to redefine its truth and its meaning in contemporary society by challenging the preestablished biotechnology narratives. The idea that the risks associated with genetic engineering are minimal and can be controlled by cautionary measures taken by the practitioners of genetic engineering were as much questioned as were the often-exaggerated scenarios (see chapter 4) of the social, economic, and cultural transformations initiated by biotechnology. Hence, the challenges concerning the meaning of biotechnology were not limited to probing the assumptions about the risks connected with genetic engineering. Instead, they included a deconstruction of the models and narratives of socio-economic development that underlay biotechnology policy.

These challenges, which came from different directions, attempted to retranslate the biotechnology problematic by deploying a policy counter-narrative that redefined the nature of the genetic engineering problematic, the interests and types of the various actors involved, and the goals of political intervention in biotechnology. Activists from animal protection groups, representatives from ministries of the environment, feminists, and professors from ecology departments—to name just some of the spokespersons—began an uncoordinated but mutually reinforcing dialogue on genetic engineering. Eventually, a process that redefined some of the key issues and goals of biotechnology policy led to a proliferation of new actors in the biotechnology networks, to new group identities, and to attempts to change the operating procedures of the biotechnology policy networks. In this chapter I will reconstruct several of the critical episodes in this attempt to restructure the political space of biotechnology.

The Impact of the Ecology Discourse

Important for the emergence of new actors in biotechnology policy and for the creation of new meanings connected with genetic engineering were a number of interrelated changes in political discourse that elevated the environment to the status of a core value in the political imaginary. With this discursive change, new actors came into existence in a number of environment-related policy fields, and they soon assumed important roles in political decisionmaking. In the early 1970s the environment had become a topic of political debate and intervention. The image of the planet as viewed from outer space by astronauts became an icon in a comprehensive political effort to address global environmental problems (Hajer 1995). After a period of framing environmental problems as incidental problems requiring ad hoc solutions, the 1980s witnessed an important modification of the metanarrative of modernization, which began to incorporate cognitive frames that were often summarized by the interchangeable slogans "ecological modernization" and "sustainable development." This new ecological metanarrative or "ecology discourse" can be understood as emerging in the interaction of academic discourses in different fields (e.g. sociology and ecology), in major policy and planning statements, and in mass media discourse. This new ecology narrative played an important role in shaping new approaches in fields such as environmental and energy policy. New policy approaches such as integrated pollution control received increased attention, and renewable energy resources emerged as an important subject in the political debate.

Sustainable development and ecological modernization became the two master signifiers of the new ecology discourse. Both of these concepts argued for the importance of integrating insights from the scientific discipline of ecology into policymaking. Ecology rejected the broadly shared model of nature as a black box that delivers energy and raw materials and absorbs waste. Ecology highlighted the complexity of the interrelation between societies and their base of sustenance, the displacement of effects on the environment in time and space, and the rapidly increasing scale of man-nature interactions (Spaargaren and Mol, 1992, pp. 328–329). Relatedly, the concept of sustainable development dealt with the developments in modern society that threatened its sustenance base. This concept, as developed in the report of the Brundtland commission,[1] attempted to integrate ecological quality with economic growth via industrialization. The precise mode of this integration, however, remained relatively vague and unspecified in terms of policy recommendations.[2]

A more specific and analytical concept had been advocated with the concept of "ecological modernization." At its core, ecological modernization stood for a major transformation: an ecological shift of the industrialization process toward a direction that would attempt to maintain the sustenance base. Without necessarily giving up modernization as a goal, ecological modernization advocated a different path of modernization: one that encompassed structural change in the economy, a preventive environmental policy, and an ecology-oriented economic policy. Ecological structural change of the economy would lead to industrial restructuring and to a de-linking of economic growth from the use of environmentally relevant inputs (e.g., through enhanced utilization of renewable resources of energy). Preventive environmental policy stood for a better balance between the anticipatory and the reactive components in policymaking—for replacing the curative post-damage strategy in environmental policymaking with a policy oriented to the prevention of damage. Finally, according to the ecological modernization narrative, the ecological orientation of economic policymaking would lead to a gradual change of the principles of economic decisionmaking through measures such as ecological accounting of individual factories and national-level accounting of the ecological costs of economic activity (Simonis 1995).[3]

The academic discourse on ecology, sustainable development, and ecological modernization was gradually disseminated by the mass media. It also found its way into policy statements by parties, interest groups, and the state. Central elements of this discourse became parts of a new political metanarrative that developed in close relationship with the emerging new environmental ethics on the cultural level (Tesier 1993; Tschannen and Hainard 1993).

Policy narratives always suggest particular goals and instruments of policymaking. In the case of the newly emerging environmental policy narratives, the following set of guiding principles developed:

- Pollution or nuisances should be prevented at the source, rather than subsequently trying to counteract their effects.
- Effects on the environment should be taken into account at the earliest possible stage in all technical planning and decisionmaking processes.
- Any exploitation of natural resources of a nature that would cause significant damage to the ecological balance must be avoided.
- The cost of preventing and eliminating nuisances must be borne by the polluter.
- Activities carried out in one state must not cause any degradation in another state.
- Administrative decisionmaking affecting the environment must be preceded by an impact assessment.

• Citizen participation and information should guide the design of environmental policymaking.[4]

Although this new framework for dealing with the environmental problems displayed some variations from country to country, it nevertheless constituted a body of new policymaking principles and tools that would become important points of orientation in a number of policy fields. Furthermore, the shaping of European Community legislation in the field of environmental policy contributed to the development of a certain uniformity of these principles among member states.

The discursive construction of the environment suggested a new model of social order. The environment had become a problem for government. But environmental discourse, like all discourses, is a social process saturated with social power relationships. Hence, social programs proposed in the name of "nature" and the "environment" are not objective programs determined by unambiguous external criteria (Haila and Heininen 1995, p. 157). By defining problems, obligatory passage points, modes of interaction, solutions to problems, and actors' identities, the new ecological discourse constituted not only a "way to solve environmental problems" but also a social technology that attempted to create a particular social order. The ecology discourse had introduced concepts that made "environmental problems" stable and manageable. If events and developments (such as environmental degradation and its relationship to economic growth) could be fixed and hence made manageable and calculable, they could become objects of government intervention. The translation of pollution into monetary units and the delimitation of procedures of scientific evaluation of environmental damage are examples of such procedures: they allowed for the displacement of highly complex processes going on in the ecosystem to administrative units and courts, where, under the heading of environmental legislation, decisions on fees and damage payments could be made. Also, the new policy narrative defined new actors of the political process, such as "the public" (which was to participate in a number of decisionmaking processes on environmental issues). In general, the discourse of ecological modernization proposed a specific model of social order. Environmental protection was represented as a positive-sum game and, hence, an administrative problem: if only every individual, every firm, and every country would cooperate, a sound society-environment interaction would be possible. According to the ecological modernization discourse, economic growth and the resolution of ecological problems could in principle be reconciled (Hajer 1995). As we shall see in this and in the next chapter, these changes in the political metanarrative would provide important conceptual tools for the rephrasing of the genetic engineering problematic. They constituted important reference points for the shaping of a policy counter-narrative with considerable impact on public policy.

The New Controversy over Deliberate Release in the United States

Equally important was the continuing scientific debate on the environmental risks related to genetic engineering. Again, the US discourse had a decisive influence on the perception of the problem in Europe. From the mid 1980s on, a scientific controversy arose over the risks of using genetically engineered organisms in the environment (Levin and Strauss 1991, pp. 1–17). In a 1985 *Science* article, the microbiologist Winston Brill argued that humans had been altering plants and animals by traditional methods of breeding for centuries without causing any serious problems (Brill 1985, pp. 381–384). Whereas traditional agricultural practices took advantage of new genetic modifications without knowing the exact nature of these modifications, genetic engineering, Brill argued, made well-characterized and specific modifications. Thus, there was no reason to expect that greater problems would arise from recombinant organisms used in agriculture than from organisms produced through traditional practices. This scientific narrative provoked a response by the ecologists Robert Colwell, Elliot Norse, David Pimentel, Francis Sharples, and Daniel Simberloff, who questioned Brill's assumption that the likelihood of negative effects of genetic engineering in agriculture was "very small." Colwell et al. (1985) argued that the phenotype of a microorganism, especially its ecological traits and population dynamics, was not fully predictable from its genotype alone, and they advocated assessing the risks of releasing genetically engineered organisms into the environment on a case-by-case basis. This debate continued in 1987 in a *Science* "Policy Forum" in which Frances Sharples continued to argue the for evaluating the deliberate release of genetically modified organisms case by case. She argued that "shifts in environmental contexts may be as important as genetic modifications in determining whether the ecological relationships of an engineered organisms will be unique relative to those of a parental form." In contrast, Bernard Davis (1987) contended that "the use of modified microbes is not entirely novel but is an extension of the old process of domestication of wild organisms-including the selection of microbial variants to make bread, wine, antibiotics, or vaccines." Furthermore, Davis (ibid.) argued that "in trying to assess the potential dangers, the experience of ecologists with transplanted higher organisms is less pertinent than are the insights of fields closer to the specific properties of engineered microorganisms: population genetics, bacterial physiology, epidemiology, and the study of pathogenesis." A 1987 pamphlet published by the US National Academy of Science came to the following conclusions regarding the risks of genetic engineering:

- There is no evidence for unique hazards either from the use of recombinant DNA techniques or from the movements of genes between unrelated organisms.
- The risks associated with the introduction of genetically modified organisms carrying recombinant DNA are the same in kind as those associated with the introduction of unmodified organisms and organisms modified by other methods.
- Assessment of the risks of introducing genetically engineered organisms carrying recombinant DNA into the environment should be based on the nature of the organism and the environment into which it is introduced, not on the method by which it was produced.[5]

On the other hand, a 1989 report of the Ecological Society of America discussed the deliberate release of genetically modified organisms. In a key passage the authors wrote:

> Genetically engineered organisms should be evaluated and regulated according to their biological properties (phenotype), rather than according to the genetic techniques used to produce them. Nonetheless, because many novel combinations of properties can be achieved only by molecular and cellular techniques, products of these techniques may often be subjected to greater scrutiny than the products of traditional techniques. Although the capability to produce precise genetic alterations increases confidence that unintended changes in the genome have not occurred, precise genetic characterization does not ensure that all ecologically important aspects of the phenotype can be predicted for the environment into which an organism will be introduced. (Tiedje et al. 1989, p. 298)

This debate was at no time "resolved," and it continued well into the 1990s.[6] However, as I will show, some of the main lines of arguments developed in this debate became textual devices for the framing of the emerging genetic engineering controversy in Europe.

New Social Movements and Green Parties

In the second half of the 1980s an important new group of actors entered the field of genetic engineering politics: "Green" parties and a variety of new social movements, social groups, and environmental organizations. Their project was to transform the hitherto passive "public"—a central black box in the biotechnology policy narratives—into an active and disruptive voice in the political process. In 1982 and 1983, mainly as a result of their influence, a number of topics—including rBST, deliberate release of genetically modified organisms into the environment, regulation of genetic engineering, and genetic engineering's social, political, economic, and ethical implications—re-entered the political debate. By 1989, an Organization for Economic Cooperation and Development report on biotechnology came to a conclusion very different from those of earlier evaluations of the public acceptance of genetic engineering: "Public acceptance of new biotechnology, or public confidence in it, has emerged as a central factor—for some, the single most important one—in the diffusion of this technology."[7] Of course, as we will see, Green parties and new social movements were not the only spokespersons for the public's perceptions of genetic engineering. But they constituted an important voice and movement in the reinterpretation of the genetic engineering problematic. How can the influence of this renewed critique of genetic engineering, which was taken so seriously by the policymakers and which inaugurated the second phase in the recombinant DNA debate in Europe, be understood?

First, the critique must be understood in the context of the changed discursive constellation of the 1980s. The environment had become a core value in the political imaginary. This discursive change provided the critics with their subject position as relevant policy actors. The new discursive situation allowed the critics to legitimately position themselves relative to the established actors in the policy field. The various social movements and groups launching the new genetic engineering controversy were not simply "against" genetic engineering. They were not just voicing grievances about potential dangers, such as the deliberate release of genetically modified organisms into the environment. Their "grievances" were not merely background factors come to the forefront; they were produced discursively (Buechler 1993, p. 222). The critics' challenge to genetic engineering materialized in a double move of dissociation and reconstruction. The critics disassembled many of the frameworks of meaning that the dominant biotechnology narratives had established. At the same time, they elaborated, in a relationship of mutual reinforcement with the discourses of ecology and sustainable development, a new narrative, which combined analyses of the shortcomings of previous biotechnology policy and descriptions of alternative fields of action for biotechnology policymaking. But the critique of biotechnology went beyond a disagreement about policymaking. If we assume that political identity is constructed in narratives (Thiebaut 1992, p. 316), then ultimately the critics' narrations constituted an attempt to disrupt the production of what had been termed "Bio-Society": a society whose identity would be simultaneously secured and transformed by the practices of genetic engineering. The new policy counter-narrative sketched a new model and a new order of policymaking.

Furthermore, the second phase of the European genetic engineering controversy must be located within a larger pattern of political mobilization. Collective political action tends to unfold in situations where the social system cannot integrate new behaviors, orientations, demands, or cultural innovations (Melucci 1985, p. 799). This was precisely the case in the during the 1970s. Most of the Western democracies experienced a wave of political mobilization and a gradual process of political and social destabilization. A multitude of new social movements focused on non-class issues emerged. In contrast to traditional social movements, such as the labor movement, the new social movements were not characterized by a shared class situation and were composed of individuals from primarily three social groups: the "new" middle class (high educational status, relative economic security, employment in personal service occupations), "decommodified" groups (middle-class housewives, high school and university students, retired people, unemployed or marginally employed youths), and the "old" middle classes (independent and self-employed individuals, such as farmers, shopowners, and artisan-producers). For these groups the notion of "collective interest" had decomposed into a variety of "life situations" and interpretations, which were no longer determined by a shared class situation. Driven by increased mobility, education, and competition, a process of individualization has been set in motion in which the individual became the reproduction unit of the social in the life world. Class and family were no longer the defining forces in individuals' lives that they had been in past decades (Beck 1992, p. 130).

In the absence of something like a "central conflict," such as class conflict, and owing to the heterogeneity of the life situations of the agents of the new social movements, the new movements might be best understood as expressions of "self-production" of the social (Touraine 1984). The new movements were responsible for the rapid diffusion of conflict across numerous social relations. The traditional movements could be seen as spearheading a process of modernization whereby a logic of differentiation and accumulation of options was articulated. The new social movements stood for strategic interventions and for reflecting on and questioning modes and directions of social differentiation (Beck 1993, pp. 176–177; Japp 1986, pp. 328–329).

The emerging social movements redefined the boundaries of the political by dissociating the notions of state and civil society. In preceding decades the conception of democracy had been based on separation of the state from civil society—a conception which postulated that the state translated the "private" interests formed in civil society into "public" institutions (Melucci 1988, p. 257). However, this distinction became unclear in the wake of the political mobilization of the 1960s.[8] The boundaries of the political and the "rules of the political game" were being rewritten. As a consequence, the institutions of civil society—including science, hitherto considered nonpolitical—came to be redefined as sites for a new type of politics (Offe 1985). The fields of conflicts broadened from political struggles in the strict sense (i.e., those that had to do with the conquest of state power) to struggles centered on the transformation of personal and collective life, which now were, to a considerable degree, subject to the intervention of new technologies (Touraine 1992, p. 183). By radically questioning the prevailing codes and descriptions of reality, social movements challenged the dominant logic on symbolic grounds. Rules of normality, spaces of difference, and frames of reality were thematized and problematized by a variety of movements, from the women's movement to ecology movements. The agendas of these movements were not necessarily rooted in their access to power; they were rooted in challenging the logic of the dominant rationalizing apparatuses (Melucci 1985, p. 810).

The critique of genetic engineering must be seen as a part of this cultural struggle over the power to name and thus shape the world. During the 1960s and the 1970s, the dominant political-economic discourse had described science and technology as contributing to progress and modernization because they promoted economic growth, agricultural expansion, and medical advances. According to this instrumental logic, molecular biology and the new biotechnology were modes of modernization. The dominant meta-narrative linked modernization with industrialization (Beck 1992)—a linkage that articulated in the problematization of the new biotechnology during the 1980s. In this reading, biotechnology constituted a differentiation of the action systems of science and industry and thereby contributed to the increasing complexity of society, which was attributed to the process of modernization. But as the industrial societies increased their ability to intervene into their own structures, the field of conflicts broadened (Touraine 1992, pp. 182–183). Biotechnology's critics rejected this linear, evolutionary logic of modernization as functional differentiation. Instead, they emphasized the contingency and multiplicity of forms and directions of modernization and the possibility that there were different paths into modernity (Japp 1986, p. 313; Beck 1993).

This counter-narrative, which evolved in an unsystematic way in Germany, France, and Britain and also on the transnational level, helped to create new public spaces for the development of political "alter identities," and it gradually extended from the sites of political actions to parties, administrative units, and the media.[9]

Organizing for Political Mobilization

The critique of genetic engineering emanated from a multitude of sites, including ecology journals, alternative think tanks, and social movements. The critical institutional sites for the creation of the new policy counter-narrative were the newly emerging Green parties at the local, state, and national levels in many European countries. In contrast to the United States, where the electoral success of Green parties was seriously hampered by an electoral system based on majority representation, in Europe proportional representation systems helped Green parties to gain seats in national or local elections. Soon they became important actors in a number of policy fields, including environmental and technology policy. This process was especially visible in Germany and at the level of the European Community. In Germany an alliance of activists from citizen groups (so-called Bunte and Alternative groups, local Green groups, and alternative-lifestyle groups) launched a national Green party in 1980. Green candidates were first elected to the Bundestag (parliament) in 1983 (when Green candidates received 5.6 percent of the votes), and in 1987 they received

8.3 percent. Die Grünen advocated a transformation of West Germany's society and economy and challenged the postwar consensus, which was based on passive citizenship, representative democracy, and a consumption. A central component of the Greens' political philosophy was *Basisdemokratie,* which meant that their party was to be decentralized, participatory, and open to continuous interaction with the activists who had been behind its creation (Frankland and Schoonmaker 1992, pp. 1–2 and 106; Pogunkte 1993; Sarkar 1993, 1994).

In France, Les Verts emerged as a unified political force in 1984 and had their greatest electoral success in 1989, when they received 10.6 percent of the votes in the European Parliament elections. In 1988 the socialist prime minister Michel Rocard appointed Brice Lalonde, leader of Les Amis de Terre, as state secretary for environment, a move that was broadly viewed as an obvious concession to ecology groups. In 1990 Lalonde founded another party, Génération Ecologie, which, together with Les Verts and independent environmentalists, received 14.7 percent of the votes in the 1992 cantonal and regional elections. Lalonde remained state secretary for the environment until 1992 (Prendville 1994, pp. 47–51; Pronier and Jacques 1992; Bennahmias and Roche 1992; Hainsworth 1990). Lalonde's position (which also meant that he would serve as France's representative in the European Community's Council of Ministers) was to play a certain role in the passage of the European Community's directives on the contained use and deliberate release of genetically modified organisms.

In Britain the potential strength of the "Green vote" became dramatically apparent in 1989 when the Green Party received 14.9 percent of the votes in the European Parliament elections. However, the majority-vote "seat" system worked against small parties in the national elections (Robinson 1992, p. 210). To a considerable extent, the "absence" of a Green party at the national level was the reason why environmental groups and movements had to channel their demands through a pattern of contacts between activists and government (Jordan and Richardson 1982, pp. 80–110). However, the Greens' success in 1989, which was the best performance of any green party in Europe, prompted much discussion of the relevance of green ideology in Thatcherite Britain (McCormick 1991, pp. 108–109).

The EC-level Rainbow Group, founded in 1984, combined three subgroups: Danish opponents of the European Common Market, the green/left alternative parties of GRAEL (Green Alternative European

Link), and the regionalist/nationalist parties of the European Free Alliance (Jacobs et al. 1992, pp. 70–75). These developments, which created a link between movement politics and institutional politics, would turn out to be of considerable importance in the second phase of the recombinant DNA controversy.

Political mobilization against genetic engineering was strongest in Germany, where the Grünen were represented in parliament and where they were thoroughly opposed to genetic engineering. Soon after the election of Grünen to the Bundestag in 1984,[10] an extra-parliamentary working group on "Bio- and genetic technology" was established. This group, financed by the Green parliamentary faction, included scientists, medical doctors, social scientists, journalists, and activists from the ecology, women's, and peace movements. Its main task was to advise the Green representatives and staff members in the Bundestag on developing a strategy for a special parliamentary commission of inquiry (Enquete-Kommission), initiated by the Grünen and the Social Democrats in 1984 to discuss the "opportunities and risks of genetic engineering." In this forum, which met 55 times between 1984 and 1986, the Grünen sought to "utilize the commission as a means to the end of fueling the public debate on genetic engineering, and to focus on the risks and the true reasons for its enforced development—and not on its opportunities."[11]

The Grünen interacted closely with various movements, working groups, discussion groups, and citizen initiatives working on genetic engineering. Some sixty of these movements were active in the second half of the 1980s. Working with the Öko-Institutes and the Grünen to provide financial support for numerous campaigns and activities,[12] they constituted an organizational infrastructure that formed the discursive space for deconstruction of the discourse of genetic engineering.

The aforementioned movements encompassed a remarkably wide variety of concerns, orientations, and backgrounds, including

- networks and documentation centers, such as the Gen-ethic Network in Berlin and the Gen Archiv in Essen, that provided information and coordination for and between activists and the public
- citizen initiatives directed against planned or active genetic engineering production sites, such as "Hoechste Schnüffel und Maagucker," which mobilized against Hoechst's insulin plant in Frankfurt
- university-based groups, such as AK Gentechnologie at the University of Erlangen

• working groups within parties, such as AK Gen-Reproduktionstechnik of the Socialist-Democratic Party of Schleswig-Holstein

• working groups within Third World movements, such as the "Congress of Development Politics Action Groups"

• working groups within ecological organizations, such as the Bund für Umwelt und Naturschutz Deutschland (Association for Protection of the German Environment and Nature)

• working groups within animal protection groups, such as the Arbeitsgemeinschaft für kritische Tiermedizin (Association for a Critical Veterinary Science)

• feminist groups, such as FINRAGE-Germany, which often focused on the relationship between new reproductive technologies such as in vitro fertilization and genetic engineering

• working groups within agricultural organizations, such as the Arbeitsgemeinschaft Bäuerliche Landwirtschaft (Peasant Farming Coordination Group)

• working groups of the Öko-Institute (ecological institutes), such as that at Freiburg

• groups representing the handicapped (often connected with university "history of eugenics" groups, which were interested in the relationship between Nazi eugenics and contemporary human genetics).

In France and in Britain the critics of genetic engineering were not as well organized, though they operated in a variety of political sites. Neither Les Verts nor the British Greens played a role comparable to that of the German Grünen in organizing and financing political campaigns against genetic engineering. In the French debate on genetic engineering, Les Amis de la Terre had assumed an important role.[13] Other groups active in the public debate on genetic engineering in France were the Confédération Paysanne (alternative farmers), Solagral (concerned with the impact of biotechnology on the Third World), and Génétique et Liberté (concerned with the ethical and social implications of the new human genetics). Recently, 29 environmental, consumer, and farmers groups created an "Alliance Paysans - Ecologistes - Consommateurs" to deal with issues of genetic engineering. Les Amis de la Terre and the important group France Nature Environment joined the alliance.[14]

In Britain the most important groups participating in the critique of genetic engineering were the UK Genetics Forum (comparable to the German Gen-ethic Network) and the Green Alliance (which

Table 5.1
Dimensions and topics of discourses critical of genetic engineering.

	Truths	Interventions	Historicity	Otherness	Transformations
Risk	Definition of risk	Probability	Retrospective assessment of risks	Discrimination as risk	Broad risk definition
Scrutiny/ monitoring	Disclosure of information	Definition of technology impact	Monitoring to counter othering	Counter othering	
Participation	Science as democracy			Parliamentary committee	
Information	Equal access	Accident information	Avoidance of discrimination	Avoidance of discrimination	
Liability		Accident case			
Technology assessment/ fourth hurdle	What is an important technology?		Unforeseen impacts		Changing agriculture
Product/process	Either/or	Is product an Intervention?			

focused on influencing parliamentary politics in the field of genetic engineering). Also important were Friends of the Earth, Greenpeace UK, the Pesticides Trust, the Women's Environmental Network, the Soil Association, and Parents for Safe Food.[15]

Although the various movements, groups, and Green parties active in Britain, Germany, and France and at the EC level hardly shared a coherent and unifying policy counter-narrative on genetic engineering, there was considerable overlap and similarity among many of the arguments, deconstructions, suggestions, and interventions they advanced. This was due to the considerable networking that was going on in Germany, France, and Britain as well as to the coordinated political interventions of various groups at the EC level. For this reason, the following discussion focuses on commonalties rather than on differences.

I suggest that we understand the critics' genetic engineering counter-narrative as unfolding along several interrelated dimensions, themes, or guiding statements (horizontal headings in the table), which were translated into and formed relationships of meaning between a set of topics that became important in political discourse (vertical headings). By borrowing freely from ecology, history of science, sociology, and philosophy, the various social movements and initiatives imploded the meaning of the politics of genetic engineering, refusing to accept disciplinary rules, scientific codes of reasoning, or existing institutional boundaries. In the next section I will examine the main themes of their discourse.

Truths

According to the Asilomar narrative, progress in molecular biology research would guarantee that the hazards associated with recombinant DNA work could be controlled. But the scientific controversy regarding the risks of genetic engineering had never been fully resolved. In the United States, it was the issue of deliberate release in particular that led to a debate between molecular biologists and ecologists over the "containability" of genetically modified organisms released into the environment. In Europe, a number of critics took this debate as an opportunity to reconsider the truth claims of molecular biology with respect to its characterization of the properties and behaviors of genetically modified organisms and their interactions with humans and the ecosystem.

In the European translation of the American debate, several important underpinnings of scientific reasoning were questioned: any scientific discipline's "monopoly on truth" or claim to be "objective," the place of science at the top of the hierarchy of knowledge, the existence of clear borders between scientific and nonscientific forms of rationality, and the superiority of scientific knowledge to nonscientific knowledge. In short, molecular biology's exclusive claims to truth were read as a strategy to manufacture a boundary separating those who know from those who don't and linking them together in a relationship of inequality and subordination. Without contesting the heuristic value of certain elements of the reasoning used in molecular biology, such as its conceptualization of DNA, other elements of the discourse, including the central dogma of molecular biology, were deconstructed as ideological textualizations of life that owed their claim to truth to the unsubstantiated representational superiority of their scientific discipline. According to the critics, life cannot be reduced to one reading which is then declared to be objective or essential.

This radical challenge revealed molecular biology's framing of life as premised by the concept of a "lisible text" (Barthes 1974) waiting for the scientist to represent it in the form of a narrative that gives an authoritative answer to the question of the meaning of the text. Instead of subscribing to the "truth" of molecular biology, critics treated scientific representations of life as something closer to an intertext (that is, a scriptible text).[16] Like a novel or a film, an intertext, far from being closed in its signification, undergoes constant inscription and reinscription in the sense that it is produced through different textualizations. By claiming priority over other views, the scientific discourse attempts to monopolize the interpretation of nature. Yet, in their deconstructive analyses, critics contested science's monopoly over interpreting nature by re-inscribing its interpretations into new discourses.

This challenge to the truth claims of molecular biology contributed to a breaching of the expert enclosure of the recombinant DNA debate constructed by the practitioners of genetic engineering. The issue of hazards came back as a central motif in the recombinant DNA debate. German critics generalized the American debate on hazards related to the deliberate release of genetically modified organisms to address not only that issue but also the issue of the contained use of such organisms. This translation was played out in the debate over "additive" versus "synergistic" risk models, a debate that played a crucial role in the shaping of German and European regulations of the contained use and deliberate release of genetically modified organisms. Critics argued that the German regulatory guidelines established after Asilomar were based on a "security philosophy" informed by an "additive" model of risk (Kollek 1989). According to this model, genetic modifications of organisms (for example, in the form of newly added or removed qualities) were seen as being incapable of creating unknown qualities in the new organisms. The critics argued that there was the alternate perspective of the "synergistic" model of risk, according to which ecological considerations must be part of the assessment of recombinant DNA's risks. Genetic manipulation is to be located in the broader context of the complex interactions of "downstream" causation and must take into account the complex dynamics of the specifics of space and time that will determine the operations of genetically modified organisms and their interactions with humans and the ecosystem. In other words, the critics rejected what they saw as the reductionist risk perception of molecular biology, a perception that was blind to specific contexts such as differences between ecosystems or seasons. One implication of this view was a more cautious and deliberate approach to certain scientific practices, such as the deliberate release of genetically modified organisms.[17]

An example of this controversy was the widely publicized year-long debate over the deliberate release of genetically modified petunias (the progeny of a particular transgenic line that contained a maize gene) in Cologne in May of 1990 and May of 1991 The experiment was to create a red pigment in the petunia's flower. The Zentrale Kommission für Biologische Sicherheit (Central Biological Security Commission), the Bundesgesundheitsamt (Federal Health Office), and the Biologische Bundesanstalt für Land und Forstwirtschaft (Federal Agriculture and Forestry Agency), which according to the 1990 Genetic Engineering Act[18] required to be involved in the authorization process, agreed with the Max-Delbrück-Laboratorium, which was planning the release, that the experiment did not carry any risks for humans and the environment.[19] A ZKBS report stated: "The transgenic plants' leukopelargonidin and the resulting pelargondin derivatives are widely existent in nature. No new substances will develop. The ZKBS sees no reason to believe that the changed pigment metabolism of the red petunias will lead to qualities that can cause a hazard for humans or the environment."[20] In contrast, an expert from the Öko-institut in Freiburg pointed at a myriad of uncertainties concerning the interaction of the transgenic plants with the environment and with humans. These uncertainties, the Öko-Institut representative argued, were not sufficiently addressed in the laboratory's description of the experiment. The lab's assessment of the risks involved in the planned experiment consisted mainly of an evaluation of the host-vector system. In the experiment, the measurement of interactions between transgenic plants and humans was based mainly on data gathered from a quarantined laboratory and from cultivation experiments, which, the Öko-Institut representative said, were no substitute for ecological research data in view of the complexity of the given problem. Hence, the Öko-Institut argued, there were good reasons to postpone the release.[21]

Related to their demand to that a multifaceted base of knowledge be applied to the evaluation of risks involved in genetic engineering was the critics' idea of a "democracy" of different scientific and so-called nonscientific interpretations. An expression of this line of thinking is their demand for public *participation* in regulatory decision-making. In 1991, in a review of the United Kingdom's attempts to implement the European Community directive on the deliberate release of genetically modified organisms,[22] the British Green Alliance commented:

> There must be specific consultation on the most useful format for the information that will go register. This is particularly important where information may be summarized. Adequate time must be allowed for the public to comment. . . . The membership of ACGM and ACRE [regulatory bodies; see chapters 3 and 6] should be revised to include a higher proportion of people with interest group backgrounds, in particular those with experience in environment and consumer groups. Some system should be found of allowing groups to nominate prospective members, rather than them being directly appointed by the regulatory authorities.[23]

In 1989, in line with this proposal, the UK Genetics Forum had demanded the creation of a "Public Biotechnology Commission" with statutory power to advise ministers on environmental, social, and ethical questions raised by biotechnology. This commission was to include consumers, animal welfare activists, representatives of religious groups, and environmentalists.[24]

Closely tied to the demand for participation was the request for access to information. This was, for example, articulated in the UK Genetics Forum's position on patenting and genetic engineering:

> What kind of information does the general public need? I would argue that the general public needs the more general kind of information that is already available, in order to have this information at the early stages of product development, in order to make a decision on whether it wants this type of

product. If the answer is "yes," then at the stage at which the product is to be marketed or released into the environment it needs more detailed technical information in order to judge whether the product is safe, which should be made available to it through the regulatory process for environmental release or product marketing. . . . Thus, it should not be the case that lack of availability of patents for plants and animals will automatically lead to a serious decrease in the type of information which the public needs in order to assess the acceptability of biotechnology products.[25]

For some of the critics, there was also a "right not to know." For example, the French group Génétique et Liberté emphasized the importance of the right not to know one's own genes and the right of privacy concerning one's genes in situations such as genetic testing in the workplace.[26]

In the critics' discourse, the determination of hazards related to genetic engineering came to be defined as an object for democratic negotiation involving disciplines and actors previously excluded from the processes of risk assessment. Such democratic negotiation would include ecological, socio-economic, and political criteria in a broadened understanding of "risk assessment," but representatives of the public would also be involved. Furthermore, some underlying assumptions of molecular biology that had been crucial in the Asilomar narrative were now being questioned, and the focus of concern had shifted away from the exclusive emphasis on ecological risks and toward the desirability of certain types of research and production and the democratic quality of decisionmaking procedures.

Interventions

A critical issue in the second phase of the recombinant DNA debate concerned the proliferation of the meanings of the intervening potential of genetic engineering. Scientists and industrialists certainly agreed with the critics on the substantial impact of genetic engineering as a technology. However, there was considerable disagreement over the meaning of this intervention. Since the Cohen-Chang experiment (1977), it had been argued that there was much similarity between recombinant DNA work and naturally occurring processes. This line of reasoning, which attempted a "normalization" of genetic engineering, continued into the 1980s. A statement by the president of the Deutsche Forschungsgemeinschaft illustrates this view:

For biologists genetic engineering is essentially an indispensable means to research life processes. If today there is talk about the expected use of genetic engineering and biotechnology, it is because hopes focus on the genetic manipulation of microorganisms, plants and animals. In doing that we use specific metabolic skills of organisms in a way which is not different from the production of food and the elimination of garbage for thousands of years. As a result many people in the life sciences following the agitated discussions on this topic or being followed by them ask: where is the problem? To be sure: this technology will also carry risks. Appropriate security measures have to be taken against this. A substantial part of research and production in genetic engineering can be done in completely contained laboratories or production systems, so that there is no question of the ethical implications of a deliberate release of genetically engineered organisms. Apart from this question, one has to judge the question of a deliberate release of genetically engineered organisms. We already have been releasing genetically modified organisms for thousands of years without any damage, if one looks at it closely, even the birth of a human being is a deliberate release experiment. To argue from an ethical perspective totally against every release of genetically modified organisms does not make sense, because all that also happens in nature in the process of sexual procreation, parasitic gene transfer, or through mutations. Concerns against release are directed against novel, non-natural risks of modern genetic engineering, and this argument can be refuted by the fact that the assumed risks are neither novel nor non-natural.[27]

In contrast to this argument, the critics read genetic engineering as a *novel, unprecedented* process of *intervention*. The Deutscher Naturschutzring (German Nature Protection Organization), for example, argued:

After physics and chemistry, biology has reached, as the third life-sciences, its "synthetic phase" and, with it, its breakthrough. Genetic engineering is one of the technologies which reduces life, in the tradition of the mechanistic world model, to a material and controllable principle; the domination of nature and technological utility are already anticipated by this methodological approach. . . . Because of its potential for disaster, genetic engineering is a technology which requires the infallibility of the human being. Past experience with other technologies has demonstrated the fallibility of human beings and that the risk factor cannot be controlled by an "infallible" technology.[28]

We can interpret this argument as suggesting that scientists present the practice of molecular biology as a supplement[29] and not fundamentally an intervention into nature—as a "nonessential" addition rather than a fundamental intervention. In this sense, genetic engineering is not doing anything different from what is already going on in nature. By fixing the meaning of the relationship between life and technology, by reducing technology to a mode of simulating life, advocates of molecular biology construct a signification of scientific practice that erases a crucial feature of modern biology: its fundamental transformation of life processes, which may result in the production of previously nonexistent forms of life. In other words, the critique of the "supplementation of life" ideology points to the contradiction between molecular biology's claim of the "primacy" of life over technology and technology's actual undermining of the "primacy" of life; it attempts to show how the dominant rhetoric is used to bracket off intervention as a constitutive element of biological practice.

This conflict of interpretations of the interventionist character of genetic engineering was played out in the "process versus product" debate, one of the central disputes in the struggle over the meaning of genetic engineering. Whereas the advocates of a process-based approach to the regulation of genetic engineering argued that any product involving processes of genetic engineering should fall under regulations of genetic engineering, the advocates of the product approach contended that regulatory decisionmaking should focus on the properties of already-engineered organisms, such as foods and pesticides.[30] Obviously, the outcome of this dispute would ultimately decide if special legislation for genetic engineering regulation would be necessary or if existing regulations would be sufficient. A staff member of the Green Party in the European Parliament commented, in a briefing for nongovernmental organizations on the European Commission's proposal for a directive on pesticides (which also included the regulation of pesticides operating with genetically modified organisms):

> The first thing to understand here is the difference between "horizontal legislation" and "vertical legislation." These are the new terms of reference for the old "process versus product" debate. Should the products of the new biotechnology be regulated on the basis of the process by which they were manufactured, requiring a new set of laws covering all GMOs [genetically modified organisms], no matter of their function (horizontal legislation)? Or is the process by which something is manufactured irrelevant as far as legislation is concerned? According to this view, regulation should only be concerned with the end product (vertical legislation) and for that, the existing laws need only be slightly adapted to cover GMOs. In this debate the promoters of genetic engineering have always pushed strongly for product-based or vertical legislation, and the critics have always argued for a process-based or horizontal regulatory approach. The crux of the matter is whether genetic engineering as such poses any unique challenges to the environment, human health, and society. We say yes, they say no—at least none requiring a unique response.[31]

This position was echoed by a variety of other reactions to the EC proposal, as in a letter from the UK Genetics Forum to the Secretary of State for the Environment (the national authority responsible for the implementation of the directive):

We are extremely concerned that this (the product-based directive) may lead to a weakening of controls on the release of GMOs by setting up a system which is both scientifically unsound and in which insufficient expertise will be available at the product release stages to assess the risk to the environment of a GMO pesticide. The fundamental reason that releases of GMO pesticides should not be regulated under the pesticides directive is that GMOs behave differently in the environment from chemical pesticides, due to the simple fact that they multiply, possibly uncontrollable. . . . [32]

The European debate that led to a ban on rBST is another example of the conflicting readings of the interventionist potentials of genetic engineering. In the middle of the heated rBST debate, an industry lobby known as FEDESA produced an "information update" for members of the European Parliament, which was soon to vote on a possible rBST ban. An excerpt follows:

What are the facts? There is well established scientific evidence that milk from BST supplemented cows is as safe and natural as other milk. The higher a cow's naturally-occurring level of BST, the higher proportionally, will be its milk yield. Without BST no milk would be forthcoming. All milk from cows contains a minute trace quantity of BST. Ultra sensitive tests in milk have shown that milk from supplemented cows is in no way different. . . . Both producers and consumers can be confident that milk from BST supplemented cows is as safe and wholesome—and as natural—as from other cows.[33]

In contrast to this position, the London Food Commission, a consumer group, insisted on the difference between the "two BSTs" and argued:

The synthetic BST hormones should be given generic names, since they are a group of similar but chemically distinct products. . . . BST milk residue test results should be available for public scrutiny. . . . Consumer rights to know and choose should be protected and enhanced. Urgent consideration should be given by MAFF and the dairy trades to setting up mechanisms for accurate labeling of dairy products.[34]

Here, again, we see two conflicting interpretations of the "nature" of rBST: that it is like what goes on in nature and that it is an intervention into and a transformation of natural processes.

The framing of genetic engineering as a largely unpredictable and difficult-to-control technology was also articulated in the critics' persistent insistence on a strengthening of liability regulations concerning genetic engineering. A Working Document of the European Parliament, drafted by a representative of the Greens, starts its framing of the liability issue by referring to a well-publicized case of damage caused by the genetically engineered drug L-Tryptophan and goes on to discuss the legal situation of product liability in the European Community:

Several thousand people in the USA and Europe became ill, some seriously, after ingesting this amino acid, produced using genetic engineering methods. . . . What is the situation as regards liability for damage cause by biotechnological products in the European Community? The Product Liability Directive of 25 July 1985 is based on the principle of liability without fault. Liability applies to damage caused by defective products of all kinds with the exception of defects arising at development state. . . . The exemption from liability for defects arising at the development state is a crucial weak point in liability for biotechnology products. It is especially important in the area of biotechnology, and particularly in the area of genetic engineering, to extend product liability to defects at development stage, since the state of scientific and technical knowledge in this field 'is still subject to considerable limitations regarding the predictability of sources of damage and harmful developments' (quote from DAMM, Gentechnikhaftungsrecht, ZRP 1989, p. 463)—as can be seen from the above example. Even where the rules governing liability apply, the injured person must as a matter of principle provide full proof that the damage was caused by the genetically modified product. Liability in the area of biotechnology is of little value to the injured person without a partial redistribution or easing of the burden of proof. The example of L-Tryptophan shows that it would be practically impossible for average consumers to prove that this drug . . . caused their illness. In addition to an easing of the burden of proof, the injured person should also be granted a right to information, since any attempt to provide evidence of the cause of the damage without information on the processes used is virtually doomed to failure.[35]

The coding of the "product versus process" debate and the discourse on liability show a deep mistrust of the practices of genetic engineering and a mistrust of the representations of molecular biology as a "self-controlled process" that is safe if only some basic rules of caution are followed. Hence, access to information and public participation became important strategies in establishing a broad-based system for the scrutiny and monitoring of genetic engineering. The president of Les Amis de la Terre put it as follows:

. . . we have to continue to be on watch to avoid an apocalypse. After the precautions decided upon at the Asilomar conference in February 1975, the security norms have been relaxed. But numerous examples, from the chemical to the nuclear industry, show that big accidents can happen. Why not in a fermenter of several hundred cubic meters? . . . It therefore is in the interest of the public as well as of industry that we have security norms. . . . An independent organization must monitor the implementation of the norms and periodically revise them. . . .[36]

But Julie Hill of the Green Alliance demanded that the monitoring of practices of genetic engineering not be limited to the proper implementation of recombinant DNA guidelines:

> We realize monitoring is difficult, but as far as we are concerned, there is no point in making accurate risk assessments as far as possible, however far sighted participative and fully informed decisions on the way releases go ahead, if we do not keep a continual watchful eye on the long-term consequences for the environment. . . . What we feel is needed is a routine long-term monitoring of the ecosystems.[37]

In other words, the critics' understanding of monitoring was not limited to short-term assessments; rather, the long-term scrutiny they advocated would go well beyond any current regulatory practice.

Historicity

Temporalization is an important strategy in the construction of political spaces. Transformist strategies of temporalization, for example, particularize ethnic identities and differentiate their contributions to and their places in nations (Alonso 1994, p. 398). Not surprisingly, an important contested field in the public controversy over genetic engineering has been its history. Both the history of eugenics and the history of the Green revolution were translated into important resources in the semantic struggle over the meaning of genetic engineering.

The women's movement in Germany saw a close relationship between reproductive and genetic technologies and subsumed the two under the heading of "population politics." Human genetics and genetic counseling (after World War II) and prenatal screening and reproductive technologies (later) were seen as continuations of Nazi eugenics and euthanasia:

> During the period of German fascism more than 100,000 disabled/women were killed and about 350,000 sterilized against their will. The search for the perfectly functioning human being quickly led to the death of those who did not correspond to the bodily and mental ideal. This highly functional and accomplished human being is today more than ever in demand. In this context the resurgence of eugenics, which we can observe today, is remarkable.[38]

The Krüppel-Initiativen[39] joined the women's groups in their effort to rewrite the history of genetic engineering in Germany, viewing modern human genetics and research efforts such as the Human

Genome Project as continued forms of Nazi eugenics. The Krüppel-Initiativen and feminist groups cooperated with new "historic" social movements, such the Anti-Eugenik Bündnis, ecological movements, and gay groups, all devoted to conducting research and to informing the broadest possible audience about the history of genetic research.

But writing the history of genetic engineering went beyond the association between eugenics and the new genetics. In a recommendation to the European Parliament, the rapporteur on the Commission's Human Genome Project (a member of the Green Party) wrote:

> This history of genetics is, finally, also the history of the grossest error and prejudice cloaked as science. Good examples of this are the grotesque story of the interpretation of treatment of mental disease, and the scientifically recognized theories once regarded as objective, such as that of the criminal chromosome (according to which an extra Y chromosome predestined men to be criminals), and that of the genetic origin of certain diseases amongst the black population which later turned out to be direct consequences of chronic undernourishment. . . . These examples have all been taken from the postwar period. The most recent example of such discriminatory theories is the interpretation of "sickle cell anemia" which even the Commission in its draft programme describes as one of the hereditary diseases which has already been established. In certain cases it may result in pathological changes, but it is completely harmless in most cases and even provides some protection against malaria.[40]

In this statement and in those linking the history of eugenics to contemporary molecular biology we see a questioning, by means of a historical argument, of a narrow notion of risk. Furthermore, it was argued that often risks can only be recognized as such in retrospect. In addition, the point was made that a narrow framing of technology assessment might fail to see the transformative potentials of technology in all its dimensions.

A related problematization of discursive framings of the past can be seen in the critique of the argument that non-Western countries would benefit from the new biotechnology. The British organization Parents for Safe Food and the UK Genetics Forum issued the following joint statement:

> The first Green Revolution began in the 1960s. It involved the replacement of traditional varieties of crops by high-yielding alternatives requiring sophisticated irrigation systems, imported machinery and expensive inputs of fertilizers and pesticides. These changes in farming methods, while they have increase food production in some places, have been at the cost of many peasant farmers' bankruptcies and environmental damage. The promise now is of a

second Green Revolution which will provide the Third World farmer with pest and drought resistant crops and crops which have the ability to fix their own nitrogen. . . . There are . . . questions about which of these technologies and products are appropriate, and, since few technologies are without drawbacks, what undesirable effects they would have in the Third World. . . . The first green revolution has already severely depleted the diversity of plant genetic material. The second revolution can only speed this process. . . .[41]

This and related assessments of the history of biotechnology highlight a particular mode of representing the past that was being advanced by the dominant biotechnology narratives, which were either ahistorical or monologically historical. Either modern biotechnology was represented as a "future," with an identity created primarily through an interpretation that offered an optimistic assessment of a potential that would be realized in an unspecified distant time, or history entered the narrative only insofar as biotechnology was represented as stemming from a long tradition, with beer and cheese production as "forerunners" of genetic engineering. Finally, traces of history can be detected in the discourse's pregnant silences (such as in Germany, where, under the shadow of Nazi eugenics, support for human genetics research proceeded with caution until the early 1990s).

These historical absences in the discourse on biotechnology provoked the writing of different histories. Given this space for alternative historical accounts, the anti-genetic-engineering movement attacked both overt ahistorical framings of medical and eugenic history and monological conceptions of the past.

Otherness

Closely related to the advancement of history as an antidote to presentistic readings of genetic engineering is what can be interpreted as the genetic engineering critics' reading of the scientific practice of modern biology as producing a new, marginalized social other[42] and a whole economy of social exclusion. The critics charged that genetic engineering could engender new types of discrimination, which could be avoided only by deliberate rejection of the idea that questions such as "What defines human life worth living?" [*lebenswertes Leben*] can actually be decided. Rather, according to the critics, what should be emphasized was the positive value of the multiplicity of different expressions of human life. Any attempt to engineer or legislate final answers to ultimate life questions, according to these critics, would necessarily lead to a construction of binary oppositions that would produce a discriminatory attitude.

Analysis of the human genome and prenatal diagnostics were thus interpreted as potential instruments of exclusion. They were not simply "neutral," and they did not just contribute to the goal of fighting disease (a goal that might or might not be reached). But they could create a new kind of class society based on an arbitrary (yet, within the context of societal organization, customary) category of "human quality." Hence, they could easily be interpreted as forms of subordination:

> Prenatal genetic screening forces one to distinguish between worthwhile life and not worthwhile life. It is impossible to draw clear borders between diseases and states of disablement, even for a commission. . . . More than ever will society ask for "healthy and pretty babies." It will not be the singularity of each individual that will be in demand, but his or her adaptability to social needs. . . . Only attempts to create and integrate quality of life for all human beings and not gene research can bring social peace and equality.[43]

In this reading, the creation of a new marginalized social other could be avoided only by building a social organization that fostered solidarity, integration, and tolerance and renounced human-related genetic technologies.

Two important preconditions of this new society were democratic dialogue and negotiation over the introduction of the new genetic technologies—that is, dialogue based on public participation and accessibility to information on planned research in human genetics. Accordingly, Génétique et Liberté criticized supporters of the Human Genome Project, who often praised the new genetic technologies as liberating on the ground that they allowed individuals to choose or alter genetic predispositions:

> . . . this is an imposed liberation. . . . The supporters of this way of handling human resources have a tendency to play down the inconveniences and dangers of the new technology: the potential for discrimination, the erosion of values of solidarity, schooling based on results of biological tests. . . . A democratic debate is indispensable to avoid that we are sold ready-made health policies under the pretext that they are based on expert advice.[44]

The critics elaborated on this point by linking the "othering" of genetic engineering to its homogenizing power. The first step of homogenization is, in the reading of a feminist network that mobilized around issues of new reproductive technologies and genetic engineering, a process of simplification:

The perspective that is used to interpret and evaluate life is the perspective of molecular biology. Life and its manifold expressions are reduced to the sum of the metabolisms and their products, to DNA sequences. The results of previous processes of selection, multiplicity of species [*Arten*], difference, and singularity of life are not taken into account. . . . [45]

This step of homogenization is followed by a generalization of simplification. In the wording of the German Green Party:

What comes increasingly into focus today is not just research on human genetics, but also manipulating human genetic composition. . . . First the agenda is to "repair defective genes." But who decides what will be defined as disease, deviation from arbitrarily created norms, or deficiencies in beauty [*Schönheitsfehler*]?[46]

Targeting "DNAism," these critics challenged the conceptual establishment of a definable, binary, fixed, and hence potentially manipulable relationship between DNA and life—a conceptual framework that conflates the borders of humanity, animals, and nature. In this context, eliminating deviants such as the disabled was seen as only one aspect of the discursive production or type of society, which was closely related to a process of homogenizing individualization that encompassed all of humanity:

The new tests, largely based on genetic analysis, don't give information on the current health status of a patient, but about his potential future status. What will be the future of this new type medical category situated between the sick and the healthy human being: the potentially sick person? Doesn't that imply the risk of attributing an exaggerated importance to the genetic makeup of a person, at the same time reducing the importance of economic and social factors?[47]

Transformations

Critics focusing on the adverse side of the socio-economic effects envisioned by the new biotechnology have called attention to the problem of transformation, and have pointed out that new risks associated with genetic engineering could lead to unanticipated and far-reaching socio-economic transformations and could require the introduction of new forms of technology assessment.

The genetically engineered hormone drug rBST, injected into cows to increase their milk production, became an important target of this type of critique. Apart from the health risks for humans and the violation of animal rights, the argument was raised that the increase in output of milk per unit of time would increase the concentration of farming and would transform it into an industry dislocated from traditional sites. As a consequence, there would be a restructuring of agriculture and a further decline in the rural population.[48] The "Rainbow Group GRAEL," represented in the European Parliament, stated the following in a leaflet supporting a Europe-wide ban on rBST:

> The authorization or banning of BST will have a decisive impact on the condition for future milk and meat production. Hormones and performance-enhancing drugs are used primarily in industrial mass production units which no longer have anything in common with traditional farming. If the use of BST is authorized throughout the EEC, it will be used immediately in large intensive farms. This will bring down the price of milk, thus forcing small farmers either to do the same thing or to go out of business. . . . BST will mean that farmers too will have to join the high-tech world, an extension of the trend towards the industrialization of agriculture.[49]

Similar arguments were brought forward in a letter addressed to the French Minister of Agriculture by the French Conféderation paysanne (Glémas 1989, p. 49) and the British group Compassion in World Farming[50]:

> BST may well have adverse effects on mastitis incidence and cause pain to treated cows from injection-site swellings. . . . We are sure you are also aware that it is widely accepted that BST will hasten the demise of the smaller dairy farm. . . . At a time of over-production in the dairy industry necessitating the use of quotas to contain surpluses, we believe that BST-use would also be an inappropriate technology to adopt. . . . [51]

The debate on rBST, in which biotechnology's critics argued for a banning of the drug based on the socio-economic reasons, was a part of the "fourth hurdle" debate. The term "fourth hurdle" was coined by a British member of the European Parliament, Ken Collins, who argued that, in addition to the traditional criteria of efficacy, quality, and safety, product evaluation should also take socio-economic impact into account:

> The real problem faced by the animal health industry is not the ban on growth promoting hormones. Rather what is important is the development of new processes and the future market authorization of new products which may be derived from these processes. . . . The use of gene technology raises ethical questions which have not hitherto been seen as central in the licensing of veterinary medicines. Demands are also increasing for consideration of the socio-economic and environmental implications of agricultural technology. It is also important to overcome public ignorance and prejudice about biotechnology. It follows from this that the traditional criteria for assessing veterinary medicines—safety, efficacy, quality—are no longer sufficient.[52]

"Fourth hurdle" quickly became a catchword for any demand requiring processes of technology assessment based on a relatively broad conception of risk.

But the concern over the transformative potential of biotechnology was not limited to socio-economic transformations in industrialized countries. Since the full scale of biotechnologies was available only in a few countries (primarily northern) and was concentrated in the hands of transnational companies, critics argued, the technology lent itself to the staging of new redistributive battles between technology haves and have-nots. In addition, in the Northern hemisphere, biotechnology would contribute to the socio-economic marginalization of certain social groups, such as small farmers.[53] A representative of Les Amis de la Terre stated:

> One has to give up the idea that [biotechnology] is a decentralized technology on the human scale. The necessary investments go into the billions. . . . One moves ahead too quickly if one says biotechnology can be of great advantage to the Third World: whereas it is possible there to develop the production of proteins and energy with traditional methods of biotechnology, the transition to the new biotechnology will increase dependence on the developed countries.[54]

The overall socio-economic impact of the new biotechnology, an important reference point in the arguments from various movements in Europe, might mean not only large-scale relocations, job losses, and newly surfaced inequalities, but also the failure to deliver its promise of growth and employment at the very core of the biotechnology industry: in chemical and pharmaceutical companies. The Hoechste Schnüffle und Maagucker, a citizen initiative opposed to Hoechst's production of genetically engineered insulin, argued that the factory planned to produce the insulin would be fully automated and hence would employ fewer people than traditional Hoechst factories. Furthermore, the traditional method of obtaining insulin from pig pancreases was highly labor intensive and involved workers in a variety of industries, from slaughterhouses to transportation. Overall, then, what could be expected was a net loss of jobs:

> Not only laymen, but even politicians assume that these new technologies create new jobs and it completely escapes their attention that this highly technical production will phase out the old insulin production and destroy jobs.[55]

This type of critique constituted a dissection of genetic engineering's interventions on the time and space axes of life and articulated a widespread feeling of displacement created by biotechnology's capacity to virtualize time and space. Critics linked the gradual disappearance and endangering of humanity and nature in its full expression and diversity to the manipulation of time and space implicit in the practice of intervention. Biotechnology, and especially agricultural biotechnology, so the argument went, gains mastery over time and space by speeding up natural processes and at the same time breaking the importance of space by separating products of living organisms from the natural environments of those organisms. What this sophisticated power technology achieves, we could interpret the critics as saying, is in fact a compression of time and space.[56]

Conclusions

During the 1980s, gradual changes in the political metanarratives in Europe—and particularly the articulation of a new discourse on modernization oriented around ideas of ecology and sustainable development along with the political mobilization of the late 1970s and the 1980s—had created a new constellation of actors and political rationalities. The various themes of the critique of genetic engineering, articulated by new social movements and Green parties, by scientists, and by historians in France, Germany, Britain, and the United States, consolidated into a narrative that challenged some of the central notions and assumptions of European biotechnology policy since the 1970s. This counter-narrative redefined the set of actors engaged in biotechnology policy and challenged the topography of the political space of biotechnology, which had mapped biotechnology as a source of wealth and progress and society as the docile and quiet recipient of the benefits of genetic engineering. What had come again into focus were the boundaries of and between science, business, society, and politics.

The critique led to a serious destabilization of the arguments and explanations that had integrated the state's biotechnology policy. Not only were central elements of policies such as suggested framings of "risk" read as contradictory and aporetic in their structure; through a deliberate effort, critics left the sphere of strictly scientific arguments and raised as issues the history of Nazi eugenics, Third World exploitation, and rBST-tortured cows in relation to biotechnology-led modernization policies. They thus created counter-discursive ruptures, blurrings, and mixtures. "The truth" of molecular biology was read as an ideology; genetic engineering was located within the history of eugenics, while agricultural biotechnology was placed within the framework of the exploitation of the non-Western world. In the critics' writing, genetic engineering was textualized as a political phenomenon that could be subjected to critique by a democratic citizenry. Molecular biology's system of signification was viewed as a strategy of cultural hegemony that could change and normalize the parameters of human self-recognition and understanding of nature—a strategy that could be countered by a politics that established spaces for the recognition of differences as alter-identities. In contrast to the idea of Bio-Society, the critics' counter-narratives outlined a different model of political order: a model of society whose identity was to be based on a novel, cautious, and ecological mode of interaction between society and nature. Their stories were to become important in the renegotiation of the economic, scientific, and political realities of biotechnology in Europe in the late 1980s.

6

Hegemonic Crisis and the Remaking of Regulatory Space in Europe

In the last chapter I discussed how during the second half of the 1980s an emerging policy counter-narrative fundamentally questioned the practices that had set up the political space of biotechnology in Europe. The practices that had defined the boundaries of biotechnology had originated in such diverse sites as laboratories, businesses, government offices, or scientific journals and had inscribed things such as the meaning of genetic engineering, its quality as the core technology of a new industrial sector, its impacts on society and economy, and social models for handling and controlling the risks of recombinant DNA technology. Within this political space, the central parameters, actors, rationalities, and institutions of biotechnology emerged and their relationships and interactions described the topography of the new field of the politics of genetic engineering. Research and industry support policies and risk regulation were two policy articulations of this new political space of biotechnology. The political challenge to biotechnology sketched in the preceding chapter was not directed specifically against either support or regulatory policies, and it cut across these two public policy fields. It offered a detailed critique of goals, values, and reality definitions of the established biotechnology support and regulatory policies and their mapping of the relationship among science, state, and society. Some of the dominant representations in biotechnology discourse, which were crucial for the specific order in the policy field, had come under considerable attack.

In this chapter I will show how from the mid 1980s on this crisis of representing biotechnology was addressed in the field of risk regulation, in particular by drafting regulations for the contained use and the deliberate release of genetically modified organism. The focus will be on developments in Germany (where in many ways this new phase of the recombinant DNA controversy originated), in France, and in

Britain. I will argue that the political conflict that characterized the preparation and drafting of these regulatory policies was more than a competition among different strategies to draw the boundaries of genetic engineering's risks. It reflected the contested, hegemonic attempt of policymaking to displace the social confrontation about genetic engineering from a multitude of different locations to the well-defined and "controllable" space of regulatory politics. As in the 1970s, the shaping of these regulatory policies must be understood against the background of highly specific, nationally varying constellations of policy narratives and specific political, economic, and scientific events and developments which differed from country to country. But during the 1980s the European Community had emerged as a player in the field of genetic engineering regulation. The negotiations and discussions about the drafting of EC directives dealing with the contained use and the deliberate release of genetically modified organism were followed with much interest by policymakers on the national level. While national specificities continued to dominate the dynamics of regulatory policymaking, the EC level added new complexity to the policymaking process and also exerted some influence on national processes of policymaking.

Policymaking is an attempt to establish a situation of fixation within a field of differences, to manage a field of discursivity. Hence, any probing of the established framings of reality, such as of the nature of risks involved in genetic engineering, could potentially lead to the proliferation of new antagonisms that might trigger a crisis of the dominant rationalities justifying the process of policymaking. Any attempt to dominate a field of discursivity is based on practices of articulation, which consist of the construction of nodal points or master signifiers (in policymaking typically presented in the form of policy narratives) that partially fix meanings. A social and political space relatively unified through the instituting of nodal points and the constitution of tendentially relational identities is what Gramsci called a historical bloc. A conjuncture where there is a generalized weakening of the relational system defining the identities of a given social and political space and where there is a proliferation of floating elements is what Gramsci called an organic crisis (Laclau and Mouffe, 1985, pp. 112, 136).

Such a situation of hegemonic crisis and overflow of meanings characterized the European politics of genetic engineering in the mid 1980s, threatening to impede further expansion of genetic engineering as promoted by industrial policies. In such situations of hegemonic crisis, policymakers tend to attempt to institute a new hegemonic articulation. In this case, they attempted to construct a new system of nodal points capable of fixing the meaning of the practices of genetic engineering. This attempt to stabilize genetic engineering's field of discursivity through policymaking sought neither to accept the critique leveled against genetic engineering policies nor to reject it. Hegemony implies domination of a social field, but not elimination of opposition. It is about re-absorption of discourses of polarity into a system of "legitimate differences" and re-articulation of systems of differences in chains of equivalence that construct social polarity. This handling of differences implies a definition of the sites where opposition, and with it a specific positioning of critics within political space, can take place.

From the mid 1980s on, the site where the discursive struggle over genetic engineering took place was the regulation of genetic engineering. Whereas in the 1970s the hazards of genetic engineering had been conceptualized as a technological problem, in the second half of the 1980s recombinant DNA's risks came increasingly to be reconfigured as a socio-ecological issue that could not be dealt with entirely by technological means. The "gene" as defined by the discourse of molecular biology ceased to serve as a boundary object that at the same time separated and linked "science" and "public" in the regulatory process. The Asilomar policy narrative, having been challenged by a number of competing narratives or counter-narratives, had lost credibility and explanatory power. In Europe, institutionalized expert enclosures such as the British Genetic Manipulation Advisory Group, the German Zentrale Kommission für Biologische Sicherheit, and the French Commission de Classement and the related administrative units lost much of their legitimizing function. In short, genetic engineering's regime of governability was confronted by a serious crisis of signification. A new regulatory policy narrative attempted to rearrange actors, events, and developments—including new social movements critical of genetic engineering, the new environmental discourse, genetically modified plants, and public opinion on genetic engineering,[1] but also the positions and interests of industry and science—into a new framework of meaning that would redefine the terrain of regulatory governmental intervention. The new system of regulation made a precautionary approach to genetic engineering an integral element of biotechnology industrial policy. At the same time, it excluded more radical positions (as articulated, for example, by the German Grünen or the UK Genetics Forum) from what was defined as politically acceptable.

As was discussed in chapter 5, conflicts about genetic engineering also took place at other sites, such as the rBST controversy, the patenting controversy, and the human genetics controversy. These various sites of controversy were not easily separable from one another and it was part of the critics' strategy to underscore the complexity of relationships such as that between the new reproductive technologies and genetic engineering. At the same time, the evolving governmental rationality articulated in the regulatory strategies developed in Germany, France, Britain and at the EC level sought to establish clear boundaries around the various sites of genetic engineering politics, such as regulation of recombinant DNA work, bioethics, and technology assessment. This boundary drawing was the expression of a hegemonic struggle to partition the political space of genetic engineering—to create a new terrain of government by resignifying the relationship among science, industry, politics, and society in the field of biotechnology policymaking.

Germany: The Drafting of the 1990 Genetic Engineering Act

At the first glance, the regulatory structure established in Germany in the second half of the 1970s seemed to display a pattern of continuity. However, from the mid 1980s on, a number of important events and critical changes in political discourse began to create a new constellation of the genetic engineering problematic. One important change in political discourse was the inscription of principles of environmental ethics into the political imaginary. As was sketched in chapter 5, during the 1980s Western societies underwent a significant change in self-perception as the environment became a central issue in the definition of collective political identity. The rise of environmental policy as an important policy field was an expression of this change. Furthermore, the rise of environmental discourse helped to position new actors, such as social movements and Green parties, in the policy field.

In Germany, environmental concerns date back to the early 1960s; however, environmental policy did not became a separate policy area until 1971, when the Social-Liberal coalition government developed a statement of principles that were to underlie an environmental program. Central to the government's strategy was an emissions-oriented approach, based on the state of technology (*Stand der Technik*), according to which environmental policy was to ensure the limitation of emissions in accordance with the state of technology even if the beneficial consequences of such limitation could not be predicted exactly.

The emissions-oriented approach was related to two other key principles of environmental policymaking: the *Vorsorgeprinzip* (principle of avoiding environmental damage through planning) and the *Verursacherprinzip* (the principle that the polluter pays). The *Vorsorgeprinzip* acknowledged that policymakers may be forced to go "beyond science" and that sometimes it is not possible to draw a sharp distinction between scientifically established dangers and suspected risks. This new understanding was to become important in the rewriting of the genetic engineering problematic.

The economic recession of 1973–74 was followed by a period in which conflict between economic and environmental priorities reduced the impetus toward more stringent environmental regulation. But by the early 1980s, with a recovering economy, an ongoing political mobilization, and the Grünen on the rise, Germany was experiencing a renewed interest in the consolidation of environmental policy. This interest culminated in the creation of the Umweltministerium (Ministry of the Environment) in 1986 (Weale et al. 1991, pp. 102–123; Hértier, et al. 1994, pp. 27–31; Hucke 1990; Malunat 1994, pp. 3–12)—a development that indicated a gradual change on the level of the political metanarrative in which, increasingly, environmental protection and economic development were conceptualized not as opposing principles (as in the mid 1970s) but as complementary.

Another important change in political discourse led to a rephrasing of the relationship between technological development and democracy. The German parliament is, like most other European parliaments, more reactive than active. Most legislative proposals are prepared by the government or the majority parties, and since the mid 1950s between 75 and 80 percent of the bills passed in each session of the Bundestag (the first chamber of the German parliament) have been government bills. This pattern is attributable largely to a system of stable coalitions and cohesive party voting, which ensures that important government proposals are hardly ever defeated (Saalfeld 1990, p. 68). But in 1969 a new instrument of parliamentary politics was established: the Enquete-Kommission (Inquiry Commission). Parliamentary representatives had complained about a trend in which ever more complicated political issues were increasingly being decided long before they reached Parliament. The purpose of Enquete-Kommissionen, which provided considerable support in terms of human and financial resources for parliamentary work, was to counter that trend. To ensure that such resources would also be available to minority parties, it was decided that an Enquete-Kommission would be instituted if one-fourth of the members of the Bundestag demanded one. The function of an Enquete-Kommission was to prepare decisions about complicated and highly contested issues. Not surprisingly, there have been several Enquete-Kommissionen dealing with potentially controversial technological topics, including nuclear power (1979–1980 and 1981–1983), information and communication technology (1981–1983), and technology assessment (1985–1986) (Hampel 1991; von Thienen 1986; Vowe 1986).

First Reconfigurations of the Regulatory Space: The Work of the Enquete-Kommission

It was in this discursive constellation, in 1983, that the Grünen were first elected to the Bundestag. When they decided to put genetic engineering high on their agenda, a group of new spokespersons for regulation of genetic engineering emerged within the German polity. In the same year, the Greens and the Social Democrats formally called for an Enquete-Kommission on genetic engineering. Soon after the Greens and the Social Democrats initiated procedures to set up an Enquete-Kommission, the parties in the coalition government (the Christlich Demokratische Union/Christlich-Soziale Union and the Freie Demokratische Partei) signaled their endorsement of this step. The government parties and the Socialists agreed that the task of the commission was to discuss the "opportunities and risks" implicit in biotechnological research with respect to its application in the areas of health, food, raw materials, energy production, and environmental protection. In particular the commission was mandated to consider the economic, ecological, legal, and social implications and the security aspects of biotechnology, and to put special emphasis on the ethical implications of the use of genetic engineering on humans (Catenhusen and Neumeister 1987, pp. 1–2).

The report of the Enquete-Kommission re-articulated the nature of the "genetic engineering question" in Germany. Far from simply reiterating genetic engineering as "a technology of the future" that should be allowed to develop without regulatory restrictions, the report of the Enquete-Kommission incorporated into its mode of reasoning a variety of master signifiers from ecological discourse, including the possibility of combining the growth of the bioindustry with precautionary regulation, the importance of risk assessment, and the importance of informing citizens and allowing them to participate in regulatory decision making. This work of problematization redefined the hazards connected with recombinant DNA as a policy challenge requiring a regulatory response that would take into account public sensibilities concerning genetic engineering and would approach new types of risk (such as deliberate releases) with extra caution. If such care were to be taken, so the narrative went, there would be no conflict among genetic engineering, economic growth, and ecological balance.

The commission's comprehensive research, data collection, interviewing, and traveling constituted an attempt to transform the hugely complex "genetic engineering question" into a narrative of 150 recommendations. It combined a framing of the "reality of genetic engineering" with the devising of policy instruments and recommendations for constructing a coherent policy approach. This inscription work was crucial for the outlining of a terrain for governmental intervention. It partitioned the political space of biotechnology in a way that established the regulation of genetic engineering as a precondition for its economic expansion. From the late 1970s on, as a result of the deregulation of genetic engineering, the boundary between "science" and "public" had become less and less permeable as regulatory decisionmaking had shifted from the regulatory commission to the laboratory. Self-regulated scientists had interacted with a public conceptualized as protected by the progress in molecular biology's understanding of genetic manipulation. The very same public had been black-boxed as a docile beneficiary of the advances of molecular biology. With the Enquete-Kommission's work, the public was re-articulated into the process of governing molecules as an entity that carefully weighed the "benefits" of genetic engineering against its "costs"—a process that transcended the narrow confines of a strictly technological risk assessment. The commission's redefinition of recombinant DNA's hazards as a socio-ecological problem constituted the core of a newly evolving and increasingly dominant policy narrative. The specification of this new discourse was left to the ensuing policymaking process.

However, this project of responding to the crisis of genetic engineering policy was fraught with difficulties from the beginning. At the beginning of the debate over the agenda of the Enquete-Kommission, the Greens argued that the goal of the commission should be to develop a framework of measures that would prohibit certain experiments involving genetic engineering but would also list experiments that were permitted. The Greens' restrictive position was not shared by the other parties. Initially the Greens had hoped to form, with some members of the SPD, an alliance powerful enough to carry through their restrictive strategy toward genetic engineering. When it became clear that the Socialist Party's leaders preferred a "more balanced" and open strategy and a dialogue with the government, the Greens began to conceptualize the Enquete-Kommission less as a forum for discussion and deliberation than as an instrument for political mobilization of the public (B. Gill 1991, p. 172). The Greens utilized the Enquete-Kommission as a highly visible site from which to launch a counter-narrative on genetic engineering policymaking—as a forum for a radical questioning of genetic engineering in particular and science and technology policy in general. They established an extra-parliamentary working group on "Bio-and genetic technology." The working group, financed by the Green parliamentary faction, consisted of scientists, medical doctors, social scientists, journalists, and activists from the ecology, women's, and peace movements. Its main task was to advise the Green representatives in the Bundestag and their staff members on how to "utilize the commission as a means to the end of fueling the public debate on genetic engineering, and to focus on the risks and the true reasons for its enforced development-and not on its chances."[2] This position differed strikingly from those held by the other parties—especially the governing CDU/CSU-FDP coalition, which viewed genetic engineering as an important economic resource to be developed. While the majority of the actors in the Enquete-Kommission sought to regulate genetic engineering as a precondition of its expansion, the Greens resisted this attempt at hegemonic re-articulation.

The Enquete-Kommission, not limited to discussing only the regulatory issues, undertook an exhaustive project of transforming genetic engineering into a domain to which government could be applied. It met 55 times between August 14, 1984, and December 1986. In addition, there were 46 meetings of the aforementioned working group and 18 hearings. The commission included four members of the CDU/CSU, four members of the SPD, one member of the FDP, and one Green. In addition, there were eight expert members: a philosopher of science, a legal scholar, a theologian, a molecular geneticist, a biochemist, a representative of the chemical union, a representative of the pharmaceutical industry, and a representative of the medical association. The work of the commission was supported by nine full-time staffers, five of whom were scientists (Catenhusen and Neumeister 1987, p. 2).

Seven working groups were established to study potential fields of application for genetic engineering: raw materials, plants, animals, the environment, health, genome analysis, and gene therapy. In addition, a working group on legal aspects was set up. The Enquete-Kommission also solicited a myriad of written expert statements and opinions from various institutions, administrative units, companies, and individuals. To collect information, members of the commission attended conferences and visited research institutes and companies inside and outside Germany, including some in Japan and the United States (Catenhusen and Neumeister 1987, p. 3). Much of the commission's work took place in the working groups, within which even further specialized groups discussed the implications of genetic engineering in the various fields. Only relatively late in its term did the commission turn to "horizontal topics," such as risks related to contained use on small and large scales, deliberate release of genetically modified organisms, changes in employment and market structures, the Third World, military applications, and current state support for research and development in genetic engineering. The last six months of the commission's term were used to develop a catalogue of some 150 recommendations for the Bundestag.

The work of the Enquete-Kommission was summarized in almost 400 pages of small print. In this text the commission tried to accomplish nothing less than to assess all potential implications of genetic engineering for Germany and to develop a comprehensive strategy of political action. But in fact the report offered two different readings or texts for a German genetic engineering policy, for the political divisions within the commission had ruled out the possibility of a common position. Of the resulting two texts, one was underwritten by the coalition parties and the Socialists and the other by the Greens, who disagreed with all the major policy assessments and recommendations made in the majority report. Apart from this split between the Greens and the other parties, the majority report itself was beset by contradictions and hesitations. In particular, there was a certain discrepancy between the analytical sections and the recommendations. In many instances, the analysis was much more critical than the recommendations, which often seemed to ignore the message of the analysis.

In part this may have been attributable to the fact that the Greens were well represented on the commission's scientific and administrative staff. When the final draft of the report was due, the members of the commission focused on the recommendations, while the staff concentrated on the analytical parts. Hence, what emerged as "The Report of the Enquete-Kommission" was in fact an intertextual framing device that could be used in a variety of different ways in interpretations of the genetic engineering problematic.

Nevertheless, the majority section of the report was an important contribution to the construction of a new dominant policy narrative on genetic engineering in Germany. In the majority report's discussion of the various fields of application, possible strategies for state support of research and development and statements of regulatory needs constituted a full endorsement of genetic engineering. The narrative anticipated a broad spectrum of potential uses of genetic engineering and endorsed state support of modern biotechnology. However, the text also incorporated elements from the discourse of the critics of genetic engineering and made ample reference to the scientific literature, including ecological studies on the hazards of genetic engineering. This was a critical move in hegemony building. As has been noted above, hegemony is not about the elimination of opposition; it focuses on the re-absorption of polarities into a system of "legitimate differences." Among the 150 suggestions were that there be no development of plants that are toxicologically problematic, that there be no mass screening of employees, and that more resources be devoted to the assessment of risks involved in the interaction between microorganisms and the ecosystem. The regulatory recommendations emphasized a cautious approach to deliberate release, with the deliberate release of microorganisms prohibited and with approval by the ZKBS and the Bundesgesundheitsamt required before the deliberate release of plants (Catenhusen and Neumeister 1987, pp. 213–239). And, most important for a regulatory framework, the report supported a legislative solution within the framework of the Bundesseuchengesetz (Contagious Disease Law). According to the Enquete-Kommission, the existing recombinant DNA guidelines should be adopted in a revised form and raised to the level of a law. This suggestion implied an extension of the validity of the existing recombinant DNA guidelines to all research, development, and production, and not only government-supported research; the introduction of liability and criminal law measures in cases of accidents and legal violations; the development of measures for the treatment and disposal of waste from recombinant DNA work; and the establishment of security measures for unintended releases in the case of "external" interruptions, as in the case of natural disasters. Furthermore, the report recommended changing the name of the Bundesseuchengesetz to "Law for the Regulation of Biological Security" and including the most important rules necessary for regulation of genetic engineering. In order to guarantee the adaptation of the law to the changing state of scientific development, the more detailed specifications would be left to administrative regulations (Catenhusen and Neumeister 1987, pp. 286–290). Since the Bundesseuchenschutzgesetz was under the jurisdiction of the Bundesministerium für Jugend, Familie, Frauen und Gesundheit and of the Bundesgesundheitsamt, this framing of the regulatory problem meant that, administratively, the agenda of recombinant DNA regulation was moved from the Bundesministerium für Forschung und Technologie to the BMJFFG. Finally, the commission proposed the creation of a genetic engineering advisory council that would include representatives of the "relevant" social groups and from the natural sciences and other disciplines. This council would reflect on the fundamental questions related to genetic engineering, advise the government, and develop guidelines for the Zentrale Kommission für Biologische Sicherheit and the ethics commissions. The council would be based either in the ministry responsible for recombinant DNA regulation (i.e., the BMJFFG) or in the parliament (Catenhusen and Neumeister 1987, p. 304).

These were far-reaching proposals which indicated important changes in the writing of biotechnology policy.

Law and Hegemony: Toward the Genetic Engineering Act

The publication of the report of the Enquete-Kommission coincided with a wave of extremely critical and negative media reporting on genetic engineering, which indicated a deepening of the crisis of genetic engineering policymaking.[3] The government sought to consolidate its control over the situation with a number of tactical and symbolic moves.

First, in 1986 (as was mentioned above) it shifted regulatory responsibility for genetic engineering from the Bundesministerium für Forschung und Technologie to the Bundesministerium für Jugend, Familie, Frauen und Gesundheit. High government officials "felt" that the responsibility of the BMFT for recombinant DNA regulation "was not appropriate any more," since genetic engineering was being increasingly applied on the industrial level. In addition, that the state's main research-supporting agency was at the same time the main regulating and, in a way, "research-inhibiting" agency was increasingly seen as a problem.[4] These changes reflected mounting concern about the credibility and legitimacy of the old system of recombinant DNA regulation (which, as was pointed out in chapter 3, constituted an expert enclosure where molecular biologists and scientists working in related fields were virtually the only legitimate spokespersons for genetic engineering regulation).

Second, a working group dealing with genetic engineering was set up in the BMJFFG. This indicated a further development toward the repositioning of the federal government as the central actor in the regulation of genetic engineering. At the same time, this move signaled an attempt to narrow down the genetic engineering problematic to a regulatory problem. Regulatory politics was becoming the site where social and political conflicts over genetic engineering would be settled. The newly emerging regulatory policy narrative emphasized that the planned recombinant DNA regulations were not inhibiting research and industry in Germany but, in contrast, were helping to support it. However, until well into the second half of the 1980s this framing was not shared by the actors representing the interest of science and industry. This became obvious in the reaction of a number of science and industry organizations to the report of the Enquete-Kommission.

The Verband Chemischer Industrie (Federation of the Chemical Industry) rejected most of the Enquete-Kommission's recommendations: "The chemical industry is convinced that the 1978 recombinant DNA use guidelines and its modifications, under participation of the ZKBS, will make it possible to provide a broad and for all acceptable protection."[5] Furthermore, the VCI renounced a moratorium on the deliberate release of genetically modified microorganisms, on the ground such a moratorium would lead to disadvantages for research in Germany against its international competitors. In a similar vein, the VCI saw a need for neither a parliamentary monitoring commission nor for changes in liability law. In general, past experiences with genetic engineering and existing legislation regulating biotechnology, according to VCI's narrative, made any further specific legislation unnecessary. Very similar positions were being articulated by research organizations such as the Deutsche Forschungsgemeinschaft[6] and the Arbeitsgemeinschaft Biotechnologie (Biotechnology Working Group).[7]

The Arbeitsgemeinschaft Biotechnologie (AG BioT), an umbrella group of scientific organizations working in biotechnology, saw no need for special legislation and considered the existing regulatory situation to be sufficient. In its evaluation, the AG BioT cited the accident-free history of recombinant DNA work since the early 1970s and highlighted the precision and predictability of genetic engineering as defined by the body of knowledge of molecular biology. Overall, the AG BioT argued that what really mattered was the education of the users of genetic engineering. Furthermore, the AG BioT agreed with the chemical industry that large-scale use of genetic engineering techniques and deliberate release of genetically modified organisms did not represent a new level of risk or necessitate any special legislation. In other words, it seemed that industry and science fully shared the position that, although industrial scale-up and deliberate release of genetically modified organisms necessitated caution and the development of modified safety guidelines, there was no need whatsoever for special legislation. This "business as usual" approach reflected the position taken by the US National Academy of Science in the debate over deliberate release.

However, industry and science found it increasingly difficult to maintain this position. They had adopted a position toward the genetic engineering problematic that was similar to industry's and science's attitude in the 1970s. But the discursive constellation of the mid 1980s was vastly different from that of the mid 1970s. Critical media reports on developments in genetic engineering, opinion polls showing a concerned public, the strong voice of the scientific discipline of ecology, and, closely related, the ecological discourse and its attendant policy recommendations had created a matrix of discursive relations which had already given rise to new legitimate actors and instruments in the policy field and which was about to contribute to the creation of new institutional designs. In contrast to the 1970s, the incorporation of prevention and public participation into policy designs had become a legitimate and expected feature in the shaping of socio-economic and technological development. Related to this change, the Bundesministerium für Forschung und Technologie had lost its centrality as an actor in the field of the politics of genetic engineering, while the Bundesministerium für Umwelt (Ministry for the Environment) and the Bundesministerium für Jugend, Familie, Frauen und Gesundheit had become the key departments. Citizens' participation in the licensing of technological projects had come to be conceptualized not only as legitimate but also as critical for the political acceptance of such projects—a view that influenced federal- and state-level reasoning in the formation of genetic engineering regulations. In other words, a policy narrative had been constructed by a number of dispersed actors, including ecologists at American universities, social scientists theorizing about "ecological modernization," actors in social movements, civil servants in state administrations, and citizens responding to opinion polls. This policy narrative evolved in a relationship of mutual reinforcement with other policy narratives (such as that of environmental policy) and, in general, with the political metanarrative of the relationship between society and the environment. This new story of recombinant DNA regulation became increasingly hegemonic as it served for more and more actors as the lens through which the reality of the genetic engineering problematic was experienced.

On May 28, 1986, the government cabinet asked the BMJFFG, with the participation of other concerned ministries, to scrutinize, first, the immediate introduction of legal means requiring a registration of laboratories working with recombinant DNA and, second, the necessity for further regulations, while considering the results of the Enquete-Kommission and the experiences of other countries.[8] However, in the context of the rather decentralized German political system the federal administration was only one among a number of somewhat interconnected locations for regulatory policymaking. Another important location was the German parliament. The report of the Enquete-Kommission was released to various subcommittees of the concerned committees of the Bundestag. Although these committees did not respond to the report before October of 1989, the release of the report had set in motion a process of parliamentary committee work that would eventually lead to the passage of the Genetic Engineering Act.

At the same time, the Länder (states) emerged as another important site for regulatory decisionmaking about recombinant DNA. In April of 1987, after the Enquete-Kommission's report was published, the Umweltministerkonferenz, a group of state environmental ministers, created a working group, which was given the task of developing the states' position on the report of the Enquete-Kommission.[9]

The Länder had their own policy goals. Confronted by critics of genetic engineering and by opposition movements at the local level, they pressured for quick action to revise their genetic engineering regulations. In the strongly federal German system, the Länder, strongly represented through the second chamber of the German parliament (the Bundesrat), traditionally play an important role in policymaking. The states' opportunity to shape genetic engineering policy came in 1987, when the government decided to regulate large-scale genetic engineering in the fourth revision of the Bundesimmissionsschutzgesetz (Federal Emissions Act), an environmental law regulating, among other things, emissions from potentially dangerous production sites (*Anlagen*). Through the reform, the government wanted to cover specifically large-scale industrial production involving recombinant DNA. According to the government, there were two possible paths to regulating facilities involving in genetic engineering: a simplified process of permitting work with "not dangerous" genetically modified microorganisms and a full procedure involving public participation and the possibility of challenging a planned production site. However, the Bundesrat rejected the government's proposal and decided that *all* large-scale work involving recombinant DNA should be subject to the regular process of approval, which included public participation. This objection found its way into the revised Bundesimmissionsschutzgesetz. This amounted to the establishment of a regulatory framework that required substantial public participation in the licensing of any production facility planned for work with recombinant DNA, no matter what the minimum level of risk associated with the genetically modified organism used would be. In other words, public participation—a core element of the ecology discourse—had become an integral part of the licensing process, independent of specific minimum measures of biological and physical containment. This provision was of crucial importance for the political confrontations to come in the following years, as it substantially extended the range of "legitimate" actors involved in the shaping of regulations for genetic engineering.

With these changes in regulatory discourse, the Bundesministerium für Umwelt, responsible for the Bundesimmissionsschutzgesetz, emerged as the second major administrative player (after the BMJFFG) in the regulation of genetic engineering. In addition, in 1987 there had been two changes in regulations explicitly referring to recombinant DNA: a regulation based on the Gefahrenstoffverordnung (BGBL. S. 2721) and the Abwasserherkunftverordnung of July 3, 1987 (BGBl I. S. 1578). Both areas were under the jurisdiction of the Bundesministerium für Umwelt (BMU). And, of course, the BMU was the department most obviously responsible for regulating genetic engineering, especially deliberate releases of modified organisms.

However, there was a disagreement between the states' induction of the BMU into the newly forming biotechnology regulation network and the government's evolving policy narrative, which attempted to reconcile new regulations with science's and industry's concerns about "overregulation." It should be remembered that policymaking is an attempt to inscribe order into an unstable discursive environment. The goal of policymaking, then, is to persuade, to establish broadly shared interpretations of reality, to define and attribute roles to actors who accept them (enrollment), and thus to create lasting networks (a precondition for governing). In this context, it was part of the government's strategy to translate the interests of industry and science into its own interests by a number of strategic moves. One of this moves was the symbolic politics of making concessions. A number of actors in the field of regulation had argued that the BMU, traditionally a strong advocate for the environment, would be more critical of and more involved in the administration and implementation of genetic engineering regulations than the BMJFFG, which was generally seen as having a friendly relationship with science and industry.[10] Which department was primarily responsible for regulating genetic engineering had turned into an issue of symbolic politics. In 1988 a special "genetic engineering" unit was set up in the BMJFFG, and that ministry was asked by the government to draft a report on the needs and possible strategies for legislation in the field of genetic engineering. Despite the 1987 reforms, which strengthened the position of the BMU, the BMJFFG increasingly became the leading department in the field of regulating genetic engineering. This responsibility would be of double importance: first, naturally, drafting a law meant political weight; second, it was clear from the beginning that the law would be a *Stammgesetz* (a very general and relatively short law that would be specified by regulations drafted and administered by the responsible ministry). A struggle for the status of the leading department began, involving particularly the BMJFFG and the BMU. This struggle concerned jurisdiction, rather than regulatory strategies. One civil servant observed:

> All departments are of the same opinion that contained use and deliberate release need to be regulated, but whether this regulation should be enacted in the form of a separate law or within existing laws, on this point there has no agreement been reached yet. . . . Concerning the contents of regulation there are hardly any disagreements . . . but there are "department-egoistic" positions as to who shall draft the regulations. This is an important field and everybody would like to have that in his house. . . .[11]

A strong indication of the consolidation of the new policy narrative was the Eckwert-Beschluss, an official position paper published by BMJFFG on the needs of and strategies for the legislation of genetic engineering. This report, the German government's first major statement on genetic engineering since the discussions of the late 1970s, presented the government's strategy for stabilizing the politics of genetic engineering. Its narrative started with the assertion that genetic engineering is indispensable for modern society. However, risks were then identified. The contained use of genetically modified organisms was said to be controllable in principle, but not completely controllable. The deliberate release of genetically modified organisms was said to be little understood or analyzed.[12] The risks and the opportunities of genetic engineering were said to require that its ethical implications be discussed intensively, that the gaps in knowledge be closed, that regulations and prohibitions be established where necessary, and that use be made of international experience.[13] Furthermore, new German regulations should, in accordance with the requests of the Enquete-Kommission and in line with the current discussions on the report in the parliamentary committees and the EC directives, combine regulation of contained use and regulation of the deliberate release of genetically modified organisms.[14] Hence, the draft law was not only an instance of national policymaking; it was also a contribution to supranational policymaking. A civil servant described this situation:

> We see on the EC level, compared with the developments in Germany relatively liberal regulations. We have to balance that out. At the end of the day we have to settle the conflict on the EC level. . . . We have to get our position through in Brussels . . . but since recently that does not work by us leaning back and saying no . . . these directives do not need unanimity. This means we need allies there and from where to get these allies, that I do not know at this moment.[15]

Finally, the Eckwert-Beschluss strongly urged the drafting of a separate law on genetic engineering. Attempting to regulate genetic engineering by changing existing laws and regulations was seen as likely to lead to inconsistencies, confusions, repetitions, and complications in implementation. The Enquete-Kommission's proposal to integrate the law into the Bundesseuchengesetz was rejected as inappropriate: according to the Eckwert-Beschluss, integrating the regulation of genetic engineering into a law focusing exclusively on the avoidance of communicable diseases such as AIDS might have negative effects on the social acceptance of genetic engineering.[16] The new law should maintain the current regulatory institutional arrangement involving the

Bundesgesundheitsamt, the Zentrale Kommission für Biologische Sicherheit, and, in the case of deliberate release, the Biologische Bundesanstalt (Federal Biology Institute). At the level of implementation of the law, the following distribution of responsibilities was outlined: in the case of the contained use of nondangerous work, permission would be given by the states; in the case of high-risk work, it would be given by federal authorities; in the case of deliberate release of genetically modified organisms, permission and definition of conditions would be given by the federal authorities. Scrutiny of the implementation of the required measures was to be the responsibility of the states.[17]

The Eckwert-Beschluss of 1988 was the first decisive moment in the intricate regulatory game around genetic engineering that had started in 1984. It tried to create a new signification of the genetic engineering problematic by integrating and translating the various positions of the federal government, interest groups, political parties, critics, administration, the Länder, and the developments at the EC and OECD level into a reasonably coherent narrative on the regulation of genetic engineering. At the same time, the position paper contained significant potential for further controversy. It stripped the Bundesministerium für Umwelt of all responsibilities for the regulation of recombinant DNA work. It indicated Germany's position for the ongoing negotiations at the EC level. And it confirmed, on the level of the political parties, the CDU-CSU-FPD-SPD axis that was shaped in the Enquete-Kommission by way of excluding the position of the Greens. Furthermore, it severely limited the role of the Länder in the authorization of recombinant DNA work. At the same time, it gave the Länder full responsibility for implementing the law. This narrative of integration would soon run into difficulties.

On December 28, 1988, a project team called Gesetz zur Regelung der Gentechnik (Genetic Engineering Act) was created in the BMJFFG to prepare a *Referentenentwurf* (ministry' s draft) for a law based on the Eckwert Beschluss to be eventually modified and sent to the parliament. Only five months later, in July of 1989, the government sent its proposal for a law to the parliament. The unusual speed with which the German genetic engineering act was drafted was to cause serious difficulties for its passage.

Speeding Up Legislation

In Germany's bicameral parliament, the Bundesrat (the second chamber) represents the Länder. Under normal circumstances, the ruling party or coalition has no trouble controlling the vote in the first chamber, the Bundestag. However, things can sometimes get tricky in the Bundesrat, where party discipline competes with the interests of the Länder. Therefore, extensive federal-states negotiations typically precede parliamentary initiatives. This was not the case with the planned genetic engineering law. According to one participant in the drafting of the law put it: "Other laws are discussed and prepared at length in all sorts of Bund-Länder committees so that things go usually relatively smooth through the Bundesrat. But not in this case."[18]

The fast pace of legislative action was attributable to a number of developments, all indicating a deepening destabilization of the German politics of genetic engineering.

The public controversy over genetic engineering had reached its peak in 1988, when a general impression that "the public" had "a highly negative attitude" toward genetic engineering had prevailed. Increasingly, industry, unions, and the government coalition began to view passing a law as a way to alleviate anxieties and rising concerns about genetic engineering. Most obvious here was the shift in industry's narrative from the rejection of legislation to its endorsement as "politically necessary":

> We depend on work in Germany and hence social acceptance is crucial. . . . If your work is based on the current guidelines I do not see any dangers. . . . Based on factual reasons we are opposed to any further regulations but from a political perspective they are unavoidable. . . . Politically a framework law is being discussed and in the negotiations we have agreed on such a law. . . . [19]

Industry had begun to accept the translation of new genetic engineering regulations as being in its very self-interest. One civil servant saw the position of industry as follows:

> Parts of industry said we are interested to have reasonable regulations which allow us to say genetic engineering must be done, is necessary, important and a blessing, and, in addition, has dangers, but these dangers can be handled by a legal measures.

Actors in industry began to conceptualize the planned law as an instrument for drawing the boundaries of industrial biotechnology, reproductive technology, and eugenics:

> We are trying to establish a clear separation between the application of genetic engineering for industrial purposes and biological aspects of reproductive technologies. . . . It is not in the interest of industry to do this kind of research. . . . How many cases of in vitro fertilization are there in Germany? Sure, these are single cases and one must help—but this is not an industrial

problem. . . . We do not want to be in one and the same law with test tube babies and surrogate mothers.[02]

The major social partner of the Verband Chemischer Industrie, the Industrie Gewerkschaft Chemie (Chemical Workers' Union), was equally supportive of legislation.[21] Like the VCI, the IG Chemie, with its tradition of close collaboration with industry, felt that, in the face of mounting social resistance, legislation was a wise step:

In the Federal Republic, at this moment, we are in a certain upheaval in the thinking on how one is to introduce new technologies into society. We are burned children concerning the introduction of nuclear technology. One has invested 130 billion Deutschmarks and one is blocked, basically because the social context has not been taken into account early enough and because there is a tendency that the ideas concerning our socio-political life lag behind the possibilities of technology, and this creates problems for democracy. In the case of biotechnology one tries to avoid the frictions experienced in the past by informing people way in advance and by making clearer the balance between risks and opportunities. We should not be just fixated like a snake on the risks, but should also look at the opportunities.[22]

At the same time, industry began to conceptualize a Genetic Engineering Act as a way to get rid of the revised Bundesimmissionsschutzgesetz, which, with its provisions concerning public participation, had created a rather rocky and delayed licensing process for large-scale industrial facilities using recombinant DNA technology. In the fall of 1988, BASF, concerned about the requirements of the Bundesimmissionsschutzgesetz, threatened to transfer a planned facility for the production of tumor necrosis factor (TNF) to a foreign country. This led to considerable reactions from the political community. The research minister assured BASF that, with the new genetic engineering law, regulation would become clearer and easier. A Socialist representative in the Bundestag, at the same time a board member of BASF and a labor union official, asked the government if it could eliminate the public participation requirements from the Fourth Amendment to the Bundesimmissionsschutzgesetz.[23] The former chairman of the Enquete-Kommission, the Socialist representative Wolf-Michael Catenhusen, attacked the government for its lack of action. He suggested that if the government did not act soon, the parliament would have to set the initiative.[42] Furthermore, in the governing coalition, concerns arose with respect to upcoming elections in the Länder. With several such elections pending, and with several Socialist-Green coalition governments a possibility, the scenario of genetic engineering legislation blocked by the Bundesrat or passed with significant concessions from the governing coalition at the federal level was becoming more likely. Hence, the government was clearly pressed by a variety of factors to move quickly on the genetic engineering law.

In February of 1989 a first draft of a Gesetz zur Regelung der Gentechnik (Genetic Engineering Act) began to circulate. On April 24 a first draft was presented by the BMJFFG to the public. This draft provided a further elaboration of the government's new regulatory policy narrative. It was designed as a law to regulate the contained use and the deliberate release of genetically modified organisms. (Unlike the EC directives, the German draft did not confine itself in the contained-use sections of the draft to microorganisms; both its contained-use section and its deliberate-release section take genetically modified organisms as the basic unit of regulatory concern.) The draft recommended that the ZKBS function as the central scientific authority and scrutinize and evaluate all questions related to security aspects of the law. The ZKBS was to cooperate closely with two other institutions involved in the implementation of the law, the Gesundheitsministerium (BMJFFG) and the Bundesgesundheitsamt. The ZKBS was to have as members eight experts from various scientific fields ranging from microbiology to ecology, and four members representing labor unions, industry, environmental organizations, and research organizations. According to §6, contained-use operations were split into four groups: group 1 (no risk and/or long experience), group 2 (pathogenic organisms implying moderate risk), group 3 (highly pathogenic organisms constituting high risks for workers and lower risks for the human population and the environment), and group 4 (work implying high risk for workers, the human population, and the environment). Work involving genetic engineering with group 1 organisms required the consent of a competent Land authority and a general licensing of the research and/or production facility. After permission was obtained, work could proceed without further consultation with the Land authorities. According to §4, work in higher groups required licensing by the Bundesgesundheitsamt, which was to consult with the ZKBS (now constituted as detailed above). New experiments or work in higher risk groups required a new license. In a similar vein, any deliberate release required the consent of the Bundesgesundheitsamt. Public participation would be required in the case of category 3 or 4 commercial contained-use work and in the case of the deliberate release of microorganisms and of such organisms, whose release might lead to an "uncontrollable distribution."[25]

The draft law signaled a major departure from the virtual self-regulation of the 1970s. At its core was a technological definition of risk. The idea that the hazards of genetic engineering could be controlled by the very technology that produced them had certainly not vanished, however. This becomes particularly clear when one looks at the contained-use parts of the draft law, which continued the well-known regulatory logic of a combined system of biological and physical containment. However, the framing of recombinant DNA's hazards had become more complex and was now, in addition, characterized by a second rationality emphasizing the socio-ecological dimensions of genetic engineering: the political importance of having in place a general licensing system applicable to research, industrial purposes, and deliberate releases; the need to allow for public participation and information, particular in cases of large-scale use of genetic technology and deliberate release; and the significance of breaching the expert enclosure of representatives from molecular biology and related fields by including ecological expertise in the licensing process. Political rationalities define the proper distribution of tasks and actions between authorities of different types and consider the ideals and principles to which government should be directed. As "the public," theories of the ecosystem, and views on the relationship between democracy and technological were inscribed in the dominant policy narrative, the dynamics of the shaping of the recombinant DNA regulations changed. In the 1970s the technological definition of recombinant DNA's hazards had led to a situation in which scientific and technological events in molecular biology and their framings (such as certain experiments or scientific conferences) had defined the shaping of recombinant DNA regulations. In the second half of the 1980s, such events in the discipline of molecular biology lost importance for the recombinant DNA regulatory debate, and other events and discourses, such as ecological arguments, perceptions of public opinion, and questions of liability in case of accidents related to recombinant DNA work, gained importance. The inscription of this new socio-ecological rationality into the regulatory policy narrative had moved into the foreground of the discursive struggle. But soon critics began to attack the centralizing aspects of the proposed law on the ground that they might undermine its declared purpose. Minister of the Environment Klaus Toepfer protested the complete elimination of his ministry from the new regulatory framework. Whereas the old regulatory framework had given the Bundesministerium für Umwelt significant influence, especially through the Bundesimmissionsschutzgesetz, the new law concentrated all regulatory matters in the hands of the Bundesgesundheitsamt, the BMJFFG, and ZKBS. In particular, Toepfer wanted to maintain the influence of the Länder in the implementation process. Furthermore, following the model of the Immissionsschutzgesetz, the BMU wanted "its" Umweltbundesamt (Environmental Agency)—the equivalent of the BMJFFG's BGA—to participate in the implementation of the Genetic Engineering Act. Joining Toepfer in his critique, the SPD argued that the ZKBS would be the power center of a regulatory framework that would be unresponsive to parliamentary influence.[26]

The government's efforts to create a commonly agreed upon framework of meaning for recombinant DNA regulation continued when Minister of Health Ursula Lehr invited representatives of industry, research, labor, consumer groups, and environmental groups to a nonpublic hearing. This hearing, held on May 24, 1989, demonstrated that negotiation of the socio-ecological rationality of the draft law had moved into the foreground of the political debate. Of the main interest groups, both IG-Chemie and the Verband Chemischer Industrie were supportive of the new law and endorsed the new system of public participation. IG-Chemie, however, supported the idea of connecting the ZKBS with the parliament. Whereas industry and the chemical union adopted something like a middle-of-the-road position, generally endorsing the draft, the scientific organizations, the environmental movements, and the Deutscher Gewerkschaftsbund articulated severe critiques. The representatives of the scientific community, in particular the Arbeitsgemeinschaft der Grossforschungseinrichtungen, the Max-Planck-Gesellschaft, the Deutscher Gewerkschaftsbund, the Gesellschaft für Genbiologische Forschung, and the Gesellschaft Deutscher Chemiker demanded considerable relaxation of the draft law. The DFG wanted group 1 work to be completely excluded from any regulations, and the GDC demanded a registration system of work without any authorization. Otherwise, the representatives argued, regression to the Morgenthau Plan[27]—deindustrialization—was unavoidable. Environmental groups and the Deutscher Gewerkschaftsbund saw too little provision for public participation and too much power and importance being granted to the ZKBS (which, in addition, they saw as dominated by representatives of the scientific establishment). Furthermore, the critics insisted on the principle of *Umkehr der Beweislast* (reverse proof of liability), according to which, in the case of an accident, it was up to the operators of a genetic engineering project to prove that they had not caused damage, not up to the victims to prove that the operators had caused the damage. Furthermore, the Deutscher Gewerkschaftsbund and environmental movements deplored the law's omission of what they saw as crucial areas of concern, such as genome analysis, human genetics, and the transport of genetically modified organisms.[28]

Finally, on July 12, 1989, the government formally submitted to the parliament a draft for a Genetic Engineering Act. Overall, this draft reflected a continuation of the middle- of-the-road strategy started with the "Eckwert-Beschluss" in 1988 and was supported by industry and the chemical union. Neither the environmental groups nor the scientific community were fully successful with their demands. On the level of administrative responsibilities, the law symbolized a defeat of the environment minister. The BMJFFG, the Bundesgesundheitsamt, and the ZKBS remained at the center of the regulatory structure. The only concession made to the Bundesministerium für Umwelt was its integration, together with the Biologische Bundesanstalt, into the process of authorization for deliberate releases, and, in case of a deliberate release of genetically modified animals, with the Bundesforschungssanstalt für Viruskrankheiten (Federal Agency for Diseases Caused by Viruses) (§ 15 (3)). With a formal license, risk group 1 work for research purposes was now allowed to proceed after a 90 days. For commercial group 1 work, a similar regulation was developed, with the difference that in this case new work had to be registered again and that the waiting period was 60 days.

The submission of this government draft proposal to the Bundesrat turned an already rocky process of legislation into a political battle with unprecedented features. The Bundesrat (the chamber representing the interests of the Länder) plays a crucial role in the passage of any law. Not only does it have to agree with the Bundestag before a law passes; it also receives the bill first, before it goes to the Bundestag, and has the right to take a position within six weeks.[29] During the consultations in the Bundesrat on the proposed Genetic Engineering Law, it quickly became obvious that the Länder had severe reservations about the draft proposal. They did not accept their role in the implementation of the planned law as envisioned by the government. To be sure, there is no unambiguous thing called "the interests of the Länder." Furthermore, the Bundesrat was divided on a variety of issues concerning on the draft, and the divisions may been ideological.

However, the one thing the Bundesrat could easily agree on was its insistence on a stronger role for the Länder in the implementation of the law. The rest of the discussion was essentially focused on the amount of protection provided by the law and on the degree of public participation allowed in the proposed regulatory structures—issues that were generally seen as important by all Länder, but whose resolution was not easy. In terms of the preventative features of the law, another important compromise within the Bundesrat was achieved by maintaining the idea of the "installation concept" in the new law. Like the Immissionsschutzgesetz, the new law would focus not only on activities involving genetic engineering but also on the sites where these activities originate and take place. Regulating the sites of production (*Anlagengenehmigung*), the states' narrative went, would provide a greater degree of prevention.

A representative of the Bundesministerium für Jugend, Familie, Frauen und Gesundheit remembered the situation as follows:

> . . . the Länder were disappointed because from their perspective they insufficiently participated in the preparation of the law. . . . We did the first draft very fast, and the Länder were not prepared for this draft and felt rolled over by the developments. Then they had very different ideas. We approached the issue with the EC draft directives in mind and our experiences here in-house with drugs, etc. . . . And in the Bundesrat it became clear that the discussion was basically shaped by environmental considerations. . . . The representatives of the Länder all came from Immissionschutzrecht and did not find in the government draft the Anlagengenehmigung and in a general decision of the Bundesrat, they demanded the integration of the Anlagenkonzept into the government draft. First they wanted the Anlagenkonzept. . . . Second they did not accept our idea of implementation which we wanted to concentrate in the [Bundesgesundheitsamt]. They wanted implementation to be concentrated in the Länder. . . . The environmental committee of the Bundesrat had had an especially strong influence on the government draft, they all came from the Immissionsschutzgesetz. . . . They thought the whole thing is nothing else than a special form of emission when it comes to contained use. As a result we need the same mechanisms of preventive control.[30]

The states' insistence on the installation concept could certainly be interpreted as a simple strategy for increasing their influence in the implementation process. Even if this was the case, it is interesting to note that this position was developed in reference to principles of ecological discourse. According to the states' narrative, the installation concept was the appropriate way to regulate large-scale use of recombinant DNA, because it was much better suited to guarantee prevention than the government's initial proposal. And prevention is, after all, the primary way to deal with environmental problems. Hence knowledge and story elements derived from ecological discourse served as the primary instrument to justify the states' position (Bruer 1991, pp. 37–79). In September the Bundesrat came up with a catalogue of 253 proposals (on 349 pages) to change the government draft—a record amount of proposed changes in the Bundesrat's history.[31] Having put forth so many partially contradictory proposals, the Bundesrat decided that it would be unable to come up with a detailed summary proposal of desired modifications; thus, it restricted itself to an "alternative" position paper asserting the states' demands. In essence, this paper demanded that permission for work related to genetic engineering be based on the installation concept and that full responsibility for the implementation of the law in research and production be given to the Länder.[32]

The Deepening of the Crisis

Besides the difficulty of getting the proposed act passed by the parliament, a series of other events created turbulence in the legislative process. In October of 1989 the Bundestag had finally come up with a decision on the Enquete-Kommission's report that essentially followed the commission's recommendations. However, the Grünen voted against this decision, and the SPD insisted on a five-year moratorium on the deliberate release of genetically modified organisms.[33] In July, the SPD in Schleswig-Holstein had decided in a special meeting on a set of stringent regulations and bans for genetic engineering, such as the prohibition of the production of drugs based on methods of genetic engineering (with strict regulations for exceptions).[34] This was a significant development, since the Schleswig-Holstein group (led by Björn Engholm, later the SPD's national leader) was, in a way, considered a model for the party's efforts to fend off the rise of the Grünen (Frankland and Schoonmaker 1992, p. 82).

On November 11, 1989, a decision of the Kassel High Administrative Court seriously questioned the government's past approval of genetic engineering regulation in Germany. In a lawsuit initiated by a citizen group against Hoechst's prospective facility for producing genetically engineered insulin, the lawyers argued that the federal Immissionsschutzgesetz could not provide a legal basis for industrial production using methods of genetic engineering. The Immissionsschutzgesetz, they argued, could not be interpreted as an expression of the will of the legislature on such a basic question as the application of genetic engineering. Just as nuclear power plants had required legislative approval, genetic engineering could not be legal without a clear articulation of legislative will:

> This need [for legislative approval] is related to the fact of possible far-reaching dangers for the environment and health . . . which are associated with the release of genetically engineered bacteria, but also to the fact that the decision on a "yes" or "no" to genetic engineering raises such far-reaching normative-ethical, social, and political questions that the answers cannot be given by the executive branch.[35]

Against this argument, Hoechst's lawyers claimed that genetic engineering was nothing new, and that hence there was no need for any special expression of legislative will.[36] But this coding of the history of biotechnology turned out to be unsuccessful in convincing the judges of the High Administrative Court of Hessen, who decided against Hoechst.

On November 6, as the Max-Planck-Institut's plans for the first deliberate release of a genetically modified organism (a new petunia) in Germany encountered massive public discussion, protest, and legal action, the High Administrative Court in Hessen decided in favor of the citizen group that there was no legal basis for industrial production involving genetically engineered organisms in Germany, and that Hoechst had to discontinue its insulin project until there was a legal basis for it. This decision was not only relevant for Hessen. Basically, the legality of industrial production involving genetic engineering was put into question for the whole of Germany. The court had clearly stated that production facilities involving the use of genetically modified organisms could be constructed only if permitted by legal rules regulating the use of genetic engineering, and that neither the Bundesimmissionsschutzgesetz nor any other existing legal rules could be regarded as sufficient legal grounds in this context. Although the Hoechst decision initially affected only the parties involved in the lawsuit, it was clear that this decision had set a precedent and thus rendered the fate of genetic engineering in Germany highly uncertain—exactly the situation that industry was most concerned about.[37] The Hoechst decision further exacerbated the crisis of signification in the German politics of genetic engineering and increased the pressure to pass the Genetic Engineering Act as quickly as possible even more.

In November, the much-debated proposal for a Genetic Engineering Act, submitted to the parliament on July 12, 1989, was sent to the Bundestag, where on December 6 the Ausschuss für Jugend, Familie,

Frauen und Gesundheit (Committee for Youth, Family, Women, and Health) created a subcommittee on genetic engineering.[38] The subcommittee met seven times to discuss the coalition proposal, and on January 17–19 it held a public hearing on the draft.[39] The three-day hearing gave an opportunity for more than 60 experts from groups ranging from the Deutsche Forschungsgemeinschaft to the Öko-Institutes to voice their positions. However, critics pointed out that this hearing was nothing more than a false display of open public discussion when, in fact, the CDU/CSU-FDP coalition in the Bundestag and the Bundesrat had already compromised on a modification of the July proposal.[40] In fact, this compromise was strongly motivated by the pending state elections in Lower Saxony, where a victory by the SPD and the Grünen seemed within reach—a victory which would mean that the CDU/CSU-FDP coalition in the Bundesrat would lose its majority and, as a result, might portend the passage of a considerably stricter Genetic Engineering Act.

The draft debated at the January hearing was revised completely; hardly a paragraph remained unchanged. The biggest changes were the addition of the *Anlagenprinzip* (facility licensing principle) and the appointing of the Länder as the main licensing authorities. In addition, the determination of risk levels was put under the jurisdiction of the Länder, which were to receive advice from the ZKBS via the Bundesgesundheitsamt. However, authorizing deliberate releases was to be the responsibility of the latter body.

The many changes proposed by the critics of the law, ranging from the SPD to the various social movements and environmental organizations, were largely ignored. The aforementioned groups, with a limited contribution from the Socialists, combined some of the main themes of the critique of genetic engineering (discussed in chapter 5) into a powerful counter-narrative that constituted an alternative framing of a socio-ecological rationality of recombinant DNA regulatory policymaking. Most important, this narrative challenged the central element of the government's policy narrative: that broad and encompassing support for, encouragement of, and development of genetic engineering were compatible with public participation, free access to information, and ecological considerations. The counter-narrative proposed a much more cautious model of development guided by ecological considerations (implying a moratorium on deliberate releases). In this model, regulatory decisionmaking was closely connected to the parliament and characterized by a committee structure where molecular biologists were in the minority.

On July 1, 1990, the Gesetz zur Regelung von Fragen der Gentechnik (Act for the Regulation of Questions of Genetic Engineering) went into force.[41] The regulatory approach of the act, first sketched in the Eckwert-Beschluss, went beyond the old regulations characterized by the rationality of a technology-based risk definition. The new act covered the preparation of genetically modified organisms (GMOs) within a manufacturing plant, their use, their multiplication, their storage, their destruction, and their transport. It applied to all biotechnology facilities and activities, including deliberate releases of genetically modified plants or animals. In principle it also covered the marketing of products containing or consisting of GMOs, and of products produced by GMOs. But at the same time, it took up some of the "classical" story lines of the old, NIH-inspired guidelines. In a clear pattern of continuity with the former recombinant DNA regulations, the Gesetz zur Regelung von Fragen der Gentechnik was said to fulfill two purposes at the same time: (1) protecting humans, animals, plants, and the environment from potential dangers of processes and products involving genetic engineering and (2) creating a legal framework for research on, development of, use of, and support of the scientific and technological potentials of genetic engineering. Furthermore, the determination of the risk levels for contained use of GMOs continued to follow a technological definition of risk. The act defined four safety levels for genetic engineering work: no risk (level 1),[24] minor risk (level 2), medium risk (level 3), and high risk (level 4). The determination of risk was based on evaluation of the donor and host organisms, on the vectors used, and on the genetically engineered organisms produced. Just as in the old guidelines, biological security measures were combined with physical containment measures in a system of "containing" hazards. But clearly there were also important elements of precautionary legislation inscribed into the act. Commercial activities and research (both academic and industrial) were differentiated. Commercial facilities required authorization both for building and for operation. A public hearing was required before the commencement of commercial activities on level 2, 3, or 4 (and, under certain circumstances, level 1). Information relevant to an application, and to security, was to be publicized and publicly displayed by the competent authority for a month. Within six weeks, any person could raise written objections that would be discussed with the authorities at a public hearing.[43] For research facilities the requirements were less stringent: Authorization would generally be necessary for facilities and activities on levels 2–4. For level 1, and sometimes even for levels 2–4, prior notification was sufficient and a continuation of the public hearing was necessary. For research activities on levels 2–4, which went beyond the initially described work, the competent authority was to be notified two months before the planned beginning of the research. In the commercial field, continuation of activity on level 1 required notification two months in advance. Continued work on levels 2–4 required authorization. Before giving formal notification or applying for an authorization, the operator had to get in touch with the appropriate authorities in the Land where the company wanted to carry out its project. The Länder, not the federal government, were to be in charge of licensing. When a project needed authorization, a licensing procedure, involving the submission of details and a comprehensive risk assessment, was required. The authority had then three months to decide. The Zentrale Kommission für Biologische Sicherheit was then be consulted on matters of risk assessment. The ZKBS, a division of the Bundesgesundheitsamt, was composed of ten experts, at least six of whom would work in recombinant DNA research and two of whom would be from the discipline of ecology. Furthermore, the ZKBS was to have one member each from organized labor, work protection, business, environmental protection, and research support. The meetings of the ZKBS would be confidential and not open to the public. The deliberate release of GMOs required advance permission from the Bundesgesundheitsamt. The application had to describe all safety-relevant features of the organism, the details of the proposed release, the surveillance, and the emergency plans. A public hearing would be mandatory, and the Bundesgesundheitsamt would have to make a decision within three months. Before a permit could be granted, the ZKBS would have to make a recommendation. Clearly, these measures constituted a breach of the expert enclosure established by molecular biologists in the 1970s. The Gesetz zur Regelung von Fragen der Gentechnik addressed the marketing of products containing or consisting of GMOs only to the extent that other legislation did not provide relevant procedures. Concerning liability, the assumption of the act was that any damage would be attributable to the genetic manipulation unless it could be shown that it was "probably" due to another feature of the organism—a response to the liability problem that clearly fell short of the burden-of-proof idea suggested by the Öko-Institutes. Liability was limited to a maximum of 160 million marks.

The 1990 law constituted an attempt fix the meaning of genetic engineering as a technology that can be safely handled by articulating a policy narrative whose intertextuality combined a variety of elements from different discourses. The law continued the framing of genetic engineering as a key technology of the future and an integral element of the German politics of modernization. At the same time, it defined the precautionary regulation of genetic engineering as compatible with its industrial exploitation. In this construction of the political space of biotechnology, the boundary between science/industry and society had become more permeable as regulatory decisionmaking had shifted back from the laboratory to the (regulatory) commission level and to other sites open for public participation. In other words, the public had been re-articulated into the regulatory system as an element of scrutiny with respect to risk issues. At the same time, this "public" or "society" remained black-boxed as the deserving beneficiary of the fruits of biotechnology. Disagreements on the ethical, social, or economic implications of biotechnology did not find sites of expression in this topography of genetic engineering.

Britain: Genetic Engineering Regulation and the Environmental Protection Act

In the second half of the 1980s, as genetic engineering policy underwent serious crisis in Germany, in France and Britain too the manipulation of genes resurfaced as a topic of political conflict to be "settled" in the regulatory field. There, as in Germany, genetic engineering emerged as a regulatory problem against the backdrop of a dramatically changed discursive constellation.

The Greening of British Politics

In the United Kingdom, it was in the early 1970s that the terms "environment" and "environmental" began to be used to refer to an important area of policy and to be taken up in the general election manifestoes of the main political parties (Robinson 1992, p. 7). Increased media attention to environmental issues went along with increasing activities by established environmental groups and by a variety of recently formed environmental groups. On the institutional level, this new discourse was reflected by the establishment of the Department of the Environment in 1970; in the same year, the Royal Commission on Environmental Pollution, a permanent advisory body to the government on environmental issues was established. In 1974 the Health and Safety Commission was set up. On the parliamentary level, the Select Committee on the Environment was organized in 1979 (Robinson 1992, p. 12).

Nevertheless, throughout the 1970s environmental policy was pigeonholed into the traditional sectoral patterns of the political agenda, and environmental law was split among a number of statutes, many of them covering material having little to do with environmental protection. The emphasis at this stage was on health and safety matters, an emphasis clearly reflected by the fact that responsibility for genetic engineering was given to the Health and Safety Executive. The trend in environmental regulation was toward reactive methods of solving environmental problems and setting standards for the discharges of emissions (Ball and Bell 1994, p. 11).

However, during the 1980s—in particular, between 1983 and 1987—the political imaginary began to change as environmental topics began to move increasingly to the center of the general political discourse. On the level of party politics, the environment became established at annual party conferences as an area of policy, and the main parties published a series of policy documents and discussion papers with the environment as their theme. And there were important new themes in the environmental discourse: First, there was an emphasis on integrating environmental policies with other policy areas. This put an end to the conceptualization of the environment as a single issue. Second, new ideas for environmental policymaking were developed in policy papers and manifestoes of the major parties, such as those that called for environmental audits or for the formation of a new Department of the Environment that would be more responsive to environmental issues. Parties gradually adopted environmental agendas that were proactive rather than reactive (Robinson 1992, pp. 26–27). This proactive approach was translated into directions in environmental law that shifted away from the more traditional reactive methods of solving environmental problems toward methods that were intended to prevent harm by planning and interlinking environmental legislation more deliberately (Ball and Bell 1994, pp. 11–12). However, in contrast to the German environmental policy narrative, the British approach focused on emissions. What mattered for environmental policymaking was the quality of the environment, not the quantity of emissions. The basic assumption, so the story went, was that the environment could bear a certain burden of pollution without suffering negative side effects. A corollary of this position was that environmental regulations cannot be justified unless negative effects on the environment could to be scientifically demonstrated. The latter was very much in contrast to the German precautionary principle, which justified preemptive regulations even in the absence of scientific proof of their necessity (Hértier et al. 1994, pp. 82–83).

Significant pressure on the Labour and Conservative parties and on the Liberal–Social Democrat alliance soon developed in the form of the radicalization of the environmental movement and the increasing electoral success of the Green Party, particularly in local and European Parliament elections. The potential strength of the "green vote" in the UK became dramatically clear in the elections for the European Parliament on June 16, 1989, in which the Green Party received 14.9 percent of the votes. However, the peculiarities of the British electoral system created a natural barrier to the success of the Greens in the national elections. The majority-vote "seat" system worked against small parties (Robinson 1992, p. 210).

The "absence" of a "green" party at the national level of policymaking was, to a considerable extent, responsible for the fact that environmental groups and movements had to channel their demands through the existing political parties. It was also responsible for the pattern of direct contacts between activists and government, which reflected to a certain extent practices of sectoralization, clientelism, consultation, and institutionalization of compromise. And the democratic metanarrative emphasized the political practices of accommodation, negotiation, and consensus, in contrast to imposed solutions (Jordan and Richardson 1982).

In his survey of 1980s environmental regulations, David Vogel (1986) sees what he calls the British style of policymaking in the absence of statutory standards: minimal use of prosecution, flexible enforcement strategy, considerable administrative discretion, decentralized implementation, close collaboration between regulators, and regulations and restrictions on the ability of non-industry constituents to participate in the regulatory process (typically taking the form of corporatist arrangements). However, the process Vogel had described changed in several significant ways in the ten years after his book was published. The British process of environmental regulation became more open, centralized, legalistic, and contentious. These changes must be partially attributed to the new discourse on the environment, which stressed the importance of a more coherent approach to environmental protection. The decentralization of environmental decision-making over a wide range of bodies, significant use of delegation, and geographical decentralization had led, over time, to a rather incoherent environmental policy with very little uniformity across the country. Even such central activities as the monitoring of the environment had been conducted in an uncoordinated way. Changes in this approach were evident in the establishment of a national body for the previously regionally coordinated regulation of water control in 1989. In a related way, there have been clear attempts to reduce administrative discretion, as exemplified by the 1991 Water Resource Act's replacement of local discretion with central prescription for water pollution. One implication of this trend toward centralizing decisions and reducing discretion is that a greater number of decisions may potentially be challenged through judicial review. This might lead to a reversal of the situation in which courts traditionally played a lesser role in the development of environmental policy in Britain than in Germany or the United States.

These domestic trends were further reinforced by Britain's membership in the European Community and by the increasing importance and number of EC directives relating to the environment. In the absence of significant input from the British local and regional bodies, these EC directives contributed to the centralizing tendencies in environmental regulation. Similarly, the imposition of EC standards contributed to the replacement of local discretion with central prescription. And the primacy of EC law had been used to develop various doctrines aimed not only at ensuring that legislation to implement EC directives would be passed but also at implementing and enforcing such legislation. This development could potentially contribute to the likelihood of courts' playing an enhanced role in environmental regulation. These significant changes in political discourse soon began to affect the conceptualization of the genetic engineering problematic.

New Challenges of Regulatory Decisionmaking

For one thing, the new environmental discourse had allowed a number of new actors to position themselves in the evolving struggle over the definition of the genetic engineering problematic. As was discussed in chapter 5, starting in the second half of the 1980s a variety of movements across Europe began to reread the meaning of biotechnology by propagating a new critical discourse that unfolded along several interrelated trajectories and created new configurations of meaning among a variety of topics and themes in political discourse. Such central notions of regulatory discourse as risk assessment, the implementation of regulations, and technology assessment began to shift in their meanings and were translated into a new regulatory narrative. Julie Hill, the director of Britain' s Green Alliance, put it this way:

> We have to try to assess risks. However, there is a major problem in making an assessment of risk in that we have very limited knowledge of the way in which our environment works. . . . If we do not know, how can we predict the effect of releasing novel organisms into it? In other words, do we have a basis for regulating at all? . . . All that means is that we need to know more about the way the environment works and we need to know it rather urgently. It is obvious that a greater proportion of research money should go into that area, that the science of ecology has very much a poor relation to the science of genetic manipulation, and it has not received nearly the funding that the latter has. (Hill 1990, p. 48)

Besides emphasizing more ecological research, the critics' evolving counter-narrative demanded a number of other important reforms, such as strict enforcement of regulations, tighter scrutiny and monitoring, a consent-based system of regulation, strict liability for originators and producers of GMOs, and greater access to information and public participation.[44] But public participation did not mean only participation in regulatory issues. By "greater public participation," critics meant more public control and involvement in all aspects of biotechnology, from decisions concerning types of research to those concerning the ethical and political implications of genetic manipulation. Hence, the Green Alliance and the UK Genetics Forum asked for the establishment of a Public Biotechnology Commission:

> The technology raises wide ranging issues. There is no mechanism with which to look at these from the point of the technology. They may be broken up into pieces by the Department of Environment or the Royal Commission, but there is no coherent central way in which they can be tackled together. What we want therefore is an official, government sponsored forum to be able to looks at these interrelated issues. . . . What we want is a government sponsored committee which would able to study all issues raised in biotechnology and to comment on those directly to ministers as well as to the public.[45]

What we find in this request is quite typical of the evolving critical discourse: suggestions to depart from a purely regulatory perspective on genetic engineering and to base regulatory decisions on broader political considerations. It is this context that we must understand the UK Genetics Forum's demand for a partial moratorium on the release of genetically modified organisms, according to which deliberate releases should be allowed only for research that enhances understanding of the behavior of genetically modified organisms in the natural ecosystem and not for commercial applications.[46]

Around the same time, in 1987, other initiatives critical of biotechnology were introduced. The London Food Commission (a consumer organization of the Greater London City Council), Compassion in World Farming (an animal protection organization), and several other organizations began a campaign against bovine somatropin (rBST). The campaign emphasized not only the potential hazards for consumers of milk from cows treated with rBST and the potential hazards to the animals themselves but also the impact of rBST on the agricultural employment structure. In the critics' counter-narrative, a new aspect was to be considered in the process of introducing new drugs: the "fourth hurdle," meaning the need to provide an assessment of the socio-economic and environmental impacts of the drugs as a precondition for their licensing.[47]

Besides the critics' efforts to reframe the genetic engineering problematic along the lines of a narrative that would transcend the established regulatory rationality, a number of other events and developments contributed to the perception that genetic engineering had run into a crisis of signification. In this context, two things that should be mentioned are an outbreak of salmonella from infected eggs in 1989 and an outbreak of bovine spongiform encephalopathy (a deadly brain disease) that by 1990 affected about 10,000 cows. These outbreaks raised the possibility of a human version of BSE and the possibility of transmission from animals to humans. Administrators began to make connections among genetic engineering, BSE, and salmonella. Asked about his perception of the public opinion on genetic technology, one civil servant responded:

> I think that the public at the moment is not day to day concerned about it, but they know so little about it. . . . In particular with food we had a number of scares, we had salmonella. . . . We had BSE in cattle. . . . People are aware that inevitably risks are taken in industry in the interest of profits and getting things done efficiently and a lot of these risks are not fully evaluated because you can't know. BSE is a classical example; we didn't think that cattle would be infected by whatever causes scabies, we relaxed rules on animal feeds and we have now a major diseases problem. I think there is a cultural problem now to be more suspicious of change, not to assume that every change is for the better. . . . When people go to the shops they need food, they can do it without a car, but you need everyday food, so they are much more sensitive to alterations in the food chain, . . . and because genetic modification is something which is so new and most people have so little understanding of

what it means, as soon as a scientist tries to explain it, they get lost after the third word, embattled. There is a resistance, a caution, if biotechnology is to get widespread acceptance in this country, somehow we have got to get over this lack of knowledge so that people can feel that they are contributing to the judgment of what we are going to eat.[48]

Another administrator described public opposition against genetic engineering this way:

I wouldn't say that all ministries are under such pressure as they are in Germany, for example, but who is to say, that may change tomorrow. It had developed here a lot in the last two or three years; definitely, the consumer voice has definitely grown. . . . The general public is concerned now; at first it may have been a group of consumer activists. What we have to do is to learn the lessons . . . carry the public with us.[49]

Another important event in development of a new regulatory space in Britain was the Thirteenth Report of the Royal Commission on Environmental Pollution (RCEP), titled "The Release of Genetically Engineered Organisms to the Environment."[50] The Royal Commission is an independent standing body that serves the Queen, Parliament, the government, and the public on matters of environmental pollution. Established in 1970, it takes a broad view of what is meant by "pollution" and its reports tend to have considerable influence on policymaking.[51] The Royal Commission is free to choose the topics for its reports, and it usually selects topics to which it feels it can make a substantial contribution. Its sixteen members are appointed by the government; half are academics and the others are from law, business, agriculture, and other fields. The typical procedure for the Royal Commission—making an announcement when it has chosen a particular topic and then inviting people to give evidence—is complemented by the gathering of written evidence, site visits, and public hearings. The procedure takes up to two years, and according to established practice the government respond to a report from the RCEP.[52] In 1986 that the RCEP announced that it would begin a study of the deliberate release of genetically modified organisms.[53] The RCEP's report on the deliberate release of genetically modified organisms into the environment was a crucial step in the shaping of a new hegemonic constellation in British regulation of genetic engineering.

Reconsidering the Regulation of Genetic Engineering

During the second half of the 1980s the regulatory situation in Britain had run into a number of difficulties. In 1984 the Genetic Manipulation Advisory Committee (GMAG) was disbanded and succeeded by the Advisory Committee on Genetic Manipulation (ACGM). The ACGM differed from GMAG insofar as its tripartite structure, with representatives of employers, employees, and "specialists" lacked representatives from the ""general public." The ACGM consisted of a chairman, five employer representatives, five employee representatives, eight specialists, and the Health and Safety Executive Secretariat. Nearly all its members were scientists.

At the core of the given regulations was a requirement to notify the Health and Safety Executive and/or the ACGM of activities involving genetic modification (although, as was pointed out in chapter 3, this requirement remained within the framework of a voluntary system of advice and control as given by the GMAG, and later the ACGM). Within this framework, the ACGM regularly issued guidelines either covering completely new areas or modifying the GMAG's guidelines from the 1970s and the 1980s. Until the "Genetic Manipulation Regulations of 1989" went into force (see below), the ACGM had issued nine guidelines.[54]

In Britain, as in Germany and France, the perception grew stronger that the existing regulatory "business as usual" framework was not sufficient to deal with the complexities of genetic engineering in the 1980s. The Genetic Manipulations Regulations of 1989 attempted to deal with some of the problems mentioned above. The new regulations defined "genetic manipulation" to cover direct insertion of heritable material as well as insertion via a vector system. In the case of a planned release into the environment, the definition was further extended to cover in vitro techniques that would enable the construction of organisms with novel combinations of genes. Centers engaging in genetic modification work were required to notify the Health and Safety Executive. For activities in containment categories 3 and 4 or in large-scale categories 2 or 3, specific notification was required; activities belonging to containment category 1 or 2 or requiring the use of good large-scale practice were subject to notification in a retrospective annual review. All intentional introductions into the environment required individual notification. Notice of intention to start genetic manipulation work and certain high-risk projects had to be given at least 30 days in advance; notice of a deliberate release into the environment had to be given 90 days in advance.[55]

The ACGM continued to update the GMAG guidance notes on the basis of the new regulations, a task that was completed with the issuing of three new guidelines in 1990.[56] The revised guidelines for deliberate releases showed a change in the terminology from "planned release" to "intentional introduction" and also gave new importance to the Intentional Introduction Sub-Committee (IISC) of the ACGM, a sub-committee created in 1986 and composed of one representative of employers, one representative of employees, and ten "specialists" from a variety of fields, including molecular biology, ecology, and medicine.[57]

Hence, the regulation of genetic engineering continued to be crafted under the Health and Safety at Work Act until 1990. But by 1987, when the revisions of the 1978 guidelines were initiated, it was becoming increasingly clear that this could be only a temporary solution. The chief weaknesses of the HASAWA were that it was designed to protect the health and safety of people in the workplace and that its duty-based approach to regulation had come under more and more local and international pressure. The criteria for classifying organisms into hazard groups made no reference whatsoever to risks for animals or plants. It was in the nature of the HASAWA to leave matters solely related to ecological effects, to protection of the environment, or to animal welfare aside unless an impact on human health could be established. This approach was evident in the way the Advisory Committee on Genetic Manipulation framed hazards related to large-scale use of genetically modified organisms in its 1987 guidelines:

> When genetic manipulation techniques were first introduced there was concern that they could give rise to potential hazards. After more than a decade of experience these hazards have remained conjectural and not based on incident. There are no health hazards specific to genetic manipulation although hazards other than those associated with non-manipulated microorganisms may be envisaged. . . . Relative to the construction of genetically modified organisms, there is nothing intrinsically more hazardous about their large-scale use.[58]

This framing was changed significantly three years later in a revised version of the guidelines:

> In arriving at a Good Large Scale Practice (GLSP) designation for work involving Genetically Manipulated Organisms (GMOs), it is essential that environmental considerations are taken into full account. It is an inherent feature of large-scale work, especially in GLSP, that microorganisms will be released incidentally at various stages of the fermentation process and at early stages of down stream processing. It is therefore important that in any large-scale use of genetically manipulated organisms, an environmental assessment be carried out before work commences.[59]

Whereas the 1987 guidelines focused on human health hazards related to large-scale use and pointed out the small likelihood of such hazards, the 1990 revisions focused on the potential impact on the ecosystem, including effects on plants, animals, and microorganisms. But the environmental considerations were still merely additions to the patchwork of existing regulations, and the other guidelines retained their focus on human health. Furthermore, keeping a strong emphasis on self-regulation, the 1989 regulations put the onus on the notifier to undertake risk assessments and draw attention to potential risks. This duty-based approach differed from a consent-based approach in that it required formal consent by the competent authority for contained use or deliberate release of genetically modified organisms. In practice, the existing system of notification turned out to be a consent system: the Health and Safety Executive had to be notified 90 days in advance of any planned release into the environment (Clifton 1990, p. 13). However, a system in which an advisory committee can react only after receiving notice of a deliberate release is qualitatively different from a system that requires prior authorization by an advisory committee.

These limitations of the 1989 regulations were also obvious to the HSE administrators. In part, the decision to continue with the 1974 regulation strategy was intended to be a strategic contribution to the evolving regulatory debate at the European Community level; in part, it was due to the perception of a developing regulatory vacuum that called for some sort of regulatory reaction. The former head of the HSE's Microbiology and Biotechnology Policy branch remembers:

> By . . . November 1989, the European Community's Directives were well advanced so we recognized that these regulations would only be a stop gap measure. But given the time scales of regulation, we decided to press on. Otherwise it would have been unlikely that any legal obligation to notify proposed releases to the environment would have been in place until 1991. The 1978 regulations were, in any case, legally defective and the new regulations incorporated some important thinking we had succeeded getting into the European Community directives. Therefore, we wished to establish this thinking in regulations in order to consolidate it. (Clifton 1990, p. 13)

The United Kingdom consistently resisted the EC's evolving regulatory directives, which differed greatly from the established HSE 1978 system in many respects. The 1989 UK regulations, which reemphasized the British position, were intended to demonstrate the success of a regulatory system based on self-regulation and an enforcement system based on negotiation. At the same time, it had become clear during EC negotiations in Brussels that the European directives would not be modeled on British approach. But the preparation of changes in the

British regulatory approach to genetic engineering was also closely related to a number of other developments.

According to the new environmental discourse, modernization and environmental precaution were increasingly defined as mutually reinforcing principles. This discourse had stabilized in the emergence of a number of powerful new actors in British environmental politics, such as the Royal Commission on Environmental Pollution and the Department of the Environment (both of which were about to become central players in the shaping of recombinant DNA regulations). And the challenge from social movements, political organizations, and consumer groups had outlined an approach to genetic engineering that went beyond demands for regulatory reform and called for the establishment of new principles of decisionmaking for biotechnology policy. As has already been discussed, the 1989 Royal Commission report was a critical move in the renegotiation of genetic engineering regulation in Germany.

The Royal Commission report made a set of comprehensive recommendations that were to be crucial in the political process (Clifton 1990, pp. 92–96):

- that statutory controls should require a license for any release of a genetically engineered organism to be issued jointly by the Secretary of State of the Environment and the Health and Safety Commission
- that it be made an offense to release a GMO without having first obtained a license or to fail to comply with any of the conditions attached to the license
- that licenses be issued on a case-by-case basis, with each proposal assessed by a committee of experts.
- that the Intentional Introduction Sub-Committee of the Advisory Committee on Genetic Manipulation be made a committee in its own right, distinct from the ACGM, and be given the task of advising both the Health and Safety Commission and the Department of the Environment
- that each release of a GMO to the environment require a license, and that the organism might then be proposed for use as or in a particular product after being assessed once more by the licensing authorities
- that the Department of the Environment and the Health and Safety Commission compile and maintain a public register of persons authorized to release GMOs

• that it might occasionally be appropriate to require registration of releases of registered products, in addition to requiring registration of trial releases

• that there be a public register of applications for release licenses and of licenses granted. Persons or organizations applying for licenses to releases GMOs should be required to place advertisements in the local press serving the areas of intended releases. Those applying for a license for the sale or supply of a GMO as or in a product should be required to place a notice in the *London Gazette* and in a national newspaper. Members of the public should have the opportunity to respond to the licensing authorities concerning any application for a release license

• that the Secretary of State for the Environment be given broad powers to ensure the implementation of the new regulations in such areas as waste disposal, the drafting of codes of practice, the handling of emergencies, the imposition of obligations on other authorities to establish emergency arrangements, and monitoring

• that any person held responsible for having carried out a deliberate release without the necessary license and registration be subject to strict liability for any damages arising from the release.

Thus, the policy narrative developed by the Royal Commission in consultation with a variety of individuals and groups—including the Green Alliance and the UK Genetics Forum (Clifton 1990, pp. 104–105)—outlined a regulatory strategy sustained by principles of what I have called socio-ecological rationality: although it did not support a moratorium on releases, the idea of a Public Biotechnology Commission, or strict liability for any release trials, the Royal Commission clearly favored a consent-based system of regulation, controls for all stages of a release (including marketing), comprehensive enforcement provisions, substantial public information, and provisions for public participation in the licensing process—all key demands of critical groups.

The Royal Commission's intervention marked the entry into the evolving regulatory struggle of an important actor in the institutional framework of environmental policy. The Department of the Environment had already, In 1988, emerged as another potentially important actor, and had added a second layer to the regulatory system by setting up an Interim Advisory Committee on Introductions, thereby creating a regulatory overlap with ACGM's subcommittee on introductions. However, the Royal Commission's strong support of the Department of the Environment's substantial involvement in the regulatory process created an important discursive alliance, and it became obvious that any new regulations on genetic engineering would entail involvement of the DoE. This constellation of mutual reinforcement of the narratives of Department of the Environment and the Royal Commission was a critical element in the discursive construction of the risks of genetic engineering.

The Struggle Over the Environmental Protection Act

Shortly before the publication of the Royal Commission's report, the Department of the Environment produced its own framing of the regulatory problematic in a consultation paper.[60] Claiming its stake in the future regulatory framework, the DoE proposed to "bring together the adaptability and economy of the Duty Based Approach with the administrative clarity of the Consent Based Approach."[61] However, the scope of the proposed legislation was not clear from the consultation paper. The paper argued that new legislation should provide exemptions for areas covered by existing legislation but did not make clear whether the new legislation should apply where no other product controls applied. In contrast, the Royal Commission had proposed extending legislation to final products, including those covered by other statues. The DoE's proposal, though it suggested that a register be available for public inspection, had little to say about the specifics of such a register; also, it did not mention public participation. Overall, the consultation paper went beyond the requirements of the existing 1989 regulations but was clearly less far-reaching than the recommendations of the Royal Commission.

Not surprisingly, the main features of the DoE's proposal met with widespread acceptance from industry. However, in various forums, industry representatives expressed ambivalence about the basis of the proposed legislation, which represented genetically modified organisms as presenting exceptional risks. But in the meantime it had become broadly accepted that public acceptance was necessary and could be secured by precautionary regulatory measures (Levidow and Tait 1992, p. 95). On the other hand, the UK Genetics Forum found many faults in the DoE paper, including the limited scope of the proposed legislation, apart from the omission of the UK Genetics Forum's two main proposals with respect to deliberate releases (imposing a partial moratorium on commercially motivated releases and narrowing regulation to purely technical issues, thus leaving out social and ethical issues).[62]

Nonetheless, what had happened was critical for the creation of a new regulatory space in Britain. The Department of the Environment's consultation paper was written in preparation for Part VI of the 1990 Environmental Protection Act, a major undertaking in British environmental law. The Environmental Protection Act reflected a new trend in environmental legal discourse that concentrated the law in a smaller number of acts focusing on sets of interrelated problems. The Environmental Protection Act focused on air pollution from stationary sources, waste management and disposal, integrated control of the most potentially polluting processes, litter, noise, statutory control of environmental nuisances, and the environmental impact of genetically modified organisms. Hence, the regulation of genetic engineering had been framed as a problem for the management of environmental pollution. The Environmental Protection Act was similar to the 1991 Water Resources Act (which included laws on water pollution and water resources) and to the 1990 Town and Country Planning Act (which including the relevant statutory laws on town and country planning and tree protection). None of these acts functions as a full code for the relevant subject matter, and much of the detailed law in any area is actually found in statutory instruments and in a wide range of other documents operating under different acts (Ball and Bell 1994, p. 11). In fact, in the preparation of the Environmental Protection Act, the government repeatedly emphasized that the act provided only enabling or primary legislation, the implementation of which would be clarified in secondary legislation.

Initially, most members of Parliament saw Part VI of the Environmental Protection Act as a largely technical issue that gave legislative force to an existing voluntary system and felt that its details could be left to the experts. But in due course the politicization of genetic engineering continued in Parliament, where deliberations on Part VI became the central site for the discursive struggle over the meaning of the genetic engineering problematic. In the House of Commons opposition members added amendments influenced by the Green Alliance and the UK Genetics Forum, and in the House of Lords similar amendments were taken up by two Labour peers who were members of the Royal Commission on Environmental Pollution. As a result of these pressures, the government introduced two major amendments to the bill: one to provide statutory backing for the newly formed Advisory Committee on Release to the Environment[63] and the other referring to the disclosure of information and guaranteeing even greater and earlier access to information than stipulated by the European Community's directives (Levidow and Tait 1992, pp. 96–97).

The Environmental Protection Act itself started out by defining the purpose of Part VI as "preventing or minimizing any damage to the environment which may arise from the escape or release from human control of genetically modified organisms."[64] The act defined "genetically modified organism" relatively broadly, including the products of "mutation-inducing agents" and not explicitly excluding any type of cell fusion. The Confederation of British Industry lobbied for amendments that would exclude mutagenesis and certain types of cell fusion as well as Good Industrial Large-Scale Practice organisms—partly because the inclusion of GILSP organisms indicated that all GMOs must be seen as potentially dangerous, a position that industry thoroughly rejected. However, the chairman of the Advisory Committee on Release to the Environment argued: "The UK position to date is that it is better to cast the net too wide at this stage than to be too lax. . . . [This] indicates the problems that will constantly arise in relation to the role of public opinion. . . . "[65]

Also contested were the meanings of "damage" and "harm" with regard to GMOs. "Damage to the environment" was defined by the Environmental Protection Act as "caused by the presence in the environment of genetically modified organisms which have (or of a single such organism which has) escaped or been released from a person's control and are (or is) capable of causing harm to the living organisms supported by the environment." "Harm" was defined as "harm to the health of humans or other living organisms or other interference with the ecological systems of which they form part and, in the case of man, includes offense caused to any of his senses or harm his property"[66] Industry complained that this definition of harm would discourage biotechnology investment; the government argued that because regulation must be anticipatory in every respect, and that a precautionary definition of environmental damage was necessary even if it might stigmatize GMOs as suspected pollutants (Levidow and Tait 1992, p. 98).

With respect to a consent-based or duty-based system of regulation, the debate in Parliament recast the diverging interpretations that developed in the report of the Royal Commission on Environmental Pollution and in the consultation paper of the Department of the Environment. The Environmental Protection Act formalized the de facto voluntary system of consent whereby agreements on the appropriate conditions of releases were reached with the advisory committee. At the same time, the act allowed for cases where consent was not required. The Green Alliance demanded that consent be required for every release. The government, however, retained the bill's original call for "a more flexible approach," although the Environment Minister guaranteed that subsequent regulations would require consent for every release.[67]

Another important discursive struggle evolved around the framing of risk in the Environmental Protection Act. Not surprisingly, the bill and the final act prohibited a range of activities that posed a risk to the environment. Industry proposed amending the characterization of some risks as "unacceptable" or "unreasonable," arguing that all activities entail risks. For environmental protection in the UK, a definitional approach to the risk question has been established by the "best practicable means" (BPM) principle. Enshrined in the 1874 Alkali Act, this principle assumed economic as well as technical criteria. The Royal Commission on Environmental Pollution's Twelfth Report (1988) extended the BPM principle to the "best practicable environmental option" (BPEO) principle, asserting that this extension would enhance the social and environmental legitimization of regulatory decisionmaking. Partly in response to critics of the BPM principle and partly to reemphasize scientific and economic criteria, British environmental regulation introduced the principle of "best available technique not entailing excessive cost" (BATNEEC), which is generally viewed as less proactive or precautionary than BPM and which today is also used in many of the European Community's documents on environmental protection. After a protracted controversy, the Environmental Protection Act extended the BATNEEC principle to all GMO releases. Since the BPM principle has come to be seen as implying an ambiguous balance between risk and cost, the implication of the introduction of BATNEEC into genetic engineering regulation may have farreaching effects. Levidow and Tait (1992, pp. 100–101) argue as follows:

> Given that different protagonists may selectively emphasize either best technology or costs, BATNEEC offers a "more contested and participatory approach," possibly challenging Britain's traditional collegiate and discretionary approach to risk regulation. . . . Although cost limitations arising from BATNEEC may preclude a truly precautionary approach, the disputes over defining risk and the duty of care illustrate a qualitative difficulty inherent in legislating against unidentified hazards, prior to any consensus on what precaution might avoid them.

The aforementioned controversies over liability continued in Parliament; however, the Environmental Protection Act followed the approach, proposed in the Department of the Environment consultation paper, that compensation should be given in the case of a violation of regulatory procedures resulting in environmental damage, but that there was no liability in a case where damage occurred but no violations against regulatory procedures were committed.[68]

Finally, the government rejected the demands of the UK Genetics Forum and the Green Alliance for a Public Biotechnology Commission. Instead, it announced, it would establish a quango-like[69] body with power to intervene in the affairs of the Secretary of State, who is answerable to Parliament and to the people. The government was concerned that a Public Biotechnology Commission would have the potential to obstruct technological progress; it also insisted on the need to separate ethical from environmental questions (Levidow and Tait 1992, pp. 102–103). Obviously, these points made by the government reflect a framing of the recombinant DNA problematic that attempted a partitioning of the political space of genetic engineering—a partitioning guided by principles of drawing boundaries between "ethics" and "the environment" or between "technological progress" and "regulatory issues," but also guided by a particular image of what constitutes democratic decisionmaking.

France: The Unpolitics of Genetic Engineering Regulation

Neither France nor the United Kingdom experienced as much political mobilization against genetic engineering as Germany. But, as the above examples show, this did not mean that there was no debate on genetic engineering in France in the second half of the 1980s. Also, there were strong indications that the French state did not intend to "settle" the evolving controversy by relying on its "strength." Just as in Germany, a policy narrative began to take shape which emphasized the importance of finding a sensitive political response to the rising concerns about genetic engineering.

The Fifth Republic was characterized by a strong concentration of power in the executive branch, which was reflected in such structural features as the government's power to legislate in certain areas by means of proposals and decrees, the weakness of the legislature, and the limited powers of the judiciary (Wilsford 1988, p. 135). However, this "strength" was not a given fact; it was a product of multifaceted processes of enrollment and the stabilization of networks. The "weakness" of the "strong" French state had always been its vulnerability to direct action. By dealing high-handedly with its opponents, the French state tended to cut them off from "normal" politics—a tactic that invited political confrontations, as in the case of nuclear power (Wilsford 1988, p. 152). But in the case of genetic engineering there seemed to be a strong determination among state administrators to reconsider the pre-established frames of political action. This change in state discourse was complicated by the fragmentation and the complexity of the French political-administrative system, with its conflicting policy objectives, its ideological competition, and its internal divisions (Mazey 1986). Institutionalized conflicts, poor vertical communication, and horizontal rivalries among the grands corps[70] forced the executive branch to form cooperative informal alliances with constituencies, rather than trying to exert administrative control by formal means. The disadvantage of such a political structure is the potential for stalemate and gridlock in decisionmaking. As we shall see, this was to a certain extent the case with genetic engineering, a highly complex field that involved a range of conflicting actors and interests. The competing strategies of key administrative actors and the reluctance to adopt regulatory measures "against" such constituencies as scientists and industry led to the establishment of a politics of genetic engineering strongly influenced by political developments in Germany and in the European Community. To an extent, this politics would also function as a substitute for national decisionmaking.

Despite the fact that France did not pass any legislation comparable to the German Gesetz zur Regelung der Gentechnik, closer inspection reveals that the French politics of genetic engineering was marked by a substantial reconfiguration of the dominant policy narrative.

The Crisis of France's Regulatory Minimalism

The reconsideration of the dominant French discourse of genetic engineering was driven by a number of events and developments, including changes in the environmental policy narrative, the evolving critical counter-narrative on genetic engineering, the public uproar that followed an unauthorized deliberate release of GMOs in Dijon in 1987, and the mounting insecurity in industry concerning the future of the French regulatory system for biotechnology.

In the second half of the 1980s, French policymakers attempted to counteract what they perceived as a rising tide of public mistrust of genetic engineering. As in Germany, this change in policy discourse was driven not by conviction but by a realization that the matrix of discursive relations had changed. One French policymaker, remembering the situation when the debate over changing the recombinant DNA regulatory structures established in the 1970s began, noted: "This was a deformation of democracy, in my view, but regulation was necessary, we had to satisfy the public opinion."[71] Another policymaker said that French government officials had learned from experience and had recognized the importance of public input into the process of environmentally sensitive decisionmaking: "The French have a very positive attitude toward new technologies. . . . Opposition against new technologies, such as against nuclear power, must be solely explained by the fact that the technocrats developed it in secrecy and it was this that caused the reactions against nuclear power."[72]

The "feeling" of negative public opinion corresponded to survey results that did not show a significant gap in opinions between, say, the German public and the French. This mounting public mistrust of genetic engineering was certainly not helped by the events surrounding the deliberate release of a GMO into the environment, conducted within the framework of the Institut National de la Recherche Scientifique, a highly respected institution that was of central importance for agricultural research in France. In March of 1987, at the INRA's soil microbiology laboratory in Dijon, a group of researchers under the direction of Noëlle Amarger conducted a deliberate-release trial with genetically modified rhizobium bacteria. The trial, which had not been submitted for prior review to either the Commission Nationale de Classement des Recombinaisions Génétiques in Vitro (set up in 1975 at the Ministère de la Recherche or the Commission du Génie Biomoléculaire (set up in 1986), was brought to the attention of the French public by Benedict Härlin, a German representative of the "Rainbow Group" in the European Parliament. In a press conference Härlin stated: "This constitutes a serious incident. The dangers of such an experiment are difficult to evaluate. And today it has not been proven that the modified bacteria cannot transmit their resistance genes to bacteria which are pathogenic for humans."[73]

Härlin's press conference inspired numerous editorials questioning the absence of binding rules for deliberate-release trials.[74] The INRA's main line of defense was to state that no laws had been violated. But this argument only made painfully clear the absence of any binding regulations in France concerning genetic engineering. Hence, the

INRA had opened up the black box of France's self-regulating system of genetic engineering work.[75] Not much more persuasive was the INRA's second line of defense: its argument that a dossier on the experiment had been submitted to the "Commission Biomoléculaire" after the trial had been started. The reason put forth for the late submission was that the commission had had no meetings scheduled before May 18, and that for technical reasons the experiment had to be started before that date.[76] In any case, the damage had been done, and genetic engineering came to be increasingly compared to the nuclear power controversy in the public discourse. An editorial in *Nouvel Observateur* stated:

> Nuclear power had its Chernobyl, the chemical industry had Bhopal, and soon biology will have a name for its future disaster. . . . Just as in the case of nuclear power, it might be only a matter of time until genetic engineering will encounter public rebellion. As long as the bacteria were only manipulated inside the laboratory, only the specialists were concerned. But now the bacteria have left the laboratory. . . . The representatives as well as the judges serving the citizens should have every interest in further devoting themselves to the mysteries of genetic engineering. This might help to avoid long and bureaucratic economic-legal battles.

Apart from the Dijon incident and the related public debate, regulatory developments abroad, particularly in Germany (see above) and in the European Community (see below), had created a feeling of "regulatory insecurity" in industrial circles. It had become increasingly clear that regulatory developments in Germany and at the EC level would affect the French regulatory situation. A representative of the French Ministry of Industry described the situation: "One of the crucial problems for the biotechnology industry is the potential risks involved in genetic engineering. At this stage the regulatory situation is in the fog, one discusses and one discusses . . . but the industrialists who are prepared to invest need to know what happens next. . . . "[77]

But, as in Germany, policymakers from the Ministry of Research and the Ministry of Industry, research organizations, and industry interest groups (particularly ORGANIBIO, the Organisation Nationale Interprofessionelle des Bio-industries) pursued a "business as usual" strategy, according to which the appropriate policy response to the rising crisis in signification consisted in the development of a regulatory system that would avoid legislation and would secure "flexibility" in the "Asilomar tradition." However, the narrative continued, though the dominant mode of state action in the field of recombinant DNA regulation had once been, in effect, non-action, a more proactive approach was now necessary. Yet such an approach should not be "too painful"—that is, too costly for industry and for research organizations.

As has already been mentioned, in the mid 1980s the Commission Nationale de Classement des Recombinaisions Génétiques in Vitro, set up by the Ministry of Research, functioned as one element of the French regulatory system. But, as was shown in chapter 3, that commission needed the voluntary cooperation of those conducting experiments. Such cooperation was not required of industry (either for research or for production) or of state research institutions.

In November of 1986, the Commission du Génie Biomoléculaire was set up within the Ministry of Agriculture. Its function was to give opinions on all questions involving hazards and concerning the use of genetically modified products, and particularly on the risks associated with deliberate releases of genetically modified organisms. The secretariat of the commission was under the jurisdiction of the Ministry of Agriculture, which also had full responsibility for choosing the commission's members.[78] Of its fifteen members, there were ten scientists, two representatives of industry, a lawyer, a representative of consumer interests, and a labor representative. Again, the legal status of this commission was mainly advisory. The establishment of this commission within the Ministry of Agriculture, an important source of funding for biotechnology research in France, corresponded structurally to the establishment of the Commission Nationale de Classement des Recombinaisions Génétiques in Vitro within the Ministry of Research. This combination of regulation and support in one institutional setting, however, was not interpreted as creating problems of credibility. One document from the Ministry of the Agriculture stated: "The role of the Commission du Génie Biomoléculaire is to accompany an important development in technological progress and thereby to assure the security of the consumers and of the population in general as well as to protect the environment."[79]

This interpretation of the function of the Commission Biomoléculaire corresponded fully to its presentation by the French Minister of Agriculture, François Guillaume, who in February 1987, after a long description of the ministry's strong interest and support for biotechnology, reasoned as follows: "It is crucial to see that our country adopts efficient and just rules which guarantee the security of the consumers, but at the same time preserve the competitiveness of our industry versus our competitors. The Commission du Génie Biomoléculaire must be very much attentive to this aspect and I count on the Commission to find the right way."[80]

Another important function of the commission was to symbolize the state's involvement in biotechnology regulation. As its president stated: "Modern biotechnology is resented as violating the laws of nature and is often difficult to understand and hence disliked by the public. It is crucial to answer to this often anxious mood for ethical and economic reasons. . . . Public anxiety, even if considered to be irrational, is an objective element which must be taken seriously. . . . Besides this, the success of biotechnology products clearly necessitates their acceptability by the public, which means the acceptance of newly used methods."[81]

The Commission du Génie Biomoléculaire reviewed some ten proposals in 1987, about fifteen in 1988 and 1989, and some forty in 1990.[82] Although it is not clear if there were more deliberate releases during this time period than were reviewed by the commission, France appears to have led Europe in deliberate-release trials, and the only comparable number is that for the United States in the period 1987–1990.[83] The commission operated without public participation, for deliberate-release trials did not require any procedures for providing access to information or soliciting such participation. Again, this system of regulation was similar to the German system before the passage of the Gesetz zur Regelung der Gentechnik.

This system of regulation was quite different from what had turned out to be another pillar of the French biotechnology regulation system: the 1976 law regulating installations that required registration for environmental reasons.[84] This environmental law provided for an extensive process of approval for installations that might have an impact on the environment—a process that included public information on planned projects. Comparable to the German Bundesimmissionsschutzgesetz, the 1976 French law was one of the cornerstones of the newly emerging integrated approach to pollution control. Also, in France environmental issues had come to occupy a critical position in the political metanarrative. The French approach to environmental policy was, like the German approach, characterized by a precautionary approach and by the "polluter pays" principle. Genetic engineering was inscribed into this piece of environmental legislation in 1985 and 1986 by decree, and the Ministry of the Environment put out a newsletter stating explicitly that installations where microorganisms and genetically modified microorganisms were handled, including industrial installations and research institutes, fell under the application of the law.[85] This discursive reconfiguration also established the Ministry of the Environment as an important player in the field of biotechnology regulation, which so far had been dominated by the Ministry of Research and the Ministry of Agriculture.

But at the same time a different regulation strategy was launched by the Ministry of Industry. This strategy focused on standardization as a substitute for legal regulations. ORGANIBIO (the French biotechnology association) suggested that AFNOR (the Association Française de Normalisation—the French standardization commission, which operated in a framework headed by the Ministry of Industry) develop "guides of good practice" for production and research in biotechnology. This strategy, intended as an alternative to legislative action, was pursued at the international level by the Organization for Economic Cooperation and Development. AFNOR's director for biotechnology described the association's approach to biotechnology regulation as follows:

> Standardization does not block innovation, but it allows the researchers to know the rules of the craft and the habits of the market. . . . Standardization takes place within a coherent framework that takes into account the points of view of the different concerned groups. . . . And the procedures guarantee that standardization develops much faster than regulation. This is a crucial advantage for companies, because what needs to be standardized is up to the scientists and industrialists. . . . Today many people talk about "deregulation." It is therefore urgent to define our national strategy and to prepare our dossiers to take into account the evolving directives of the European Community and of the OECD.[86]

In AFNOR's and ORGANIBIO's policy narrative, standardization was a "middle-of-the-road strategy" between "irresponsible abstinence and unnecessary constraints."[87] At the core of this strategy, however, was essentially the old idea of biotechnology regulation as a form of self-regulation configured around the principle of a technological definition of risk. For example, one AFNOR publication asserts: "The security of industrial or agricultural recombinant DNA work is based on a sense of responsibility of the researchers and their concern to define acceptable risks. Guides of Good Practice are an expression of the specialists in the field of recombinant DNA research and application."[88]

The specification of this narrative was left to a variety of AFNOR commissions and subcommittees, a series of guides such topics as research with genetically modified organisms and on deliberate-release trials and large-scale industrial production, which, in sum, were designed as substitute measures for stringent regulations and as contributions to France's increasingly internationalized biotechnology regulation strategy. Specification of this narrative was left to a variety of AFNOR commission and subcommittees. The guides these working groups produced, which addressed large-scale genetic engineering, deliberate releases of genetically altered organisms, and other topics, were seen as substitutes for legal regulations. It was also hoped that the standardization of biotechnology à la française would establish a model for other countries and, in particular, for the emerging EC directives on genetic engineering

Another Ministry of Industry initiative was to set up a Groupe Interministeriel des Produits Chimiques to bring together representatives from a variety of ministries to discuss regulatory issues affecting the chemical industry.[98] Although this group was officially charged with coordinating activities among the ministries, its main function was to safeguard the chemical industry from "overregulation."[90] In 1987, it began to study the regulatory situation in the field of recombinant DNA regulation. After a survey of the various regulatory commissions and initiatives launched since the mid 1980s, it came to the following unsurprising conclusion: " . . . France has a regulatory 'puzzle' which is in need of restructuring in order to create an efficient system with respect to the economic development of biotechnology and which assures appropriate security for humans and the environment."[91] In the same report, the Groupe Interministeriel des Produits Chimiques proposed that one pillar of the new system be the Commission de Génie Génetique. That commission was set up in 1989 to replace the old Commission Nationale de Classement des recombinaisons génétiques in vitro (originally established in 1975). The newly established Commission de Génie Génetique was attached to the Ministère de la Recherche et la Technologie, but it now had nineteen members, nine of them nominated by other ministries.[92] This indicated a certain "opening up," especially since the old commission had been exclusively composed of scientists. According to GIPC, the new commission would be responsible for classifying any genetically modified organisms used in research, development, or production. This was a far-reaching proposal, for it implied that even work with class 1 microorganisms would require notifying the commission—a step justified by GIPC as being necessary to appease public opinion and to comply with similar regulatory tendencies in the emerging EC legislation.[93]

The second pillar of the new regulatory system would be the Commission du Génie Biomoleculaire, which would have the primary responsibility of scrutinizing the risks associated with products involving genetic engineering, such as organisms used in deliberate release, production of transgenic plants and animals, and environmental applications of the new biotechnology. This extension of the tasks of the Commission du Génie Biomoleculaire would also require an extension of its membership to include representatives of such institutions as the Ministry of the Environment in order to reflect the new agenda.[94] Furthermore, just as classification of recombinant DNA work by the Commission de Génie Génetique would become obligatory, the submission of planned deliberate releases and other high-risk experiments to an oversight commission would be required by law. AFNOR would continue to develop its Guides of Good Practice, which in the future could become a valuable element within the new regulatory framework. The GIPC document concluded: "It seems that all in all the proposals developed in this report are compatible with the currently negotiated EC directives on the protection of workers, contained use and deliberate release."[95]

The GIPC's policy narrative was a watershed in the French regulatory approach to genetic engineering. For the first time, a coherent system of regulation with binding legal impact was proposed. Furthermore, the Ministry of the Environment was acknowledged as an important actor in the newly designed system of regulation. Again, as in the case of the AFNOR standardization strategy, the GIPC report outlined a political strategy designed to serve as a model for the evolving European Community directives. In the course of the GIPC's research and discussions it had become clear that the EC's regulatory framework would go beyond the adoption of a system of standardizations. This was reflected in the GIPC's proposal for binding legislation. At the same time, the GIPC's policy proposals remained relatively vague and did not incorporate important elements of the planned EC directives, such as provisions for public information and participation. Unlike the German and British governments, the French government did not pass any legally binding regulations for biotechnology. This refusal to pass such regulations probably reflected a situation in which the government had to refrain from any impositions on resistant constituencies, particularly since a regulatory decision was soon expected on the EC level. Despite the apparent absence of policymaking activities, France was undergoing a transition from the Asilomar regime of signification to a new regulatory narrative that had been in preparation since the mid 1980s.

Conclusions

In this chapter I have discussed how policymaking responds to a hegemonic crisis—a fundamental questioning of its premises, world-views, rationalities, and reality definitions. In the wake of the second phase of the recombinant DNA controversy of the 1980s, the risks of recombinant DNA technology reemerged as a controversial topic and as a site for the social conflict over genetic engineering. This conflict was not limited to risk; it also included a reconsideration of the very desirability of many of the practices and potentials of genetic engineering.

After Asilomar, in the United States and in Europe, the regulation of genetic engineering had been inscribed as a problem that was best handled through a system of self-regulation. In practice, this meant that those engaged in genetic engineering would police themselves. The answers to major regulatory questions were sought in the disciplines that had initially raised the regulatory questions. From the early 1980s on, genetic engineering expanded from a laboratory setting to large-scale production sites. Other new areas included field trials of genetically modified plants and other organisms and the sale of biotechnology products on the markets. To cope with such different developments, a disjointed system of regulation evolved in Germany, France, and Britain. That system was increasingly perceived as beset by "regulatory gaps" and inconsistencies. But these regulatory loopholes were not the "cause" of the emergence of a new regulatory discourse. Nor was the development of strong opposition to genetic engineering or the emergence of the "ecological modernization discourse" the "reason" why new regulatory approaches developed. The breaching of molecular biologists' self-enclosed domain of expertise was the result of a situation in which a number of mutually reinforcing events and developments had created a new discursive constellation, which was critical for the shaping of new modes of perception in which genetic engineering's risks were visualized by a variety of actors. Furthermore, the changed discursive constellation had given rise to new actors and institutions, and they got a voice in the regulatory debates. Hence, resistance to genetic engineering and regulatory gaps were intertwined with processes of co-articulation within a more complex discursive configuration. Other important elements in this discursive field were events such as the deliberate-release scandal in Dijon, the outbreak of BSE in Britain, the "rediscovery" of the history of eugenics in Germany, and the German Greens' focus on genetic engineering.

The robustness of the established frameworks of meaning defining the risks of genetic engineering underwent a tough endurance test. Ecology's discussion of the deliberate release of genetically modified organisms, the Dijon "test" of the French voluntary system of recombinant DNA regulation, the reasoning of the German administrative court in the Hoechst case, and the arguments of the Royal Commission on Environmental Pollution increasingly raised questions about the ability of the old regulatory narratives to fix the meaning of genetic engineering and the boundaries of the political space of biotechnology. This confluence of events led to a hegemonic crisis in which the legitimacy of the existing regulatory system was challenged and in which new actors' positions, arguments, meanings, and identities were being articulated and institutionalized. The response to this crisis was the articulation of a new hegemony through a restructuring of the political space of biotechnology. The regulation of the risks of genetic engineering became the site where the attempt to bring stability and order to the increasingly messy field of genetic engineering policy took place.

A number of structural similarities in the risk-regulation policy responses in Germany, France, and Britain have been considered. The evolving policy narratives intermediated between the changed contexts of regulatory policymaking and the transformed discursive constellation by establishing a new system of regulation that, at its core, translated a precautionary approach to genetic engineering as an integral element of biotechnology industrial policy and, at the same time, marginalized more radical positions (as articulated, for example, by the German Greens or the UK Genetics Forum) from the field of social positivity. This demonstrates that policymaking attempts to establish hegemonic reality definitions not by eliminating opposition but by re-absorbing discourses of polarity into a system of "legitimate differences" and by defining the locations where differences can be articulated. The demands of the critics were not rejected; they were partially absorbed and thus transformed from statements of antagonism into "legitimate differences." The Gesetz zur Regelung der Gentechnik and the British Environmental Protection Act document this strategy of splitting the resistance by separating an "acceptable" critique from forms of resistance that could be removed from the field of social positivity and (potentially) vilified as expressions of irrationalism.

Despite ongoing deliberations about the drafting of European Community's genetic engineering regulations, the various discursive constellations in Germany, Britain, and in France had led to rather different regulatory responses. For example, in Germany and Britain the specific discursive constellation had helped the social movements to position themselves in a way that increased political pressure for the establishment of new genetic engineering legislation. By contrast, in France the absence of such pressure had somehow created less of a sense of an urgent need to pass recombinant DNA regulations. On the other hand, in Germany the highly antagonistic style of the genetic engineering controversy had led to substantial mistrust between the opponents of genetic engineering and the government. In Britain there was a much better base for negotiations between governmental agencies and critics. Such cultural differences referred to different meta-narratives about modes and traditions of state-society interactions and had substantial impacts on policy negotiations and passed regulations.

In all three countries, the attempt to stabilize a political crisis led to the creation of a new system to deal with the risks of genetic engineering. Complex new configurations were shaped around the master signifier of precautionary regulation—relationships that, in the United Kingdom, comprised of such disparate elements as the Royal Commission, the concept of integrated pollution control, BATNEEC, ecological models of the deliberate release of GMOs into the environment, and the Confederation of British Industry. By establishing equivalencies between the deliberate release of a GMO and many other important agents, such as British Industry and the Department of Environment, chains of significations or networks of meaning were shaped which attempted to redraw the boundaries of the political space of biotechnology by redefining the relationship among the state, science, industry, and society in the context of deliberate-release field trials. In such situations, the boundary between experts and non-experts had become more permeable and thus an object for negotiation.

However, as in the risk regulation of the 1970s, the rephrasing of the "genetic engineering risk" problematic in the second half of the 1980s was also a critical contribution to the expansion of biotechnology. Initially questioned on safety grounds, with the establishment of the new regulations such trials became legitimate scientific projects, provided certain information was given to the public and particular procedures were followed under the scrutiny of regulatory agencies. Consequently, the evolving recombinant DNA regulations in Britain, Germany, and France hardly curbed the expansion of genetic engineering. On the contrary, the new regulations contributed considerably to the stabilization of a policy field in crisis by facilitating a confluence of events that were stable and calculable. The limitation of liability provided by the Gesetz zur Regelung der Gentechnik, for example, constituted for German companies an asset that was the envy of many US companies and which certainly fostered investment.[96] The existence of commissions scrutinizing deliberate-release trials transformed projects that were potentially highly controversial on socio-economic grounds into problems of ecological risk assessment.

Finally, another critical aspect of the genetic engineering controversy of the 1980s was the construction of recombinant DNA regulation as the exclusive site for the political negotiation of genetic engineering. It was through this definitional work that certain topics (such as different definitions of "genetically modified organism") became legitimate concerns for the process of regulatory policymaking whereas other topics (such as the socio-economic implications of genetic technologies) could not legitimately be spoken of. Society continued to be black-boxed as greatly benefiting from genetic engineering, while industry and research were framed as locations of scientific and social progress. However, like all other hegemonic attempts, this particular partitioning of the political space of biotechnology remained contested, as was evident from the continuation of the recombinant DNA controversy into the 1990s.

7

Genetic Engineering, Identity Politics, and Poststructuralist Policy Analysis

Beginnings and endings are elusive phenomena. Narratives tend to begin and end *in medias res*—in the middle of things— presupposing as a future anterior some parts of itself outside itself (Martin 1986). The narrative of this book opened by tracking the politics of genetic engineering back to major socio-political transformations in US biology since the 1930s. It is not easy to argue for a convincing closure for my story. In chapter 6 I discussed how in Germany, France, and Britain an attempt was made to bring closure to the genetic engineering debate by constructing the field of risk regulation as a site for carrying out the social controversy over biotechnology. Shortly thereafter, a similar attempt was made at the level of the European Community, which passed legislation in the form of two directives regulating the contained use and the deliberate release of genetically modified organisms for all member states.[1] But genetic engineering did not disappear from the center of political interest at the level of the individual states or at the level of the European Community. Likewise, genetic engineering continued to be debated in the United States. Regulated under the so-called coordinated framework for recombinant DNA regulation, which attempted to avoid special legislation for genetic engineering and to deal with genetically modified organisms under existing laws, biotechnology and its new methods continued to raise many highly controversial questions (Auchincloss 1993, pp. 37–63). More recently, discussions of the patenting of genetically manipulated organisms and demands for a ban on cloning experiments involving humans have been prominent in political debates in the United States and in Europe.[2]

Fortunately, I did not start this book with the intention of offering a comprehensive or final analysis of the complicated relationship between genetic technologies and policymaking in Europe and in the

United States. Rather, I intended to discuss, from a poststructuralist perspective, several important episodes of genetic engineering policymaking in a number of countries as a case study for policymaking under postmodern conditions. Policymaking today is characterized by (among other things) increasing uncertainty concerning central dimensions of society and politics. Recourse to "grand narratives" and to belief guarantees such as science or ideology has become highly problematic. Not only are there rising doubts about the very possibility of completing the project of modernity, with its central assumption of a close correspondence between increasing rationalization and increases in freedom, autonomy, and happiness (Lyotard 1984; Smart 1992a, p. 219), but many previously accepted distinctions between sectors and fields such as nature and technology, health and disease, economics and politics, and the national and the international or supranational field have become increasingly difficult to maintain. Today processes of mapping science, politics, and society tend to be characterized by contestation and controversy. This development has important power implications.

Boundaries of phenomena such as animals, the human body, or society and their constitution as knowledge always depend on techniques of social control, and thus on mechanisms of power. This micropolitics of boundary construction goes on at a number of different sites. One central site of boundary construction is the institutions of the democratic state, which are often called upon to address problems and challenges involved in boundary work. As I noted at the start of this book, for example, there is a relatively short road between the genetic engineering of bovine growth hormones in the laboratory and this new technology's turning into a controversial policy issue. Today government agencies and parliaments have become prominent locations for clarifying the nature and the meaning of complicated scientific and technological issues. Genetic engineering is one among many such topics, including the state of the earth's ozone layer and the relationship between bovine spongiform encephalopathy (a fatal brain disease affecting cattle) and Creutzfeld-Jacob disease (a similar brain disease affecting humans).

The shift of attention in policymaking to questions of boundary work and meaning construction can be observed in a number of locations. It is, for example, not an accident that in the past 15 years the concept of difference seems to have displaced inequality as the central concern of political and social theory. The question of how to achieve equality while still recognizing difference has moved into the foreground not only of political reflection but also of political struggle. Since the early 1970s, new social movements have engaged in social conflicts about the prevailing codes that describe reality and the rules that define normality and subject identities. The movements have reconfigured democracy as not only the place for struggles to eliminate inequality but also the medium through which difference can establish space for itself. Many of the political issues of recent decades, including those of race and those of sexuality and gender, have addressed concerns of difference and have underscored the paramount importance of these concerns for the affected individuals and groups. But it would be mistaken to equate this "politics of difference" with "minority politics." A number of deep-going transformations, from the globalization of economy and culture to ongoing technological changes, have also raised topics of identity and difference. The issue of what it means to be "French" or "German" in the European Community, the debate about depletion of the ozone layer, and the impact of the dramatic changes in ultrasound technology on our concept of unborn life all involve boundary questions, such as how nation, environment, and person are constituted as knowledge and how they gain specific meanings in particular semiotic contexts. Processes of constructing the meanings of social, political, and economic phenomena are gaining more and more importance because they can no longer be settled by mobilizing "grand narratives" such as those of science and ideology. At the same time, policymakers and people affected by policies are increasingly understanding the importance of signs and codes in the politics of ordering and structuring life. Technology policymaking in general and the genetic engineering in particular are good examples of the importance of boundary work and the micropolitics of knowledge in the constitution of policy fields.

Genealogies of Policymaking

Genetic engineering and molecular biology certainly are technologies that are connected to strong economic interests. But, as I have attempted to show, it would be an oversimplification to portray the politics of regulating and supporting genetic engineering as expressing a conflict between economic interests and risk concerns or a "technology race" among the United States, Europe, and Japan. The story is much more complicated. In order to come to terms with the difficult cultural, economic, and social ramifications of genetic engineering politics, we need to reconsider the analytical tools and strategies of policy analysis.

From an epistemological standpoint, we need to acknowledge that economic and other interests cannot be conceived outside the discourses within which they are generated. It is for this reason that I have focused on reconstructing the complex discursive matrix within which genetic engineering policies originated. The analytic strategy I have pursued has traced the emergence of genes and their boundaries as a legitimate topic for policymaking to a number of sites and locations inside and outside the political realm. I have shown that practices that were concerned with the study and the manipulation of genes on the molecular level originated in a space together with other sets of practices that targeted problems such as the support and the regulation of science or the control of populations. I have called this space a regime of governability—a structure of interrelated sites or fields of practice where the boundaries of the objects and subjects of policymaking such as the risks of genetic engineering and biotechnology industry were constituted. It is at this intersection between various "political" and "nonpolitical" forces that operate in a regime of governability that policymaking comes into existence as an attempt to create structure and order in a discursive field.

One important implication of my analytic strategy was to conceptualize science and power as two strongly interconnected phenomena. Hence, I suggested that manifestations of genetic engineering such as the Human Genome Projects or genetically modified agricultural products are insufficiently explained as natural outcomes of scientific developments (such as the elucidation of the structure of DNA or advances in protein engineering). Rather than interpret practices and impacts of genetic engineering as inevitable results of the progress of science, we should view them as products of a variety of contingent-accidental events of micro-level power struggles that could have ended differently than they did (Foucault 1984, p. 76; Kusch 1991, p. 167).

The inseparable relationship connecting power, knowledge, and politics becomes clear when we look at the constitution of molecular biology as a policy field. Molecular biology played a central role in an ongoing and contested process of re-inscribing the frontier that constitutes life. In its discourse, properties of life, health, and disease were defined in terms of genetically directed macromolecules operating on a subcellular level. As we saw in chapter 2, this scientific discourse on life and other discourses such as modernization discourse contributed to the shaping of the objects, actors, and institutions of the policy field of molecular biology. The new definition of life in the discourse of molecular biology expressed a specific operation of power. Foucault has pointed to a significant historical transition contemporaneous with the emergence of industrial capitalism, in which a shift of emphasis occurred from the primacy of sovereignty, law, and coercion or force "to take life" to the emergence of new forms of power constitutive of life (Smart 1992b), p. 161). This power over life evolved in two forms: disciplining the body and regulating populations. Whereas the former had as its object the individual, the latter addressed itself explicitly to the "ensemble of the population" as a field of shaping and forging (Gordon 1991, p. 20). These two strategies constituted the two poles around which the power over life was organized (Foucault 1980, p. 139). When bio-politics was deployed throughout the population, state racism was born in many countries. The life of the species became a political issue and gave rise to such projects as eugenics (Simons 1995, p. 34). As I argued in chapter 3, eugenics—discredited by historical consequences—vanished as a state project. But there was a critical link of continuity between eugenics and molecular biology. The object of intervention had changed from the populations (the gene pool) to the subcellular level, but geneticization—the framing of central dimensions of humans in terms of genetic factors—continued to inform scientific practice. The genetic determinism and the eugenic politics of force had given way to the subtle cultural politics of molecular reductionism of the new biology, but the control of life processes on the genetic level remained the goal.

During the 1970s, with the development of recombinant DNA technology, the interventionist possibilities that had been inscribed into molecular biology from the very beginning turned into veritable devices for transforming the structure of life on the molecular level. It was at this point that molecular biology and its representations and interventions became controversial. Genetic engineering's risk emerged as a major topic for policymaking. But neither the "risks of genetic engineering" nor the actors and institutions involved in its handling were "out there." In complicated processes of boundary drawing, a technological definition of risk was developed on the basis of the representation of the recombinant DNA hazard as controlled by the technologies that produced their possibility. Furthermore, boundaries that related scientific expertise to politics and the public were constructed. Although in most of the countries discussed in this book the boundaries between experts and public were at least somewhat permeable for a few years, eventually the regulation of recombinant DNA became an expert enclosure. Since recombinant DNA's hazards were defined as a function of the availability of knowledge in molecular biology, the advancement of the field and the experience that had been accumulated were used to argue for deregulation. The normalization of genetic technologies in terms of the equivalence "genetic engineering = other, well-known technologies as used in biochemistry etc. = processes going on in nature" and the construction of regulation as a negation of modernization constituted a temporarily successful attempt to remove genetic engineering from the field of political conflict.

Research laboratories, field-test areas, hospitals, and farms became terrains for practices geared toward interventions on the subcellular level. The integration of these various sites of government was made possible by what I have called "discourse of deficiency" and the "myth of biotechnology industry." These narratives constructed the boundaries of phenomena such as "biotechnology industry" and explained the transformative practices of genetic engineering, such as deliberate releases, as essential to socio-economic development (Rheinberger 1993, pp. 9–10). In Europe, biotechnology policies began to establish a semantic relationship between the collective identity of Europe in general and of France, Germany, and Britain in particular and the project of rewriting life through the new genetic technologies. In this construction of a "bio-society," the identity of society was at once secured and modified by the practices of molecular biology. In these significations, the futures of France, Germany, and Britain became dependent on the expansion and diffusion of molecular biology.

Beginning in the early 1980s, this construction of collective identity and redefinition of what constitutes a human being became the subject of a multi-faceted political struggle carried out in a number of locations. In the mid 1980s critics mobilized against genetic engineering and its underlying cultural strategies. Furthermore, in the political debate the frames of ecological modernization and sustainable development gained importance. The weakening of the relational system defining the identities of the political space of genetic engineering gave rise to a sequence of attempts at hegemonic re-articulation. In Europe, the drafting of regulations for the contained use and the deliberate release of genetically modified organisms constituted the central site for the social negotiation of genetic engineering. This move constituted an attempt to manufacture a closure of the debate by re-articulating the identities of the critics of genetic engineering by selectively absorbing their demands (such for public participation) and by transforming these demands into positivities within the system. At the same time, more radical demands (such as for the banning of deliberate releases, for public participation in the decisionmaking on research support for genetic engineering) and the critique of genetic engineering's redefinition of central human dimensions had their legitimacy removed and were cast as undermining Europe's competitiveness. Whereas the regulation of genetic engineering had been articulated as an acceptable topic for public deliberation in political institutions, there was no corresponding political space within the polity in which to debate biotechnology's desirability, its acceptability, and its economic, social, political, and ethical implications. In other words, despite the fact that genetic engineering was in the center of public controversy, in many respects decisionmaking concerning its applications, its diffusion into research, medicine, industry, and agriculture, and its support by public and private resources was effectively removed from the scrutiny of democratic institutions and became a form of subpolitics (Beck 1993). This demonstrates that hegemonic articulation can dramatically narrow the available space for subject positions in a policy field and thus establish exclusionary patterns of decisionmaking. As I argue, the price of such a construction of the political space of biotechnology might be a high one.

Theoretical Implications for Policy Analysis

At the core of the poststructuralist approach used in this book is a theory of language that conceptualizes language not as a neutral medium of transmission but as a performative practice constitutive of the phenomena it seems to describe. This perspective has profound implications for the study of science and technology policy and for that of policy analysis in general; it also helps to overcome some of the shortcomings of conventional policy approaches that center on the concepts of actors, structures, or institutions. We cannot simply assume the "prediscursive existence" of entities such as institutions, "unified" subjects of policymaking, and political identities. There is, for example, no reason to deny the importance of the structure of the German parliament for the shaping of the 1990 Genetic Engineering Act. But how can we explain the strong involvement of the German parliament in the shaping of the legislation in the first place? What created the dynamics for a policymaking process that led to legislation? Here purely neo-institutionalist arguments seem insufficient. Thus, poststructuralist policy analysis pays strong attention not only to the organization of politics, but also to the politics of organization, not only to the actors of politics but also to the politics of actors—in other words, to the semantic struggles and discursive constructions that define who counts as an actor in a particular policy setting and who does not, and which institutions are legitimized and authorized to take part in the shaping or the implementation of policymaking and which are not. My argument here is not that actors don't have and pursue goals or that institutions don't matter in policymaking. Actors do things in politics, and institutions do shape policymaking. But these processes must be understood within the discourses where actors are constituted and institutions framed as relevant in a given policy field. These are extremely important analytic considerations that address dimensions of power widely ignored in conventional policy analysis. Power is not only articulated in interactions between actors, in institutional biases, or in ideologies. In addition, discourses, representations, scientific statements, and "public philosophies" are critical articulations of power that construct subjectivity and position individual or institutional actors in the socio-political field; thus, they deserve a prominent place in policy analysis.

The poststructuralist focus on language implies a reconceptualization of what constitutes the proper *domain* of policy analysis and what the objects of such analysis are. Since ideas, scientific theories, representations of certain political, economic, and social or scientific realities, and, in general, knowledge are seen as shaping the process of policymaking, policy analysis views the various fields where these communications originate as important sites of power which are in a relationship of exchange and interaction with other more traditional centers of power, such as parliaments or ministries.

This leads to an important redefinition of the *nature* of policymaking. I defined policymaking as situated at the intersection between forces and institutions deemed "political" and apparatuses that shape and manage individual conduct in relation to norms and objectives yet are constituted as "nonpolitical," such as science and education. Policies are often guided and informed by ideas, theories, and knowledge that originate outside the political realm. This implies that policy analysis needs to extend its focus of interest from what is traditionally considered to constitute a political phenomenon to other areas, such as laboratories or the complexities of scientific expertise. Here, for example, I am not arguing that everything that occurs in a genetics lab deserves the attention of policy analysis. But certainly, if we are to understand the regulation of genetic engineering or the controversy over nuclear power safety, we have to pay considerable attention to arguments originating in laboratories, to their histories, to their discursive constructions, and to their underlying strategic goals. I have shown how the discursive production of genetic engineering, its hazards, and its industrial potentials were central to the shaping of policies that regulated and supported biotechnology in Europe. This "micropolitics" of meaning—which originated in diverse sites, from laboratories to parliaments—was instrumental in creating, structuring, and prioritizing the range of possibilities considered open to the government agencies involved in the policymaking process. It was predicated on representations of nature, technology, medicine, economy, and society in scientific and political discourse. Furthermore, this micropolitics of meaning relied on policy narratives that established relationships among economic growth, molecular biology, recombinant DNA's hazards, technologies, collective identity, and international economic competitiveness.

This analytical perspective requires that policy analysis *take seriously* the knowledge base underlying many policies, from health policy to environmental policy. Public policy questions such as "What is responsible for the return of tuberculosis in many Western countries?" and "What causes global warming?" should not be black-boxed as issues of scientific opinion or controversy; they should be addressed in policy analysis as issues in which scientific arguments have become inseparable from expressions and structures of power. Today, from economic to social and agricultural policy, there is hardly a policy field whose operations and rationales are not in some way justified, explained, and legitimized by reference to knowledge-based arguments. This is not to say that power is causally related to knowledge. For example, Peter Haas (1992), in his approach to "epistemic community," conceptualizes the power of experts as based on their access to information. In this reading, knowledge and interests are understood as distinct phenomena. As a result, science is conceptualized as potentially transcending politics and thus helping to make politics more rational. Poststructuralist analysis is less optimistic about science as some sort of neutral judge in policy disputes. It underscores that today the legitimacy of political power relies increasingly on support by scientific or technical expertise. From this insight follows the important analytical suggestion to conceptualize power and knowledge as intertwined rather than causally related and to study precisely how knowledge operates as it makes up actors and as it guides, orients, and structures social and economic practices.

Another important theoretical tendency in poststructuralist policy analysis is to pay attention to the forms and figures of the language of politics, such as myths, tropes, and narratives. Of special significance in this context is the recovery of "narrative" from being mainly a phenomenon of fiction to being designated as a mode of organizing political, scientific, and economic reality. Narratives connect centers of domination (such as the state) with peripheral sources of power (such as laboratories). Policy narratives are constructed and used to make sense out of events and, at the same time, to position actors in a policy field. For policy analysis, this means that identifying a narrative's structure or constellation and elucidation the interplay among narratives, counter-narratives, and metanarratives in a policy field are important goals when it comes to explaining a policy dynamic. Such narrative constellations will vary from policy field to policy field and from country to country and must be reconstructed by careful analysis of the written and oral sources in a policy field. Of particular interest for policy analysis is why certain narratives become dominant or hegemonic at certain times in a policy field, whereas others remain marginal. I have described the success of a policy narrative as depending on its success in intermediating between discursive context (i.e. certain economic, political, or scientific realities) and discursive constellation (the other dominant narratives in a policy field). The analytical challenge that stems from this kind of analysis comes from balancing analytic sensitivity for "the discursive" without resorting to the naive argument that "everything is discourse," and from acknowledging the "realities" of politics without naively believing that there is a thing called "reality" that presents itself without the intervention of language.

The strong focus on language in poststructuralist policy analysis allows for the reconceptualization of one of the classical questions in social theory: How does social and political order become possible? If language only reflects reality, its contribution to the creation of order can only be marginal. However, if language is constitutive of politics, the picture changes. Policymaking then comes across as a performative process that uses and mobilizes complex systems of representations to fix the meaning of transient events (such as an economic recession). In doing so, it is possible to move them in space and time and make them susceptible to evaluation, calculation, and intervention. An event such as a recession enters a government document in the form of statistical data and can be used, for example, to raise taxes or to distribute resources for investment. The risks of genetic engineering are expressed in tables and categories, which become the basis for regulatory decisionmaking. This ordering activity of language is precisely what creates governability—what, for example, transforms the anarchic and multiple activities of an economy or the risks of genetic engineering into a body of knowledge and markers which create the basis and orientation for political intervention. Instead of assuming governability and practices of policymaking, one must see them as a question: How do things and phenomena become objects of government and policymaking? Such a question cannot be answered by pointing to actors, groups, or other social forces; it requires attention to representations, computations, and procedures of assessment. Policymaking, then, must be studied as a material practice that does not simply react to its environment but, rather, inscribes itself into the texture of the social and creates or rewrites order by drawing from a multitude of discursively available narratives and modes of representations—as a process that entails intermediation and translation between different social worlds and realms of perception. In this analytical perspective, the focus is on textual strategies of policymaking. Such strategies are articulated in memoranda, in programs, in expertise, and in speeches, which provide frameworks for the deployment of important resources or technologies of government (including calculations, statistical computations, procedures of assessment, and standardizations of evaluation practices). Textual manifestations of policymaking have long been of interest for political science analysis, but they used to be studied mainly to decipher the intentions and goals of government. Poststructuralist policy analysis, in contrast, looks at the texts of government not only as declarations of interest or as statements of strategy but also as material practices and as strategies to create order—as representations and interventions that actually shape politics.

Genetic Engineering, Democracy, and the Politics of Identity

What policy lessons can be drawn from this book? First, this book has emphasized the importance of narratives or stories for the perception of policy problems. Politics is an "empty space" until demarcated and partitioned by struggles of boundary drawing. Policy narratives and related political metanarratives serve as interpretations of the topography of a policy field. Certain narratives have a good chance of becoming hegemonic for a particular period of time and of defining the nature of a policy problem and the actors and institutions that should be involved in its handling. But there is also ample evidence that, under postmodern conditions, such definitions of political reality tend to be short-lived and rather quickly become objects of contestation and destabilization.

The enormous difficulty of developing broadly accepted policy solutions is certainly not restricted to genetic engineering policymaking; indeed, it seems to be prevalent today in many policy areas and in many countries. One might deplore this as indicating the increasing ungovernability of modern democracies, or one might decide that this political instability reflects the working of the postmetaphysical culture of our time—a culture that has given up the search for general ordering principles, laws of history, fixed notions of who holds rights, and ultimate foundations of knowledge (Daly 1994, pp. 180–181). In this new culture, the idea that there is one truth, which can be revealed by means of the "grand narratives" of science or ideology, is rejected in favor of the idea that society is an ongoing process of re-invention that should be able to draw on the plurality of the culturally available stories that deal with social and political problems. Business organizations are conceptualized not as monolithic structures but as texts written by a plurality of authors, ranging from consumers to producers and regulators (Linstead 1993, pp. 59–60; Boje et al. 1996). Democracy is not understood as an already-given narrative that must be "read" or revealed by certain "specialists" so that people may discover their objective interests and what they were really intended for. Instead, democracy is something "written" by the democratic subject, who is not a passive consumer of a given narrative but rather an unstable locus for identities, and who is responsible for the constitution of democracy (Daly 1994, p. 182). Nor is science an unquestionable authority whose narratives can settle policy conflicts. In this reading, democracy is configured as a Janus-faced phenomenon that can create the space for the expression of alter-identities or can be a medium by which the dogmatization of identity and the elimination of dissensus are politically legitimized (Connolly 1991, p. x).

One lesson that can be drawn from this interpretation of current political conflicts as reflecting today's postmetaphysical culture concerns the importance of tolerance for the multiplicity of the stories available in a policy field. Accepting the elusiveness of the search to define the "true" nature of a policy problem and coming up with a generally acceptable solution for it might be a crucial step in the direction of what ultimately could turn out to be a more efficient style of policymaking—a postmodern style that acknowledges modernity as a fundamentally incompletable project. But this does not imply that policymaking is impossible or that it cannot find points of orientation. Postmodern policymaking takes up themes and logics of the modern, but without recourse to a telos or an ultimate foundation (Daly 1994, p. 183). Critical elements of this style of policymaking are more tolerance for the multiple ways to tell the story of a policy problem and the openness to negotiate frames of meaning and political realities. Hence, democracy is understood not as an expression of a social totality but as a way of responding to politics and its many uncertainties and anxieties. It is a strategy to articulate and negotiate political issues and to keep them contestable, not to deny their controversial character or to vilify certain groups and their positions (Warren 1996, pp. 244–250).

There are a number of different techniques available to keep problems, practices, anxieties, and issues contestable in a democracy (Daly 1994, p. 266). One way is to create deliberative spaces throughout the institutions of state, economy, and civil society for the social negotiation of concerns and grievances. Various authors, including Ulrich Beck (1993) and Frank Fischer (1995),[3] have argued for the importance of institutionalizing new public spaces that facilitate communication among different semiotic universes dealing with alternative descriptions of the realities of policymaking. Certainly, such new institutional platforms for the public negotiation of policy problems would be useful improvements of the democratic process. Political groups such as the UK Genetics Forum have already proposed such deliberative institutions as a new strategy for dealing with controversial technologies (for example, the Biotechnology Commission proposed in the United Kingdom).

Deliberative spaces could also be created by reforming existing institutions—in particular, by empowering hitherto excluded groups and individuals to participate in the negotiation of policies. The diligence and competence displayed by the German Parliament and by the Royal Commission in the shaping of regulations for genetic engineering raises the questions why such political institutions should not be in a position to address other complex issues relating to genetic engineering, such as the desirability of certain product developments in the field of biotechnology or of certain practices of research support. The reason that the patterns of decisionmaking concerning genetic engineering discussed in this book took on features of a subpolitics of science and technology was not that, in principle, no institutional sites were available for the social negotiation of genetic engineering. Rather, what was often missing were public spaces within the confines of institutional politics that would allow for the articulation of alternative readings of genetic engineering. In other words, it was not so much a lack of institutions per se that impeded the better communication between opposing conceptualizations of policy problems as a lack empowerment of important groups and actors to engage in such a dispute between cognitive worlds.

Hence, democratic policymakers would be well advised to create spaces within existing institutions and in institutions yet to be designed to keep issues of great public interest, such as genetic engineering, politically contestable and negotiable. Failure to do so may not only create political alienation; it may also create excessive anxieties concerning the risks of genetic engineering or lead to consumer boycotts against products of the biotechnology industry. Such a public space of reflection and communication about genetic engineering would be characterized by more permeable boundaries between science and nonscience—between experts and non-experts. Science would *not* be conceptualized as some sort of meta-narrative for the solution of policy disputes. This more realistic understanding of science would probably make many processes of political decisionmaking more complicated and time consuming. But it would also liberate policymaking from mobilizing science into endless and delegitimizing cycles of expertise and counter-expertise. Without the burden of the "truth claim," and as one story next to others, the often elegant and elaborate narratives of science might acquire new value in the orientation of policymaking. Furthermore, the negotiation of these boundaries would be a public process with a specified place in the political-institutional framework.

In the case of genetic engineering politics, such political spaces are, for example, constructed through a variety of resistances against genetic engineering. These new political spaces—occupied by a variety of actors, including members of the Green Party in Germany, members of Génétique et Liberté in France, and socialist members of the European Parliament— were autonomous from the constructed boundaries of the state, science, industry, and society, extended beyond the state borders, and were not constrained by the state system. These resistances point to the fact that political identity tends to be a field of contestation. In the genetic engineering controversy, the critics' policy counter-narrative rejected the reduction of the genetic engineering problematic to case of regulatory policymaking. Their deconstructions re-inscribed genetic engineering as a practice of identity politics and as a new strategy of subjectivation, of defining what constitutes a human being. Furthermore, the critics outlined a model of a society whose identity should be based on a novel, cautious, and democratic mode of interaction between society and nature. If democracy is an imagination that introduces a logic of equality and self-determination, then it involves the interactive constitution of new identities through an invitation to complete this logic. The various resistances against genetic engineering have already begun to outline such a political space for the production of political alter-identities on the regional and national levels and at sites that transcend the territorial state. But certainly the creation of public spaces within the confines of institutional politics for the social negotiation of genetic engineering (on the regional, national, and supranational levels) would be an equally important expression of an interactive democratic logic.

Finally, policymakers should acknowledge that such standard adversarial procedures as voting and litigation have a place beside deliberative processes in which an irreducible pluralism of interests limits the capacities of deliberation (Churchill et al. 1991, p. 266). All these steps require a better understanding on the part of institutionally entrenched policymakers that we are living in a "writerly democracy" where any attempt to impose a narrative carries the seeds of its own deconstruction. Without such tolerance for dissenting voices, institutional reforms would be no more than an empty politics of symbolic democratization.

The postmodern style of policymaking would not only enhance democracy; it would also help to avoid policy failures. The comparison between the politics of genetic engineering and that of nuclear power points to the close relationship between processes of narrowing political spaces of contestability and failure of policy.

There was a time when nuclear power energy generation was framed as the cornerstone of the economic development of industrialized countries. That interpretation underwrote a particular political strategy for improving the economic situation of the nuclear power industry. During the late 1950s and the early 1960s, when reactor manufacturers and the US government realized that unanticipated cost increases were undermining initial attempts to sell nuclear power plants to utilities, they took two steps to promote their commercial acceptance and to cut costs. First, to gain economies of scale, reactor manufactures began to offer plants four and five times larger than those previously produced. Second, also driven by economic concerns, utilities began to build nuclear power plants closer to metropolitan centers, in order to reduce the costs of electricity transmission. Both decisions eventually led to the mounting public controversy in the 1960s, which was exacerbated by the Atomic Energy Commission's relaxed standards of licensing and safety. In addition, rising costs and capital requirements made it increasingly difficult for the utilities to generate capital. In the 1980s, when pension funds, insurance companies, and other institutional investors began to pull out of the electricity bond market, the US nuclear industry collapsed (Campbell 1988, pp. 52, 61, 101). The Three Mile Island incident was only a link in a chain of developments in the demise of the industry.

History does not repeat itself. The biotechnology industry is very different from the nuclear power industry, and so far genetic engineering has not had a Three Mile Island or a Chernobyl. But if my analysis is correct there is no reason to assume that genetic engineering's promises, projects, and products enjoy broad public acceptance. While policymakers preferred to assign to the "public" the role of a happy beneficiary of biotechnology developments instead of listening to the many voices of concern, consumers in many countries began to boycott products of the biotechnology industry. Though it is difficult for consumers to choose between electricity produced by nuclear power plants and electricity from other sources, most products and services of the biotechnology industry allow for choice. The considerable public mistrust of genetic engineering might create serious difficulties for the biotechnology industry and, by implication, question the rationality of its extensive support by the state. Moreover, there seems to be a serious mismatch between the initial promises of biotechnology and the products and services it has yielded so far. Some authors have related this gap to the theoretical limitations of reductionist molecular biology and the genetic research paradigm. Two of the many questions that arise in this context are "Why did the scientific voices that have developed a comprehensive critique of the dogmas of molecular biology not receive more attention in policy deliberation?" and "Why were the voices of consumer protest who warned so passionately about the dangers of genetically modified foods ignored in the policymaking process?" These questions need to be raised in the policy process. If they are, it will seem that more tolerance for dissenting policy stories about biotechnology and the creation of deliberative spaces that make genetic engineering not less but more contestable not only might improve the democratic quality of decisionmaking; it might even avoid currently unanticipated policy failures.

Notes

Chapter 1

1. For a general outline of this argument in the social sciences, see Smart 1982.

2. See, e.g., Evans et al. 1985; Nordlinger 1981.

3. For good overviews, see Cammack 1992; Hall and Taylor 1994.

4. See, e.g., Mayntz 1994; Hértier 1993, p. 16; Bressers et al. 1994, p. 6; Dowding 1995.

5. See also Dobbin 1994, pp. 5–10; Grafstein 1992.

6. See, e.g., Thelen and Steinmo 1992, p. 27.

7. For a systematic study of the interplay between paradigms and institutions in a country' s most important policy fields, see the chapters on economic and financial policy, industrial policy, social policy, legal policy, educational policy, environmental policy, energy policy, minority policy, agricultural policy, and technology policy in Dachs et al. 1991.

8. For a discussion of structuralism and Derrida's critique, see Frank 1989, pp. 20–33; Welsch 1996, pp. 245–255.

9. For important examples of the "linguistic turn" in political science, see Jobert and Muller 1987; Patzelt 1987a; Fischer and Forester 1993.

10. For key contributions to post-structuralist political theory, see Laclau and Mouffe 1985; Connolly 1991.

11. For pioneering work on the importance of boundary drawing in science politics, see Jasanoff 1990.

12. For important contributions to this perspective from organization sociology, see Cooper 1989; Reed and Hughes 1992; Hassard and Parker 1993.

13. The importance of "micropolitical practices" in policymaking has also been pointed out in ethnomethodology and in political science studies

influenced by ethnomethodological theory. See Battershill 1990, p. 164; Goffman 1974. For the applications of Goffman's approach and ethnomethodology in political science, see Fenno 1978; Patzelt 1987b; Patzelt 1993; Gottweis 1988. For a social constructivist view of the state, see Jessop 1990, p. 342.

14. See, e.g., Streeck and Schmitter 1985; Hollingworth et al. 1994.

15. For a most illuminating discussion of a "Foucauldian government analysis," see Rose and Miller 1992, p. 183.

16. See also Wells 1996.

17. For the definition of "frame" as used here, see Goffman 1974, pp. 10–11. I follow in my usage of the term "plot" the distinction made in literary theory between story, the preliterary sequence of events providing the writer with the raw material, and plot, the literary reordering of this sequence.

18. For a good discussion of the relationship between inscription and representation, see Rheinberger 1994.

Chapter 2

1. Eukaryotes are cells or organisms whose DNA is organized into chromosomes with a protein coat and which have cell nuclei with nuclear envelopes. All living organisms except viruses, bacteria, and blue algae are eukaryotic. Organisms whose genetic material is not sequestered in a well-defined nucleus (including bacteria and blue-green algae) are called prokaryotes.

2. On the history of molecular biology, see Kohler 1976; Yoxen 1982; Abir-Am 1982; Kay 1993.

3. Quoted on p. 45 of Kay 1993.

4. On the history of eugenics, see Adams 1990; Weingart et al. 1988; Kevles 1985; Ludmerer 1972.

5. On the continuity between eugenics and molecular biology, see Kay 1993, pp. 34–39; Proctor 1992; Weingart et al. 1988.

6. See, e.g., Jacob 1988, p. 231.

7. See Stiftung Volkswagenwerk, *Bericht 1965*. The Stiftung Volkswagenwerk (Volkswagen Works Foundation) was founded in 1961 by the Federal Republic of Germany and the state of Niedersachsen (Lower Saxony) as the result of a dispute concerning the property relationships of the Volkswagen company after 1945. In a state treaty, Volkswagen was transformed into an Aktiengesellschaft (i.e., a corporation), and 60 percent of its equity became the object of a public offering. The returns from this sale, about a billion marks, became the capital of the Stiftung Volkswagenwerk. The board of the foundation has fourteen members, seven of whom are nominated by the federal government and seven by the state of Niedersachsen. See Stiftung Volkswagenwerk, *1962–1987: 25 Jahre Stiftungsarbeit* (1987), pp. 11–12.

8. Interview, Volkswagen-Stiftung, Hannover, September 27, 1991. (On the interviews conducted for this book, see the appendix.)

9. Of course, science and technology received state attention early on. Whereas in the eighteenth and nineteenth centuries it was primarily educational policy that fostered scientific and technological development, the "age of science and policy" is closely linked to the two world wars. However, not until the nineteenth century did the education of scientists begin to be seen as calling for more systematic state intervention. Prussia's was the first national government to address the task of a science policy by investing enormous sums of money in modern forms of education, including technical and vocational training at the secondary level. Prussia also revitalized its universities through the establishment of new research facilities and the expansion of slots. Other European countries, including France and England, rallied behind and made efforts to catch up with Germany' s leading position in the advancement of technology and innovation. However, the French and English states were not successful in establishing research institutions comparable to those in Germany, nor were their industries on a level of development that allowed for substantial industrial research comparable with Germany. In most European countries, World War I created a situation of increased state intervention in the fields of science and technology. In Britain, a Department of Scientific and Industrial Research was set up after the war. In Germany, an "Emergency Association of German Science" was created to improve German science's position after the war. In France, however, it was not until 1939 that the Centre National de Recherche Scientifique was created, signaling the new agenda for state action in science and technology. After World War II, state intervention was qualitatively and quantitatively different. See Herman Ros, *The European Scientific Community* (Longman, 1986).

10. The socio-political importance of the identification of modernization and industrialization has been extensively explored in the work of Ulrich Beck. See Beck 1992, 1993.

11. See chapters 1 and 2 of Kay 1995.

12. I use "Fordism" in the sense of the "regulationist approach" in political economy. The key concepts of this approach are regime of accumulation and mode of regulation. Regime of accumulation means the whole set of regularities that allow a general and more or less consistent evolution of capital formation. "Mode of regulation" refers to the institutional ensemble and the complex of cultural habits and norms that secure capitalist reproduction as such. Fordism can be described in terms of an "ideal type" on different levels: e.g., as a distinctive type of labor process involving mass production based on assembly-line techniques, and as a stable mode of macroeconomic growth involving a virtuous circle of growth based on mass production, rising productivity based on economies of scale, rising incomes linked to productivity, increased mass demand due to rising wages, increased profits based on full utilization of capacity, and increased investment in improved mass production equipment and techniques. As a mode of social and economic regulation, Fordism involves the separation of ownership and control in large

corporations with distinctive multi-divisonal, decentralized organizations subject to central controls; monopoly pricing, union recognition, and collective bargaining; wages indexed to productivity growth and retail price inflation; and monetary emission and credit policies orientated to securing effective aggregate demand. See Nielsen 1991, p. 22; Jessop 1991, p. 136; Boyer 1991, p. 107; Leborgne and Lipietz 1988.

13. Quoted on p. 41 of Salomon 1991.

14. Organization for Economic Cooperation and Development), *Innovation Policy in France* (1986), p. 205.

15. Deutsche Forschungsgemeinschaft/Arwed H. Meyl, *Denkschrift zur Lage der Biologie* (Franz Steiner, 1958), p. 9.

16. Deutsche Forschungsgemeinschaft, *Stand und Rückstand der Forschung in Deutschland in den Naturwissenschaften und in den Ingenieurwissenschaften aufgrund einer Umfrage verfaßt von Dr. Richard Clausen* (Franz Steiner, 1964), p. 12.

17. Ibid., p. 37.

18. Even in 1967, the annual support money of the Stiftung Volkswagenwerk (DM 127 million) was not much less than the DFG's budget (DM 184 million). See Deutsche Forschungsgemeinschaft, *Aufgaben und Finanzierung III* (Franz Steiner, 1968), p. 23; Hirsch 1970, p. 159.

19. Stiftung Volkswagenwerk, *Bericht 1966,* pp. 44–45; *Bericht 1968,* p. 42.

20. Interview, Stiftung Volkswagenwerk, Hannover, September 27, 1991.

21. Stiftung Volkswagenwerk, *Bericht 1974,* p. 44.

22. Interview, Bundesministerium für Forschung und Technologie, Bonn, July 5, 1988.

Chapter 3

1. In this characterization of molecular biology and genetic engineering I follow Rheinberger 1993 (p. 8). See also Rheinberger 1992.

2. For the general argument, see Yoxen 1978.

3. See also Wright 1994, p. 194.

4. Interview with Paul Berg, May 17, 1975, box 7, Institute Archives and Special Collection, Recombinant DNA History Collection, Massachusetts Institute of Technology (hereafter cited as "MIT Recombinant DNA History Collection").

5. Interview with Syndey Brenner, May 21, 1975, MIT Recombinant DNA History Collection, box 26.

6. Report of the Working Party on the Experimental Manipulation of the Genetic Composition of Micro-Organisms (HMSO Cmnd. 5880, 1975).

7. "Les manipulations génétiques: des risques, des réalités, des fictions," *Le Monde,* September 18, 1974.

8. Ministère de la Recherche et de la Technologie, *Normes et sécurité pour les recombinaisons génétiques in vitro* (Paris: 1985), p. 24.

9. Interview with François Jacob, March 22, 1976, MIT Recombinant DNA History Collection, box 9.

10. Interview with Hans Zachau, August 29, 1977, MIT Recombinant DNA History Collection, box 15.

11. Ibid.

12. My analysis will focus primarily on the regulatory narratives developed at the Asilomar conference and in the NIH guidelines. The purpose of this interlude is not to give a political sociology of the development of recombinant DNA regulation in the US; this has been accomplished in an exemplatory way by Krimsky (1982) and Wright (1994). Rather, this interlude is limited to a close textual analysis in the strict sense of the phrase, an analysis which will be broadened into a political sociology of the recombinant DNA debate when we look at the European situation.

13. US Department of Health, Education, and Welfare). National Institutes of Health, Recombinant DNA Research Guidelines, *Federal Register,* July 7, 1976.

14. Ibid. p. 27902.

15. Ibid., p. 27903.

16. Ibid., p. 27902.

17. Ibid., p. 27904.

18. Ibid., p. 27911.

19. Ibid., p. 27912.

20. Ibid., p. 27908.

21. Ibid., p. 27908.

22. *E. coli* K12, one of the variants of *E. coli,* was first isolated in 1922 and was soon widely used in laboratories.

23. Ibid., p. 27912.

24. An autoclave is a pressurized, steam-heated device used for sterilization.

25. For an excellent exposition of the "logic" of the NIH guidelines, see pp. 184–191 of Krimsky 1982.

26. According to the "central dogma" of molecular biology, three types of processes are responsible for the inheritance of genetic information and for its conversion from one form to another: (1) Information is perpetuated by

replication, a double-stranded nucleic acid is duplicated to give identical copies. Information is expressed by a two stage process: (2) Transcription generates a single-stranded RNA identical in sequence with one of the strands of the duplex DNA. (3) Translation concerts the nucleotide sequence of the RNA into the sequence of amino acids comprising a protein. See Lewin 1990, pp. 109–113.

27. All three meetings were funded by the NIH. The Ascot meeting (the US-EMBO Workshop) was co-sponsored by NIH and EMBO. For an excellent discussion of these three meetings, see pp. 228–243 of Wright 1994.

28. US Department of Health, Education, and Welfare. 1978. National Institutes of Health, Recombinant DNA Research. Proposed Revised Guidelines, National Institues of Health, Environmental Impact Assessment of a Proposal to Release Revised NIH Guidelines for Research Involving Recombinant DNA Molecules, p. 33124.

29. Ibid., pp. 60080–60131.

30. See ibid., pp. 33042–33044.

31. This committee, known as COGENE, was an offspring of the International Committee of Scientific Unions.

32. US Department of Health, Education, and Welfare). 1979, 1980. Recombinant DNA Advisory Committee, Meeting, Recombinant DNA Research, Proposed Guidlines and Actions, Federal Register, Vol. 44, No. 232, Nov. 30, 1979, 69235, and Recombinant DNA Research, Actions under Guidelines, January 29, 1980, *Federal Register* 45, no. 20, 6732.

33. For a rich, in-depth study of this process, see pp. 256–311 and 337–405 of Wright 1994.

34. On the concept of the "obligatory passage point," see pp. 196–233 of Callon 1986.

35. On the concept of the "spokesperson," see p. 71 of Latour 1987.

36. Interview, Charles Weissmann, May 27, 1975, MIT Recombinant DNA History Collection, box 14.

37. Report of the EMBO Delegation that attended Asilomar (memo, no date), MIT Recombinant DNA History Collection.

38. Report on the First Meeting of the EMBO Standing Advisory Committee on Recombinant DNA, London, February 14 and 15, 1976 (minutes), MIT Recombinant DNA History Collection.

39. John Tooze, Emerging Attitudes and Policies in Europe (memo, no date), MIT Recombinant DNA History Collection.

40. Council of the European Communities, Council Recommendation of 30 June 1982 concerning the registration of work involving recombinant deoxyribonucleic acid (82/472/EEC).

41. Report of Robens Committee on Safety and Health at Work (1972, CMnd.5034), 28, paragraph 28, quoted in Drake and Wright 1983.

42. Report of Working Party on Experimental Manipulation of Genetic Composition of Micro-Organisms, p. 3.

43. Report of the Working Party on the Practice of Genetic Manipulation, Presented to Parliament by the Secretary of State for Education and Science by Command of Her Majesty, August 1976.

44. Report of Working Party 1976, p. 3.

45. Such, for example, was the case with Sidney Brenner, one of the key participants in the second Asilomar conference.

46. Ibid., p. 5.

47. Ibid., p. 15

48. Ibid., p. 4.

49. The term "intertext" is used to designate the complex ways in which a given text is related to other texts. Just as a sign only gets meaning in relationship to other signs, there are no texts apart from other texts.

50. Report of the Working Party 1976, p. 6.

51. First Report of the Genetic Manipulation Advisory Group. Presented to Parliament by the Secretary of State for Education and Science, May 1978, Cmnd 7215 (London: HMSO, 1978).

52. *Health and Safety at Work, Genetic Manipulation* (London: HMSO, 1978).

53. Interview, Association of Scientific, Technical, and Managerial Staffs (ASTMS), London, July 14, 1989.

54. *Health and Safety at Work.*

55. Ibid., p. 2.

56. Second Report of the Genetic Manipulation Advisory Group, Presented to Parliament by the Secretary for Education and Science (London: HMSO, Cmnd. 7785, 1979), p. 4.

57. GMAG Note 14, Revised Guidelines for the Categorisation of Recombinant DNA Experiments; Third Report of the Genetic Manipulation Advisory Group, Presented to Parliament by the Secretary of State for Education and Science (London: HMSO, Cmdn. 8665, 1982), p. 54.

58. Third Report of the Genetic Manipulation Advisory Group, Presented to Parliament by the Secretary of State for Education and Science (London: HMSO, Cmdn. 8665, 1982), p. 46.

59. Ibid., pp. 46–49.

60. Ibid., p. 3.

61. Interview, Deutsche Forschungsgemeinschaft, Bonn, September 1991.

62. Interview with Hans Zachau, August 29, 1977, MIT Recombinant DNA History Collection, box 15.

63. Interview, DFG.

64. Interview with Hans Zachau.

65. Antwort der Bundesregierung auf die Kleine Anfrage der Abgeordneten Duerr, Prinz zu Sayn-Wittenstein-Hohenstein, Spitzmueller und Genossen-Drucksache 8/880-Risken bei der Neukombination von Genen, Deutscher Bundestag, 8. Wahlperiode, Drucksache 8/924, September 22, 1977, p. 1.

66. For example, experiments to be submitted to the ZKBS for approval started in the initial draft with (in NIH terminology) experiments P3/B1. In the final guidelines they started with P2/B1.

67. Bundesminister für Forschung und Technologie, Bekanntmachung der Richtlinien zum Schutz vor Gefahren durch in-vitro neukombinierte Nukleinsäuren vom Februar 1978, Bonn.

68. "Richtlinien zum Schutz vor Gefahren durch die in-vitro neukombinierte Nukleinsäuren (Fassung vom 26. 7. 1979)," in Klingmüller 1980 (pp. 145–146).

69. For a discussion of deconstructivist approaches to the law, see Culler 1988.

70. Bekanntmachung der Richtlinien zum Schutz vor Gefahren durch in-vitro neukombinierte Nukleinsäuren vom Februar 1978, E. II. 8., 9.

71. Ibid., G. 17.

72. Ibid., G. 15.e.

73. Interview, ASTMS, London, July 14, 1989.

74. For reprints of the two drafts, see pp. 142–155 of Lukes and Scholz 1986.

75. Deutscher Bundestag, 9. Wahlperiode, Antwort der Bundesregierung auf die Anfrage de Abgeordneten Dr. Riesenhuber et al., Drucksache 9/418, Gesetzliche Regelungen auf dem Gebiet der Gen-Forschung (Gen-Technologie-Gesetz). Drucksache 9/682, July 21, 1981, 2.

76. Bericht über die zurückliegende Amtsperiode der Zentralen Kommission für die Biologische Sicherheit, ZKBS (September 29, 1981 to June 30, 1988), Berlin 1988, 4–5.

77. Interview, Umweltministerium Nordrhein-Westfalen, Düsseldorf, September 26, 1991.

78. Interview, Institut für Genetik, Universität Köln, September 24, 1991.

79. Bericht, ZKBS, 2.

80. Interview, François Jacob, March 22, 1976, MIT Recombinant DNA History Collection, box 9.

81. Interview, Philippe Kourilsky, March 20, 1976, MIT Recombinant DNA History Collection, box 9.

82. Ibid.

83. Interview, Kourilsky.

84. Rapport d'activité 1975–1976 de la Commission Nationale de classement. Reprinted in: Ministère de la Recherche et de la Technologie, Normes de Sécurité pour les recombinaisons génétiques in vitro (Paris, 1985), 25.

85. Tooze, "Emerging Attitudes and Policies in Europe."

86. See, e.g., Rapport d'activitée, 27.

87. Ibid., 26; Rapport au Secretariat d' état à la recherche sur le contrôle des expériences comportant des recombinaisons génétiques "in vitro" établi par un groupe de travail interministériel présidé par Monsieur Poignant, Conseiller d' Etat, Février 1978, 7.

88. Rapport au Secrétariat d' état.

89. Ibid., 9.

90. Rapport Royer, Mai 1981, La sécurité des applications industrielles des biotechnologies (Paris: La Documentation Française, 1981), 75.

91. Ibid., 74.

92. Ibid., 285. Interesting in this context is that the answers to the same question with respect to nuclear energy yielded almost opposite figures with respect to an increase or a decrease in support: 36 percent supported an increase in support, 16 percent a decrease.

93. Rapport Poignant, 1.

94. Ibid., 17.

95. Ibid., 22–25.

96. Rapport Royer, 88.

97. Ibid., 44.

98. Minstère de la Recherche et de la technologie, Normes de Sécurité pour les recombinaisons génétiques in vitro, Paris 1985.

99. On the concept of the "obligatory passage point," see pp. 205–206 of Callon 1986.

100. On the concept of the "spokesperson," see p. 71 of Latour 1987.

Chapter 4

1. For example, an economic crisis that brings high unemployment will have an impact on social and economic policy discourse. But the really interesting question is not whether the crisis had some impact on discourse, but how the crisis was coded, signified, and translated in policy narrative and if a hegemonic definition of a policy problem could be achieved.

2. An allele is one of several alternative forms of a gene occupying a given locus on a chromosome.

3. For a general discussion of this political strategy, see p. 207 of Connolly 1991.

4. "First-generation biotechnology" refers to beer brewing and similar activities, which goes back as far as the ancient Egyptians. "Second-generation biotechnology" describes disciplinary and industrial developments, going back to the time around World War I, which were based on systematic screening of microorganisms existing in the natural environment and their industrial exploitation and which led in the ensuing decades to the growth of the antibiotic industry, enzyme technologies, and attempts at continuous fermentation. See Bud 1993.

5. For overviews of the evolution of science and technology policy, see Ronayne 1984 and Herman 1986.

6. The political construction of the global high-tech race has not been studied enough. Traces of the "race theory" can be found in innumerable documents and public statements. For a good example of this rhetoric, see Narjes 1988.

7. See also Bio-Industry in Europe (unpublished manuscript, Club de Bruxelles, 1990).

8. For a critique of a structural explanation in the shaping of European technology policy, see Sandholtz 1992.

9. Commission of the European Community, DG XII, FAST, FAST Work Programme, Memo, Brussels, 1979.

10. Interview, DG XII, Brussels, February 25, 1991.

11. At the administrative core of the Commission of the European Community are "mini-ministries" called Directorates General. DG XII is the Directorate General for Science, Research, and Development.

12. Ibid.

13. Commission of the European Communities, DG XII, Directorate Biology, Radiation Protection and Medical Research, The Biomolecular Engineering Programme (BEP) of the European Communities. Implementation Activities and Preliminary Evaluation of Achievements, Brussels, 1986, 3.

14. Commission of the European Communities, COM (83) 672 final/2-ANNEX, Biotechnology in the Community (Communication from the Commission to the Council, Brussels, October 3, 1983), p. 33.

15. Interview, DG XII, February 25, 1991.

16. DG XII - FAST Programme, Biotechnology and the European Community. The Strategic Challenges. A discussion note on current action required (Internal Working Document by Mark Cantley, Riccardo Petrella, and Ken Sargent, Brussels, 1981), p. 7.

17. *Tomorrow's Bio-Society* (European File, Commission of the European Communities, Brussels, 1980), p. 2.

18. Commission of the European Communities, DG XII-FAST, The Social Dimensions of Biotechnology in the European Community (report of a multidisciplinary working group, Brussels, 1982), p. 5.

19. Ibid., pp. 5–6.

20. DG XII - FAST, The Strategic Challenges, p. 9.

21. Ibid., pp. 8–15.

22. Ibid., p. 22.

23. Commission des Communautés Européennes, *Europe 1995. Mutations Technologiques & Enjeux Sociaux. Rapport FAST* (Paris: Association Internationale Futuribles, 1983).

24. Interview, DG XII, February 25, 1991.

25. Biotechnology in the Community, pp. 38–51.

26. Ibid., pp. 52–74.

27. Ibid., p. 56.

28. Ibid., p. 69.

29. Quoted on p. 312 of Elizalde 1987.

30. Senior Advisory Group in Biotechnology, Community Policy for Biotechnology. Priorities and Actions, Brussels; Greenwood and Ronit 1991, pp. 480–481.

31. Commission of the European Communities, FAST Occasional Papers, Bio-Feedback. Contributions from the FAST Bio-Society Network toward a Community Strategy for European Biotechnology (Brussels 1983), p. 53.

32. Fifteen million ECUs, 1982–1986.

33. Fifty-five million ECUs, 1985–1989, plus a 20-million-ECU "revision" for 1987–1989.

34. See also United Nations Centre on Transnational Corporations, *Transnational Corporations in Biotechnology* (United Nations Publications, 1988); G. Steven Burill "Strategic Alliances in Biotechnology," *International Industrial Biotechnology* 9 (1989): 13–16.

35. Sandor L. Boyson, Market Revolution: the Explosion of Biotechnology Strategic Alliances (Paper prepared for the United Nations Center for Science and Technology for Development, September 14, 1989), pp. 36–41; "Accords: L'Années des grandes manoeuvres," Biofuture-Hors Série (1990).

36. Commission of the European Community, Evaluation of the Biomolecular Engineering Programme-BEP (1982–1986) and the Biotechnology Action Programe-BAP (1985–1989). Research Evaluation, Report No. 32 (Brussels 1988), pp. 2–3.

37. One hundred million ECUs, 1990–1993.

38. Eighty million ECUs, 1988–1993.

39. Twenty-five million ECUs, 1989–1993.

40. Entscheidung des Rates, vom 29. Juni 1990 zur Annahme eines spezifischen Programms für Forschung und technologische Entwicklung auf dem Gebiet des Gesundheitswesens: Analyse des menschlichen Genoms (1990–1991), July 26, 1990, Brussels.

41. Commission of the European Communities COM (88) 424 final—SYN 146, Proposal for a Council Decision adopting a specific research Programme in the field of health: Predictive Medicine: Human Genome Analysis (1989–1991), Brussels, July 20, 1988.

42. Commission des Communauteés Européennes, COM (92) 60 final—SYN 265, Proposition reéxaminé de Decision du Conseil en vue de l'adoption d'un programme spécifique de recherche et de dévelopment technologique dans la domaine de la Biotechnologie (1990–1994), February 28, 1992.

43. DECHEMA: Deutsche Gesellschaft für chemisches Apparatenwesen.

44. Interview, Bundesministerium für Forschung und Technologie, Bonn, July 5, 1988.

45. Interview, BMFT, July 5, 1988; Buchholz 1979.

46. Interview, BMFT, July 5, 1988.

47. Interview, BMFT, July 5, 1989.

48. Deutsche Gesellschaft für chemisches Apparatewesen e.V. 1974. *Biotechnologie. Studie über Forschung und Entwicklung. Möglichkeiten, Aufgaben und Schwerpunkte der Förderung—ausgearbeitet im Auftrag des Bundesminsteriums für Forschung und Technologie,* p. 129.

49. Ibid.

50. Interview, BMFT, July 5, 1988.

51. Sozial-Demokratische Partei and Freie Demokratische Partei.

52. For the classical framing of this concept, see Scharpf and Hauff 1975.

53. Bundesministerium für Forschung und Technologie, *Biotechnologie 2000* (Bonn, 1990), pp. 50, 151.

54. BMFT, *BMFT Leistungsplan 04—Biotechnologie, 1979–1983,* Bonn 1980; Deutscher Bundestag, Unterrichtung durch die Bundesregierung. Programm der Bundesregierung "Angewandte Biologie und Biotechnologie," 10. Wahlperiode, Drucksache 10/3724, August 14, 1985, Bonn; BMFT, *Biotechnologie 2000.*

55. BMFT, *Leistungsplan 04,* 12–14; Deutscher Bundestag, "Angewandte Biologie," p. 5; BMFT, *Biotechnologie 2000,* 12; Interview, BMFT, July 5, 1988; B. Gill 1991, pp. 125–141.

56. Interview, BMFT, Bonn, September 26, 1991.

57. Deutscher Bundestag, "Angewandte Biologie," pp. 8–12.

58. BMFT, *1979–1984,* 4.

59. Ibid., 5.

60. BMFT, *Biotechnologie 2000,* 85 and 108.

61. GBF, *Wissenschaftlicher Ergebnisbericht 1989* (Braunschweig 1989).

62. Deutscher Bundestag, "Angewandte Biologie," pp. 17–19.

63. Deutsche Forschungsgemeinschaft, *Tätigkeitsbericht 1989* (1988), pp. 86 and 93; Deutsche Forschungsgemeinschaft,*Tätigkeitsbericht 1990* (1991), p. 83.

64. Max-Planck-Gesellschaft, *Jahrbuch 1988,* p. 87; *Jahrbuch 1990,* p. 81.

65. RAUCON Biotechnologieberatung GmbH. *Biotechnology in the Federal Republic. Policies,* pp. 44–45; B. Gill 1991, p. 95.

66. "100 Millionen Mark für Kreativität und Innovationen in Ost und West," *BioEngineering* 7 (1991), p. 6.

67. Michael Wortmann, *Country Study on the Federal Republic of Germany. Prospective Dossier No. 2: Globalisation of Economy and Technology, FAST FOP 289* (Brussels 1991).

68. Interview, Laboratory of the Government Chemist, London, July 9, 1992.

69. Interview, Department of Trade and Industry, Teddington, July 8, 1992.

70. Advisory Council for Applied Research and Development, Advisory Board of the Research Councils, and The Royal Society, *Biotechnology: Report of a Joint Working Party,* March 1980 (HMSO, 1980).

71. Ibid.

72. Ibid., 7.

73. Advisory Council, *Biotechnology,* p. 8.

74. Ibid., pp. 37–38.

75. Ibid., pp. 18–22.

76. Ibid., p. 41.

77. *Biotechnology.* Presented to Parliament by the Secretary of State for Industry by Command of Her Majesty, March 1981 (HMSO, 1981), p. 3.

78. See ibid.

79. National Economic Development Council, Government Support for Biotechnology: A Summary of a Memorandum by the Secretary of State for Industry (London, 1983).

80. See ACOST, *Developments in Biotechnology;* EC Commission/CUBE, National Biotech Policy, EC Member States Review, May 1992, Brussels, 1992; Lex 1990, p. 39. All figures given are estimates based on different definitions of biotechnology and should be interpreted as rough indictors.

81. "Stock-Exchange Proposals Disappoint UK Firms," *Biotechnology* 11 (June 1993): 146–147.

82. The latter company then opted for listings in both London and the United States. See "UK Set For Blockbuster Year," *Biotechnology* 12 (October 1994): 21.

83. "Does UK Biotech Offer Better Opportunties?" *Biotechnology* 11 (June 1993): 106.

84. National Economic Development Office, New Life for Industry. Biotechnology, Industry and the Community in the 1990s and Beyond (London, 1990); Oakey et al. 1990; Dodgson 1991.

85. This effectively led to the abandonment of the Délégation Generale de la Recherche Scientifique et Technique.

86. *Innovation Policy. France* (OECD, 1986), p. 189.

87. Interview, Institut Pasteur, February 14, 1991.

88. Ministère de la Recherche et de l'Industrie. Mission des Biotechnologies. Programme Mobilisateur: l'Essor des Biotechnologies (Paris, 1982).

89. Ibid., p. 12.

90. *Financial Times,* "600 m francs plan over three years," August 3, 1982.

91. Ministère de la Recherche, Mission des Biotechnolgies, Avant-Propos.

92. Ibid., 12–13.

93. D. Thomas and P. Printz, Le Programme Français en Biotechnologie, July 1991, memo, Ministère de la Recherche et de la Technologie, Paris 1991, 2.

94. Robert T. Yuan, Biotechnology in Western Europe (International Trade Administration, US Department of Commerce, 1987), p. A-7.

95. OECD, *Innovation Policy,* 174.

96. Thomas and Printz, Le Programme Français, p. 2.

97. Ministère de l'Agriculture, Le Programme Aliment 2000. Développement de la recherche et de la technologie dans les industries agricoles et alimentaires, Paris 1985; "Financing of Biotechnology in France," *BFE* 9 (1991), p. 93.

98. Lipietz 1991, pp. 33–34; Messerlin 1987, p. 105; Gilbert Wasserman, "Au Nom de la Modernisation," *L'Evenement,* December 14, 1984.

99. Jean-François Augereau, "La relance de l'innovation industrielle: priorité du nouveau gouvernment," *Le Monde,* May 18, 1988.

100. Thomas and Printz, Le Programme Française, p. 2; "Financing of Biotechnology," p. 92.

101. Aline Richard, "La politique de l'innovation en panne," *Tribune de l'Expansion,* January 14, 1988; Michel Cahier, "ANVAR: Budget en légère baisse en 1987," *Tribune de L'Economie,* January 30, 1987; Lipietz 1991, p. 38.

102. Interview, Institut Pasteur, February 14, 1991.

103. René Sautier, Les Biotechnologies. Rapport établi pour le Premier Ministre, April 1988, pp. 109–111; Yuan, *Biotechnology.*

104. Sautier, Biotechnologies.

105. Yuan, Biotechnology in Western Europe, p. 24.

106. Thomas and Printz, Le Programme Français, pp. 3–6.

107. "Le Financement des Enerprises de Biotechnologies. Entretien avec Marie-Christine Candellé-Gardneq, Sofinnova," in *La Mosaïque Biotech. Une Etude d'Ernst & Young Conseil sur le Secteur des Biotechnologie en France* (Paris: Elsevier, 1991), pp. 47–50; "Entretien avec Olivier de Pelet, Agrinova," ibid., pp. 50–54.

108. Gardener and Molyneux 1990, p. 116; *La Mosaïque Biotech,* p. 21.

109. "Entretien avec Didier Masson, Transgène," *La Mosaïque Biotech,* 55–62.

110. Raugel 1992, pp. 206–211; "French Biotechnology Relies on Large and Small Firms and Government Help," *Genetic Engineering News,* May 15, 1993.

111. Ward 1993, pp. 799–801; "Biotechnologies: En attendant Bruxelles," *Le Nouvel Economiste,* November 1, 1991.

112. For this interpretation, see p. 42 of Orsenigo 1989.

113. See, e.g., *The Economist,* February 25, 1995, pp. 5–6; *Fortune,* October 14, 1996, pp. 293–294.

114. According to the "dogma," three types of processes are responsible for the inheritance of genetic information and for its conversion from one form to another: (1) Information is perpetuated by replication, a double-stranded nucleic acid is duplicated to give identical copies. Information is expressed by a two stage process: (2) Transcription generates a single-stranded RNA identical in sequence with one of the strands o the duplex DNA. (3) Translation concerts the nucleotide sequence of the RNA into the sequence of amino acids comprising a protein. See Lewin 1990, pp. 109–113.

115. Ernst & Young, *Biotech' 95. Reform, Restructure, Renewal. Industry Annual Report, 1994.*

116. Ibid.

117. See also Teitelman 1994, pp. 192–195.

118. For a related study of policy failure, see p. 276 of Ferguson 1994.

119. A. Düvell, Überblick über Freisetzungsexperimente betreffende Richtlinien verschiedener Länder und bekanntgewordene Freisetzungsexperimente mit gentechnisch veränderten Organismen (manuscript, Braunschweig, 1990). Öko-Institut, Freiburg and Gen- ethisches Netzwerk, Freisetzungsorientierte Forschungsprojekte in der Bundesrepublik. Gen- und Biotechnologie. Schadstoffabbau. Schädlingsbekämpfung. Pflanzen. Sicherheitsforschung (Berlin: 1988).

Chapter 5

1. World Commission on Environment and Development, *Our Common Future* (Oxford University Press, 1987).

2. Ibid. See also Spaargaren and Mol 1992, p. 333.

3. On the concept of ecological modernization, see also Huber 1985; Spaargaren and Mol 1992, p. 337.

4. See, e.g., Prier 1984; Krämer 1990, pp. 59–86; Ball and Bell 1994, pp. 3–175; Kloepfer 1989, pp. 71–96.

5. Quoted after Levin and Strauss 1991 (p. 15).

6. See, e.g., Hagerdorn 1989; van Elsas and Trevors 1991; Regal 1994; Abbott 1994; Ryder 1994.

7. OECD, *Biotechnology. Economic and Wider Impacts* (1989), p. 56.

8. The study of social movements as expressions and agents of political transformation was radically renewed by the movements of the 1960s and has become a major subfield in political science and sociology since then. For a

review of the different approaches, see Tarrow 1988; on the analytical shift in research from more structurally oriented theories of collective action (e.g. resource mobilization) to more culturally oriented ones (with a strong focus on the study of discourse and language for social movement organization and practices), see pp. 102–110 of Ellington 1995. Ses also Morris and Mueller 1992. For discussions of the German case, see Brand et al. 1986; Roth and Rucht 1987; Dalton and Kuchler 1990; Roth 1991.

9. On the importance of narratives for social movements, see p. 61 of Couto 1993.

10. On the Grünen, see Kitschelt 1989; Zeuner 1991; Raschke 1991.

11. Bundesarbeitsgemeinschaft Gentechnologie der Grünen, *Rechenschaftsbericht—BAG Gentechnologie,* Memo (1987), Bonn.

12. Interview, Grüne-Nordrhein-Westfalen, Düsseldorf, 1991.

13. See Amis de la Terre, *Problèmes de l'Environment: Les Biotechnologies. Dossier Environment* (Paris, 1990).

14. Institut für angewandte Ökologie, Darmstadt, *Einschätzungen des nationalen Gentechnikrechts* (Darmstadt, 1992), 23–25.

15. Ibid., 22–23.

16. According to Barthes (1974), a scriptible or writerly text is one in which the reader is invited to play a part in the construction or fabrication of the text's meaning.

17. See, e.g., Deutscher Naturschutzring and Bundesverband für Umweltschutz, *Memorandum zum Gentechnikgesetz* (Bonn, 1989), pp. 16–18.

18. See chapter 6 below.

19. Max-Planck-Institut für Züchtungsforschung, Köln, Antrag an die Kommission für Biologische Sicherheit: Anbau der gentechnologische veränderten Petunienlinien RL 01–17–3 im Freiland, June 24, 1988; Bundesgesundheitsamt, Bescheid zum Antrag des Max-Delbrück-Laboratoriums, Köln, auf Durchführung eines Freilandexperimentes mit gentechnisch veränderten Petunien in der Vegetationsperiode 1991, April 29, 1991; Stellungnahme der "Zentralen Kommission für die Biologische Sicherheit" zum Antrag auf Durchführung eines Freilandexperimentes mit gentechnische veränderten Petunien in der Vegetationsperiode 1991, April 24, 1994.

20. Stellungnahme der "Zentralen Kommission für die Biologische Sicherheit" zum Antrag auf Durchführung eines Freilandexperimentes mit gentechnische veränderten Petunien in der Vegetationsperiode 1991, April 24, 1994, p. 4.

21. Öko-Institut Freiburg, Gutachten zu der wissenschaftlichen Zielsetzung und dem wissenschaftlichen Sinn des Freisetzungsexperimentes mit transgenen Petunien, Freiburg, Februar 1980.

22. See chapter 7 below.

23. Julie Hill (Green Alliance), "Genetically Modified Organisms: Implementation of the EC Directives on Deliberate Release and Contained Use. The UK Approach to Information Provision and Participation. An Assessment, Position Paper, London, October 1991, pp. 4–5.

24. UK Genetics Forum, "Proposal for Additional Legislation on the Intentional Release of Genetically Manipulated Organisms. A Response to the Department of Environment Consultation Paper," London, August 1989, p. 1.

25. David King (Genetics Forum), The Patenting of Plants and Animals: Legal, Socio-Economic and Ethical Issues, Briefing Paper, London (undated).

26. See "Genomes," *Politis,* December 6, 1990.

27. Quoted after Hubert Markl, "Vom Sinn des Wissens. Auch die Genetik ist keine Wissenschaft im "wertfreien Raum," *Die Zeit,* September 8, 1989. (Markl, an evolutionary biologist, was president of the Deutsche Forschungsgemeinschaft.)

28. See Deutscher Naturschutzring and Bundesverband für Umweltschutz, *Memorandum zum Gentechnikgesetz* (1989).

29. For the notion of "supplement," see pp. 244–282 of Derrida 1983.

30. See also pp. 223–226 of Tait and Levidow 1992.

31. Linda Bullard, Briefing for NGOs. Some Lines of Argumentation on Legislation of Pesticides Containing or Consisting of Genetically Modified Organisms (GMO-Pesticides), Memo, Brussels, June 18, 1991, p. 1.

32. David King (Genetics Forum), letter to Michael Heseltine, Secretary of State for the Environment, London, March 6, 1991.

33. FEDESA, Bovin Somatropin. Information Update for Members of the European Parliament, Brussels, November 1987, p. 3.

34. Eric Brunner (London Food Commission), Bovine Somatropin: A Product in Search of a Market, London, April 1988, p. 9.

35. European Parliament, Committee on Energy, Research and Technology, Working Document on the Commission Communication on Promoting the Competetive Environment for the Industrial Activities Based on Biotechnology within the Community, SEC (91) 629 final, Draftsman: Hiltrud Breyer, March 12, 1992, pp. 4–5.

36. "Exposé d'ouverture," Biotechnologies: Quels Risques? Quels Choix? Pur Quelle Societé? (Paris, 1987).

37. Julie Hill (Green Alliance), "Influences on Specific Sectors: Environment," in The Impact of New and Impending Regulations on UK Biotechnology (Report of a Meeting Sponsored by the Department of the Environment, the

Health and Safety Executive and the BioIndustry Association held at the Royal Over-Seas League, London, 23 April, 1990), pp. 48–51.

38. Quoted after "Behinderte" Neugeborene werden liegengelassen. Aussonderung und Sterilisation von Krüppeln," in Anschlag auf die Schere am Gen und die Schere im Kopf. Texte zu Gentechnologie, Frauenbewegung, Faschismus und Bevölkerungspolitik. Ausgesucht von Ulla Penselin und Ingrid Strobl, Hamburg 1988, p. 35.

39. The usage of the word *Krüppel,* which in German has connotations similar to those of "cripple" in English, was meant to demonstrate a positive identification with a situation through the assertive usage of a pejorative term.

40. European Parliament, II Recommendation of the Committees on Energy, Research and Technology on the Common Position of the Council with a View to the Adoption of a Decision adopting a Specific Research and Technological Development Programme in the Field of Health A3–89/90 SYN 136: Human Genome Analysis (1990–1991), Rapporteur Hiltrud Breyer, April 20, 1990, p. 25.

41. Parents for Safe Food and Genetics Forum, Genetic Engineering, London (undated).

42. For the notion of the Other, see Foucault 1967.

43. Quoted after SPD Schleswig-Holstein, Zur Sache: 28. Gen- und Reproduktionstechniken in der Diskussion, Kiel (1988), p. 13.

44. André Klarsfeld (Génétique et Liberté), "Notre avenir entre les mains des experts," *Le Monde,* March 6, 1991.

45. Gen-Archiv Essen, Selbstdarstellung (Memo, Essen, 1988).

46. Die Grünen, Frauen und Fortpflanzungstechniken (leaflet, 1988).

47. Thierry Damerval (Génétique et Liberté),"Génétique et Liberté," *La Recherche,* October 1990, p. 1173.

48. See, e.g., Antrag der Abgeordneten Flinner, Kreutzeder, Schmidt-Bott und der Fraktion Die Grünen, Verbot der Produktion und Anwendung und des Inverkehrbringens von gentechnologisch erzeugten leistungssteigernden Hormonen und Verbindungen, Deutscher Bundestag, 11. Wahlperiode, Drucksache 11–1507, December 10, 1987.

49. Rainbow Group GRAEL, With or without genetically engineered hormones? How do you like your milk? What you always wanted to ask about gene technology but were afraid to find out the answer (undated leaflet, Brussels).

50. Compassion in World Farming was an important animal welfare group that linked traditional animal protection arguments with socio-economic reasoning.

51. Compassion in World Farming, Open Letter to the Minister of Agriculture, Petersfield, October 25, 1991.

52. Ken Collins, Biotechnology in the European Community: Redressing the Uncertainties Caused by the Hormone Ban (memo, Brussels, June 1988).

53. See, e.g., Altner et al. 1988.

54. Pierre Samuel, "Colloque Biotechnologies. Exposé d'ouverture," "Les Amis de la Terre: Biotechnologies: Quels Risques? Quels Choix? Pour quelle Societé?" Paris, April 24–25, 1987.

55. Aktionsbündnis gegen Gentechnik Hoechst, "Nr. 5: Gen-Insulin und Arbeitsplätze," Frankfurt (undated).

56. On the importance of space and time as power technologies, see Lefebre 1991 and Harvey 1989.

Chapter 6

1. In a number of public opinion polls on biotechnology (the most extensive one commissioned by the Commission of the European Community, the image of a rather concerned and critical public attitude toward genetic engineering across Europe gained shape. A variety of surveys conducted in the 1980s showed that the public, on the one hand, believed that "genetic engineering will improve our way of life" (1991: Germany: 40.9 percent; France: 53.7 percent; UK.: 51.2 percent); on the other hand, the same respondents had a strong perception of risk. When asked, for example, if research involving genetic engineering on plants might involve risks to human health and the environment, the following percentages of respondents definitely agreed or tended to agree: Germany 53 percent, France 75.5 percent, UK 53.8 percent. Concerning the risk perception of genetically engineered foods, the following percentages of respondents definitely agreed or tended to agree that there were risks involved: Germany 68.6 percent, France 67.9 percent, UK 62.7 percent. A more favorable picture of risk perception was reflected in answers to the question concerning risks involved in genetically engineered medicines and vaccines. Here the following percentages of respondents agreed that risks were involved: Germany 61 percent, France 49.6 percent, UK 55. 8 percent. Taking into account this pattern of risk perception, the vast percentage of the respondents tended to believe, not surprisingly, that research in genetic engineering "needs to be controlled by the government." (Especially with respect to medicines and vaccines, the pattern of agreement was Germany 90.9 percent, France 86.7 percent, UK 88.5 percent.) At the same time, according to the polls, trust in public authorities in regard to genetic engineering seemed to be low. When asked about which sources of information they had confidence in telling the truth about biotechnology and genetic engineering, respondents gave the following picture. Germany: public authorities 7.8 percent; consumer, environmental, and animal welfare organizations 60.9 percent. France: public authorities 4.7 percent; consumer, environmental, and animal welfare organi-

zations 65.3 percent. UK: public authorities 7.5 percent; consumer, environmental, and animal welfare organizations 50.9 percent. Furthermore, a variety of polls showed quite clearly a strong correlation among respondents between knowledge about genetic engineering and the presence of concerns about it; this correlation flatly contradicted the commonly held belief that more information about science would alleviate concerns. See Commission of the European Communities, Eurobarometer 35.1 Biotechnology, June 1991; Bernhard Zechendorf, Public Opinion on Biotechnology, Manuscript, Brussels 1992; Büro für Technikfolgenabschätzung beim Deutschen Bundestag (TAB), Gentechnologie und Genomanalyse aus der Sicht der Bevölkerung. Ergebnisse einer Bevölkerungsumfrage des TAB. TAB-Diskussionspapier Nr. 3, Bonn 1992.

2. Bundesarbeitsgemeinschaft Gentechnologie der Grünen, Rechenschaftsbericht-BAG Gentechnologie, 1987, Bonn, Memo.

3. For an extensive and detailed study of the reporting of German media on genetic engieering, see Hans Mathias Kepplinger, Simone Christine Ehmig, and Christine Ahlheim, Gentechnik im Widerstreit: Zum Verhältnis von Wissenschaft und Journalismus, Manuscript, Mainz 1991.

4. Interview, Bundesministerium für Jugend, Familie, Frauen und Gesundheit, Bonn, July 5, 1988.

5. Verband Chemischer Industrie e. V. Stellungnahme der chemischen Industrie zum Bericht der Enquete-Kommission "Chancen und Risken der Gentechnologie," May 13, 1987, Frankfurt, Memo.

6. Deutsche Forschungsgemeinschaft, Stellungnahme zum Bericht der Enquete-Kommission "Chancen und Risken der Gentechnologie" des 10. Deutschen Bundestages, May 8, 1987.

7. See "Chancen und Risken der Gentechnologie." Stellungnahme und konstruktive Vorschläge der "Arbeitsgemeinschaft Biotechnologie" zum Bericht der Enquete-Kommission des Deutschen Bundestages, Forum Mikrobiologie, December 1987, pp. 472–474.

8. Bundesminister für Jugend, Familie, Frauen und Gesundheit, Bericht des BMJFFG über gesetzliche Regelungen zur Gentechnik. Unterrichtung durch die Bundesregierung. Eckwert-Beschluss der Bundesregierung, Bundestags-Drucksache. 11/3908), Bonn 1988, 2.

9. Interview, Ministry of the Environment, Nordrhein-Westfalen, Düsseldorf, September 26, 1991.

10. Interview, Ministry of the Environment.

11. Interview, BMJJF, July 5, 1989.

12. BJFFG, Eckwert-Beschluss, 1988, p. 3.

13. Ibid., 4.

14. Ibid., 5.

15. Interview, BMJJF, July 5, 1989.

16. BJFFG, Eckwert-Beschluss, 7.

17. Ibid., 7–11.

18. Interview, Department of Environment, September 26, 1991.

19. Interview, Verband Chemischer Industrie, Frankfurt/Main, June 29, 1989.

20. Interview, VCI.

21. See, e.g., 13. DGB-Bundeskongress, Hamburg, May 25–31, 1986, Antrag 159, IG-Chemie-Papier-Keramik. Betrifft Biotechnologie und Genforschung.

22. Interview, Industrie Gewerkschaft Chemie (IG Chemie—Chemical Industry Union), Düsseldorf, June 30, 1988.

23. *Frankfurter Allgemeine Zeitung,* "Die BASF-Pläne für Gentechnik-Zentrum schaffen Unruhe," October 26, 1988; ibid., "Die Gentechnik braucht nicht auszuwandern," November 12, 1988.

24. *Frankfurter Rundschau,* SPD prescht bei Gentechnik vor, September 23, 1988.

25. BMJFFG, Entwurf eines Ersten Gesetzes zur Regelung von Fragen der Gentechnik, April 24, 1989, Bonn.

26. "Auch das Umweltministerium meldet Bedenken an," *Frankfurter Rundschau,* March 18, 1989; "Die Labors werden abgedichtet," *Die Zeit,* April 7, 1989.

27. The Morgenthau Plan was "a radical plan developed by the US Secretary of the Treasury from 1934–45, Henry Morgenthau Jr., to turn Germany into 'a country primarily agricultural and pastoral' [and] without 'war-making industries.'" Source: *Encyclopaedia Brittanica,* CD 97 version.

28. Gen-ethischer Informationsdienst, "Das Gengesetz-ein Freifahrschein," Nr. 45, July 1989, pp. 5–6.

29. According to article 76(2) of the basic law.

30. Interview, BMJFFG, Bad Godersdorf, September 25, 1991.

31. Bundesratsdrucksache 387/1/89 from September 12, 1989. See also Wurzel and Merz 1991, p. 15.

32. Bundesratsdrucksache 387/89.

33. Bundestagsdrucksache 11/5320 and 11/5468. Since these were minority opposition parties, these dissenting votes had no consequences.

34. Gen-Ethischer Informationsdienst, Nr. 46, August 1989, 3–4.

35. Lutz Meissner and Hans Neumann. Lawyers, "Schreiben im Widerspruchsverfahren gegen die Hoechst Fermtec Produktionsanlage," December 8, 1987.

36. Konrad Redeker, Kurt Schön, Hans Dahs, and Dieter Sellner. Lawyers, Stellungnahme zur Widerspruchsbegründung im Verfahren Fermtec und Chemtec, March 14, 1988.

37. *Neue Juristische Wochenschrift,* 1990, 336; *Deutsche Verwaltungsblätter,* 1990, 63. See also Bruer 1991, p. 16.

38. Deutscher Bundestag. 11. Wahlperiode. Drucksache 11/5622. Gesetzesentwurf der Bundesregierung. Entwurf eines Gesetzes zur Regelung von Fragen der Gentechnik. November 11, 1989.

39. Deutscher Bundestag. 11. Wahlperiode. Drucksache 11/6778. Beschlussempfehlung des Ausschusses für Jugend, Familie, Frauen und Gesundheit. Zum Entwurf eines Gesetzes zur Regelung von Fragen der Gentechnik. Bericht des Abgeordneten Catenhusen, 23.

40. "Wer schützt uns vor dem Schutzgesetz?" *TAZ,* January 19, 1990.

41. Gesetz zur Regelung von Fragen der Gentechnik, June 20, 1990, Bundesgesetzblatt 1080–1095.

42. At the time of the passage of the act, all ongoing commercial work was at this level.

43. Since at the time all commercial work was limited to level 1 activity, not many public hearings were expected.

44. For examples of this position, see Wheale and McNally 1990 and Hill 1990.

45. Interview, UK Genetics Forum, London, July 11, 1989.

46. UK Genetics Forum, The Deliberate Release of Genetically Modified Organisms to the Environment. Submission of Evidence to the Royal Commission on Environmental Pollution, March 1989, p. 12.

47. Compassion in World Farming, "BST—Politics, Persuasion and the Poor Cow," Agscene, No. 98, Spring 90, 5; Compassion in World Farming, BST—the Dairy Cow Growth Hormone, Memo, Petersfield, August 1988; Eric Brunner/The London Food Commission, Bovine Somatropin: A Product in Search of a Market, London 1988.

48. Interview, Ministry of Agriculture, Fisheries, and Food (MAFF), London, July 10, 1992.

49. Interview, MAFF, London, July 27, 1989.

50. Royal Commmission on Environmental Pollution. Chairman Rt Hon The Lord of Newnham, Thirteenth Report. The Release of Genetically Engineered Organism to the Environment. Presented to Parliament by Command of Her Majesty, July 1989 (London: HMSO, 1989).

51. Royal Commission on Environmental Pollution. A Short Guide to the Commission and its Activities (London, 1989), Ball and Bell 1994, p. 34.

52. Royal Commission, Short Guide, Interview, Royal Commission of Environmental Pollution, London, July 28, 1989.

53. Royal Commission, *Report,* 107.

54. AGCM/HSE/Note 1, Advisory Committee on Genetic Manipulation, Guidance on Construction of Recombinants Containing Potentially Oncogenic Sequences, London 1985; ACGM/HSE/Note 2, Disabled Host/Vector Systems, London 1986; ACGM/HSE/Note 3, The Planned Release of Genetically Manipulated Organisms for Agricultural and Environmental Purposes. Gudidelines for Risk Assessment and for the Notification of Proposals for Such Work," London 1986; ACGM/HSE/Note 4, Guidelines for the Health Surveillance of those involved in Genetic Manipulation at Laboratory and Large-Scale, London 1986; ACGM/HSE/Note 5, Guidance on the Use of Eukaryotic Viral Vectors in Genetic Manipulation, London 1986; ACGM/HSE/Note 6, Guidelines for the Large-Scale Use of Genetically Manipulated Organisms, London 1987; ACGM/HSE/Note 7, Guidelines for the Categorizaton of Genetic Manipulation Experiments, London, 1988; ACGM/HSE/Note 8, Laboratory Containment Facilities for Genetic Manipulation, London 1988; ACGM/HSE/Note 9, Guidelines on Work with Transgenic Animals, London 1989.

55. Health and Safety Executive, Genetic Manipulation Regulations 989. Guidance on Regulations (London: HMSO, 1989).

56. Advisory Committee on Genetic Manipulation, ACGM/HSE/Note 9, Guidelines on Work Involving the Genetic Manipulation of Plants and Plant Pests, London, 1990; Advisory Committee on Genetic Manipulation, ACGM/HSE/Note 11, Genetic Manipulation Safety Committees, London 1990; Advisory Committe on Genetic Manipulation, ACGM/HSE/Note 3 (revised), The Intentional Introduction of Genetically Modified Organisms into the Environment. Guidelines for Risk Assessment and for the Notification of Proposals for such Work, London 1990.

57. ACGM, The Intentional Introduction, 25–28.

58. ACGM, Large Scale Use, 3.

59. ACGM, Appendix IV of ACGM/HSE/Note 6—Large Scale Use of Genetically Manipulated Organisms, August 1990.

60. Department of the Environment, "Environmental Protection. Proposals for Additional Legislation on the Intentional Release of Genetically Manipulated Organisms. A Consultation Paper, London, June 1989. See also Shackley and Sharp 1989.

61. DoE, Intentional Release, 8.

62. UK Genetics Forum, Proposals for Additional Legislation on the Intentional Release of Genetically Manipulated Organisms. A Response to the Department of Environment Consultation Paper by the UK Genetics Forum, August 1989.

63. The ACRE was the successor to ACGM's Intentional Introduction Subcommittee and to the Department of the Environment's Interim Advisory Committee on Introductions.

64. Environmental Protection Act, Part VI, Clause 106, 115 (London, HMSO, 1990)

65. Quoted after Levidow and Tait 1992, p. 98.

66. EPA, Clause 107, 117.

67. Ibid., 99, EPA, Clauses 111–113, 121–124.

68. EPA, Clause 118, 128–130.

69. "Quango" is British political jargon for "quasi-governmental organization."

70. Ibid., 419–420, Michel Crozier, *La Sociéte bloquée* (Paris: Seuil, 1970).

71. Interview, Organisation Nationale Interprofessionelle des Bioindustries (ORGANIBIO), Paris, February 19, 1991.

72. Interview, Institut National de la Recherche Agronomique (INRA), Paris, February 15, 1991.

73. *Le Monde,* July 9, 1987.

74. *Le Monde,* July 9, 1987; *Liberation,* July 9, 1989.

75. It had been one of the central assumptions of the French system of recombinant DNA regulation that self-regulation was as efficient as regulation by law.

76. *Liberation,* July 9, 1987.

77. Interview. Ministère de l'Industrie, Paris, September 28, 1988.

78. Arréte du 4 novembre 1986 instituant la commission d'étude de l'utilisation de produits issus du génie biomoléculaire, *Journal Officiel de la République Française,* 25 novembre 1986, 14199.

79. Ministère de l'Agriculture de la Forêt, La Commission du Génie Biomoléculaire. Activité en 1989, Paris 1990, Memo, 4.

80. François Guillaume, Missions et Objectives de la Commmission du Génie Biomoléculaire Institutée auprés du Minstre de l'Agriculture, Memo, Paris 1987, 5.

81. Axel Kahn, Biotechologies: Encadrere sans Entraver, Memo, Paris (undated).

82. Axel Kahn, La Commission du Génie Biomoleculaire. Son Role, son Action et ses But, Paris 1990, Memo.

83. A. Düvell, Überblick über Freisetzungsexperimente betreffende Richtlinien verschiedener Länder und bekanntgewordene Freisetzungsexperimente mit gentechnisch veränderten Organismen. Stand 12.04 1990 (manuscript, 1990).

84. "Loi No 76–663 du 19 juillet 1976 relative aux installations classées pour la protection de l'environment," *Journal Officiel de la Republique Française.* Annexe VI, 4320–4323.

85. Ministère de l'Environnement, "Décret no 85–822 du 30 juillet 1985 modifiant la nomenclature des installations classées, *Journal de la République Française,* 2 aout 1989, 8826–8828; Ministère de l'Environnement, "Circulaire no 86–32 du 19 Septembre 1986 relative aux installations classées pour la protection de l'environment (décret de nomenclature du 30 juillet 1985, rubrique 58–11: installations mettant en oeuvre des micro-organisms."

86. Jocelyne Gantois, L'AFNOR s'interesse aux Biotechnologies, memo, Paris 1986, 1,12.

87. See Association Française de Normalisation (AFNOR), Compte Rendu de la Deuxième Reunion du Groupe de Travail, Memo, 6 novembre 1987, Paris, ANNEX II and AFNOR, Compte Rendu de la Premiere Reunion du Groupe de Travail, memo, 21 October 1987, Paris, ANNEX II.

88. AFNOR, Deuxième Reunion, ANNEX II.

89. This commission was initially set up in 1981. See Ministère de l'Industrie, Décret no 81–278 du 25 mars 1981 portant création d'un groupe interministeriél des produits chimiques, *Journal Officiel de la République Française,* 27 mars 1981, 881.

90. In 1986 the premier minister writes to the president of the Commission: "GIPC has to make sure that in matters concerning the protection for users, workers and the environment, no "excess" legislation or regulation takes place." Letter Premier ministre a Pierre Creyssel, 29 Mai 1986.

91. République Francaise. Premier Ministre. Groupe Interministeriel des Produits Chimiques, Reglementation des Biotechnologies. Etat actuel et propositions soumise au premier ministre, 7 juin 1989, Paris, Memo, 10.

92. Ministère de la Recherche et la Technologie, "Décret no 89–306 du 11 mai 1989 portant création d'une commission de génie génétique," *Journal Officiel de la République Française,* 13 mai 1989, 6089.

93. GIPC, Reglementation, 12–13.

94. Ibid., 13–14.

95. Ibid., 20.

96. For empirical evidence, see Deutscher Bundestag. 12. Wahlperiode, Drucksache 12/7095, Bericht des Ausschusses für Forschung, Technologie und Technikfolgenabschätzung, Biologische Sicherheit bei der Nutzung der Gentechnik (Bonn: March 16, 1994), p. 124.

Chapter 7

1. European Council, Council Directive of April 23, 1990, on the contained use of genetically modified microorganisms; and Council Directive of 23 April on the deliberate release into the environment of genetically modified organisms, *Official Journal of the European Communities* L: 117–127.

2. "Cloning Report Leaves Loophole," *New Scientist,* June 14, 1997, p. 7.

3. See also Hajer 1995.

Interviews Conducted, 1988–1992

In-depth interviews conducted in Britain, France, Germany, and Belgium were of central importance in the writing of this book. These interviews gave me a first impression of what was going on in the field and helped me to explore the various readings of the genetic engineering problematic. Before I started my field work, I certainly had a number of assumptions about the dynamics of genetic engineering policymaking in Europe. Interviewing helped me to gain data that either supported or contradicted my initial thoughts and assumptions. Furthermore, interviewing allowed me to explore the structure of the biotechnology policy networks. In more or less classical fashion, one interview led to the next. Either I asked my interview partner for other important actors in the biotechnology network to which he or she belonged or names simply happened to be mentioned in the course of the interviews. And the interviews gave me access to written sources. When I started my project in 1988, hardly anything had been published about European biotechnology policymaking. In the absence of accessible archives on the topic (with the exception of the European Community's BIODOC archive), interviewing turned out to be the most valuable way to get hold of written sources.

Almost all of the nearly eighty interviews I conducted were recorded on audio tape.

Here and in the notes to the chapters, I give the dates and the locations of my interviews and the institutional affiliations—but not the names—of my interviewees.

Britain

Agricultural and Food Research Council, Swindon, July 8, 1992

Association of Scientific, Technical, and Managerial Staffs, London, July 14, 1989

Borough of Brighton, Brighton, July 29, 1989

British Technology Group, London, July 11, 1989

Compassion for World Farming, Petersfield, July 14, 1992

Department of Health and Security, London, August 8 and 9, 1989

FINRAGE, London, July 28, 1989

Greenpeace UK, London, July 19, 1992

Green Alliance, London, August 4, 1989 and July 7, 1992

Health and Safety Executive, London, July 13, 1989 and July 9, 1992

Laboratory of the Government Chemist, Teddington, July 8, 1992

Laboratory of the Government Chemist, London, July 9, 1992

Linacre Center, London, August 3, 1989

London Food Commission, London, July 21, 1989

Ministry of Agriculture, Fisheries, and Food, London, July 27, 1989 and July 10, 1992

National Farmers Union, London, August 8, 1989

Royal Commission of Environmental Pollution, London, July 28, 1989

Science and Engineering Research Council, Swindon, July 18, 1989

Transport Union, London, August 10, 1989

UK Genetic Forum, London, July 11, 1989 and July 9, 1992

France

Biofuture, Paris, September 29, 1988 and February 14, 1991.

Confédération Générale du Travail, Paris, September 28, 1988

Confédération Paysanne, Paris, September 16, 1988

Fédération Nationale des Syndicates d'Exploitants Agricoles, Paris, September 26, 1988

Génétique et Liberté, Paris, February 19, 1991

Institut Monod, Université Paris 7, Paris, February 15, 1991

Institut National de la Recherche Agronomique, Paris, September 22, 1988 and February 15, 1991

Institut National de la Santé et de la Recherche Médicale, Paris, February 18 and 19, 1991

Institut Pasteur, September 21, 1988 and February 14, 1991.

Ministère de L'Agriculture, September 23, 1988

Ministère de l'Environment et du cadre de vie, Neuilly/Seine, February 14, 1991

Ministère de l'Industrie et de lámanagement du territoire, Paris, September 28, 1988

Ministère de la Recherche et de la technologie, Compiènge, September 12, 1988

Nature et Progrès, October 3, 1989

Organisation National Interprofessionelle des Bioindustries (ORGANIBIO), Paris, September 15, 1988 and February 19, 1991

Sanofi, Paris, February 20, 1991

Germany

Bundesministerium für Forschung und Technologie, Bonn, July 1 and 8, 1988, and September 26, 1991

Bundesministerium für Jugend, Familie, Frauen und Gesundheit, Bonn July 5, 1988 and September 25, 1991

Bundesministerium für Wirtschaft, Bonn, July 1, 1988

Bundesumweltamt, Berlin, October 2, 1991

Bürgerinitiative Bürger gegen Petunien, Köln, September 23, 1991.

Deutsche Forschungsgemeinschaft, Bonn, September 25, 1991

Deutscher Gerwerkschaftsbund, Düsseldorf, June 30, 1988

Die Grünen, Bonn, July 4, 1988

Die Grünen, Nordrhein-Westfalen, Düsseldorf, September 26, 1991

European Molecular Biology Organization, Heidelberg, September 24, 1991

Gen-Ethisches Netzwerk, Berlin, September 30, 1991

Hoechste Schnüffle, Frankfurt/Main, September 23, 1991

Industrie Gewerkschaft Chemie, Düsseldorf, June 30, 1988

Institut für Genetik, Universität Köln, July 2, 1988 and September 24, 1991

Sozial-Demokratische Partei Deutschlands, Bonn, July 4, 1988

Umweltministerium Nordrhein-Westfalen, Düsseldorf, September 26, 1991

Verband Chemischer Industrie, Frankfurt, June 29, 1988

Stiftung Volkswagenwerk, Hannover, September 27, 1991

Brussels

DG XII, February 14, 1989, June 29, 1990, and February 25, 1991

DG XI, February 25 and March 1, 1991

DG III, February 26 and March 1, 1991

DG VI, February 27, 1991

Greens, European Parliament, March 1, 1991

Bibliography

Abbott, Richard J. 1994. Ecological Risks of Transgenic Crops. *Ecology and Evolution* 9: 280–282.

Abir-Am, Pnina. 1982. The Discourse of Physical Power and Biological Knowledge in the 1930s: A Reappraisal of the Rockefeller Foundation's "Policy" in Molecular Biology. *Social Studies of Science* 12: 341–382.

Abir-Am, Pnina. 1993. From Multidisciplinary Collaboration to Transnational Objectivity: International Space as Constitutive of Molecular Biology, 1930–1970. In *Denationalizing Science,* ed. E. Crawford et al. Kluwer.

Abir-Am, Pnina. 1992. The Politics of Macromolecules: Molecular Biologists, Biochemists, and Rhetoric. *Osiris* 7: 164–191

Adams, Mark B., ed. 1990. *The Wellborn Science: Eugenics in Germany, France, Brazil, and Russia.* Oxford University Press.

Adams, Pamela. 1991. State Policy and the Chemical Industry in Western Europe. In *International Markets and Global Firms,* ed. A. Martinelli. Sage.

Adams, William James. 1989. *Restructuring the French Economy: Government and the Rise of Market Competition since World War II.* Brookings Institution.

Alonso, Ana María. 1994. The Politics of Space, Time and Substance: State Formation, Nationalism, and Ethnicity. In *Annual Review of Anthropology,* ed. W. Durhamn et al. Annual Reviews Inc.

Altner, Günter, Wanda Krauth, Immo Lünzer, and Hartmut Vogtmann, eds. 1988. *Gentechnik in der Landwirtschaft. Folgen für die Umwelt und Lebensmittelerzeugung.* C. F. Müller.

Altvater, Elmar, and Kurt Hübner. 1988. Das Geld einer mittleren Hegemonialmacht—Ein kleiner Streifzug durch die ökonomische Geschichte der BRD. *Prokla* 73: 6–36.

Andrain, Charles F. 1985. Social Policies in Western Industrial Societies. Institute of International Studies, Berkeley.

Atkinson, Michael M., and William D. Coleman. 1989. Strong States and Weak States: Sectoral Policy Networks in Advanced Capitalist Economies. *British Journal of Political Science* 19: 47–67.

Auchincloss, Stuart. 1993. Does Genetic Engineering Need Genetic Engineers? Should the Regulation of Genetic Engineering Include a New Professional Discipline? *Boston College Environmental Affairs Law Review* 20: 37–63.

Aujac, Henri. 1986. An Introduction to French Industrial Policy. In *French Industrial Policy,* ed. W. Adams and C. Stoffaës. Brookings Institution.

Bachrach, Peter, and Morton S. Baratz. 1963. Decisions and Nondecisions: An Analytical Framework. *American Political Science Review* 57: 632–642.

Ball, Simon, and Stuart Bell. 1994. *Environmental Law: On the Law and Policy Relating to the Protection of the Environment.* Blackstone.

Barthes, Roland. 1972. *Mythologies.* Hill and Wang.

Barthes, Roland. 1974. *S/Z.* Hill and Wang.

Battershill, Charles D. 1990. Erving Goffman as a Precursor to Postmodern Sociology. In *Beyond Goffman,* ed. S. Riggins. Mouton de Gruyter.

Baudrillard, Jean. 1976. *L'échange symbolique et la mort.* Gallimard.

Beck, Ulrich. 1992. *Risk Society: Towards a New Modernity.* Sage.

Beck, Ulrich. 1993. *Die Erfindung des Politischen. Zu einer Theorie reflexiver Modernisierung.* Suhrkamp.

Bennahmias, Jean-Lu and Agnés Roche. 1992. *Des vertes de toutes les couleus.* Albin Michel.

Bennet, David, Peter Glasner, and David Travis. 1986. *The Politics of Uncertainty: Regulating Recombinant DNA Research in Britain.* Routledge & Kegan Paul.

Bentley, Arthur F. 1908. *The Process of Government.* Harvard University Press, 1967.

Berg, Paul, et al. 1974. Du danger potentiel des manipulations génétiques. *Biochimie* 56: X–XII.

Berg, Paul, David Baltimore, Sydney Brenner, Richard O. Roblin III, and Maxine F. Singer. 1975. Summary Statement of the Asilomar Conference on Recombinant DNA Molecules. *Proceedings of the National Academy of Sciences* 72, no. 6: 1981–1984.

Bill, James A., and Robert L. Hardgrave Jr. 1973. *Comparative Politics: The Quest for Theory.* Charles E. Merrill.

Bimber, Bruce, and David H. Guston. 1995. Politics by the Same Means: Government and Science in the United States. In *Handbook of Science and Technology Studies,* ed. S. Jasanoff et al. Sage.

Binder, Norbert. 1980. Richtlinien für die Genforschung im Spannungsfeld zwischen Gefahrenschutz und Forschungsfreiheit. In *Genforschung im Widerstreit,* ed. W. Klingmueller. Wissenschaftliche Verlagsgesellschaft.

Boje, David M., Robert P. Gephart Jr., and Tojo Joseph Thatchenkery, eds. 1996. *Postmodern Management and Organization Theory.* Sage.

Boyer, Robert. 1991. The Eighties: The Search for Alternatives to Fordism. In *The Politics of Flexibility,* ed. B. Jessop et al. Edward Elgar.

Brand, Werner, Detlef Büsser, and Dieter Rucht. 1986. *Aufbruch in eine andere Gesellschaft. Neue soziale Bewegungen in der Bundesrepublik.* Campus.

Bressers, Hans, Laurence J. OToole Jr., and Jeremy Richardson. 1994. Networks as Models of Analysis: Water Policy in Comparative Perspective. *Environmental Politics* 3: 1–23.

Brill, Winston J. 1985. Safety Concerns and Genetic Engineering in Agriculture. *Science* 227: 381–384.

Bruer, Rüdiger. 1991. Ansätze für ein Gentechnikrecht in der Bundesrepublik Deutschland. In *6. Trier Kolloquium zum Umwelt- und Technikrecht vom 26. bis 28. September 1990, Gentechnik und Umweltrecht.* Werner.

Buchholz, Klaus. 1979. Die gezielte Förderung und Entwicklung der Biotechnologie. In *Geplante Forschung,* ed. W. van der Daele et al. Suhrkamp.

Bud, Robert. 1993. *The Uses of Life: A History of Biotechnology.* Cambridge University Press.

Budge, Ian, and David McKay. 1993. Turning Britain Around? In *The Developing British Political System in the 1990s,* third edition, ed. I. Budge and D. McKay. Longman.

Buechler, Steven M. 1993. Beyond Resource Mobilization? Emerging Trends in Social Movement Theory. *Sociological Quarterly* 34: 217–235.

Burchell, Graham. 1993. Liberal Government and Techniques of the Self. *Economy and Society* 22: 267–282.

Butler, Judith. 1993. *Bodies That Matter: On the Discursive Limits of "Sex."* Routledge.

Callon, Michel. 1986. Some Elements of a Sociology of Translation: Domestication of the Scallops and the Fishermen of St. Brieuc Bay. In *Power, Action and Belief,* ed. J. Law. Routledge & Kegan Paul.

Callon, Michel, and Bruno Latour. 1981. Unscrewing the Big Leviathan: How Actors Macro-Structure Reality and How Sociologists Help Them to Do So. In *Advances in Social Theory and Methodology,* ed. K. Knorr-Cetina and A. Cicourel. Routledge & Kegan Paul.

Camiller, Patrick. 1989. Beyond 1992: The Left and Europe. *New Left Review* 175: 5–17.

Cammack, Paul. 1992. The New Institutionalism: Predatory Rule, Institutional Persistence, and Macro-Social Change. *Economy and Society* 21: 397–429.

Campbell, David. 1992. *Writing Security: United States Foreign Policy and the Politics of Identity*. University of Minnesota Press.

Campbell, John L. 1988. *Collapse of an Industry: Nuclear Power and the Contradictions of US Policy*. Cornell University Press.

Canals, Jordi. 1993. *Competetive Strategies in European Banking*. Oxford University Press.

Catenhusen, Wolf-Michael, and Hanna Neumeister, eds. 1987. *Chancen und Risken der Gentechnologie. Dokumentation des Berichts an den Deutschen Bundestag*. Schweitzer.

Celarier, Michelle. 1993. A New Leader for Europe. *Global Finance* 7: 52–59.

Chafer, Tony. 1984. The Greens in France: an Emerging Social Movement? *Journal of Area Studies* 14: 36–43.

Chataway, Joanna. 1991. Biotechnology and Business Blues. *AgBiotech News and Information* 3: 1003–1005.

Chilcote, Ronald H. 1981. *Theories of Comparative Politics: The Search for a Paradigm*. Westview.

Churchill, Robin, John Gibson, and Lynda M, Warren, eds. 1991. *Law, Policy and the Environment*. Blackwell.

Clifton, Richard. 1990. The UK Regulatory Structures; Development and Current Concerns. In *The Impact of New and Impending Regulations on UK Biotechnology*, ed. D. Bennett and B. Kirsop. Cambridge Biomedical Consultants.

Colwell, Robert Elliot Norse, David Pimentel, Frances Sharples, and Daniel Simberloff. 1985. Genetic Engineering in Agriculture. *Science* 29: 111–112.

Connolly, William E. 1991. *Identity/Difference: Democratic Negotiations of Political Paradox*. Cornell University Press.

Coombs, Rod, Paolo Saviotti, and Vivien Walsh. 1987. *Economics and Technological Change*. Rowman and Littlefield.

Cooper, Robert. 1989. Modernism, Post Modernism and Organizational Analysis 3: The Contribution of Jacques Derrida. *Organizational Studies* 10: 479–502.

Couto, Richard A. 1993. Narrative, Free Space, and Political Leadership in Social Movements. *Journal of Politics* 55: 57–79.

Cox, Andrew. 1986. The State, Finance and Industry Relationship in Comparative Perspective. In *State, Finance and Industry*, ed. A. Cox. Wheatsheaf.

Cox, Robert W. 1987. *Production, Power, and World Order: Social Forces in the Making of History*. Columbia University Press.

Crozier, Michel. 1970. *La Sociéte bloquée*. Seuil.

Culler, Jonathan. 1988. *Framing the Sign*. Blackwell.

Dachs, Herbert, Peter Gerlich, Herbert Gottweis, et al., eds. 1991. *Handbuch des politischen Systems Österreichs*. Manz.

Dahl, Robert A. 1957. The Concept of Power. *Behavioural Science* 2: 201–205.

Dahl, Robert A. 1961. *Who Governs? Democracy and Power in an American City*. Yale University Press.

Dalton, Russel J., and Manfred Kuchler, eds. 1990. *Challenging the Political Order: New Social Movements in Western Democracies*. Polity Press.

Daly, Glyn. 1991. The Discursive Construction of Economic Space: Logics of Organization and Disorganization. *Economy and Society* 20: 79–102.

Daly, Glyn. 1994. Post-metaphysical Culture and Politics: Richard Rorty and Laclau and Mouffe. *Economy and Society* 23: 173–200.

Davis, Bernard D. 1987. Bacterial Domestication: Underlying Assumptions. *Science* 235: 1329–1335.

de Chadarevian, Soraya. 1994. Building Molecular Biology in Post War Britain. Paper given at Fourth Mellon Workshop, "Institutional and Disciplinary Contexts of the Life Sciences in the Late Twentieth Century." Massachusetts Institute of Technology.

de Meyer, Arnoud, and Atsuo Mizushima. 1992. Global R&D Management. *R&D Management* 19: 135–146.

de Rosnay, Joel. 1979. *Biotechnologies et Bio-Industrie*. Seuil/La Documentation Francaise.

Derrida, Jacques. 1983. *Grammatologie*. Suhrkamp.

Deutsch, Erwin. 1986. Zur Arbeit der Enquete-Kommissionscholz "Chancen und Risken der Gentechnologie. In *Rechtsfragen der Gentechnologie,* ed. R. Lukes and R. Scholz. Carl Heymanns.

de Woot, Philippe. 1990. *High Technology Europe: Strategic Issues for Global Competitivness*. Blackwell.

Dickson, David. 1989. Genome Project gets Rough Ride in Europe. *Science* 243: 399.

Digeser, Peter. 1992. The Fourth Face of Power. *Journal of Politics* 54: 977–1007.

DiMaggio, Paul J., and Walter W. Powell. 1991. Introduction. In *The New Institutionalism in Organizational Analysis,* ed. P. DiMaggio and W. Powell. University of Chicago Press.

Dobbin, Frank. 1994. *Forging Industrial Policy: The United States, Britain, and France in the Railway Age.* Cambridge University Press.

Dodgson, Mark. 1991. Strategic Alignment and Organizational Options in Biotechnology Firms. *Technology Analysis and Strategic Management* 3: 115–125.

Dowding, Keith. 1995. Model or Metaphor? A Critical Review of the Policy Network Approach. *Political Studies* 43: 136–158.

Drake, Charles D., and Frank B. Wright. 1983. *Law of Health and Safety at Work: The New Approach.* Sweet & Maxwell.

Dreyfus, Hubert L., and Paul Rabinow. 1983. *Michel Foucault: Beyond Structuralism and Hermeneutics.* University of Chicago Press.

Duclos, Denis H., and Jocelyne J. Smadja. 1985. Culture and Environment in France. *Environmental Management* 9: 135–140.

Duster, Troy. 1990. *Backdoor to Eugenics.* Routledge.

Elizalde, Jose. 1987. Legal Aspects of Community Policy on Research and Technological Development (RTD). *Common Market Law Review* 30: 309–345.

Ellington, Stephen. 1995. Understanding the Dialectic of Discourse and Collective Action: Public Debate and Rioting in Antebellum Cincinatti. *American Journal of Sociology* 101: 100–144.

Elzinga, Aant, and Andrew Jamison. 1995. Changing Policy Agendas in Science and Technology. In *Handbook of Science and Technology Studies,* ed. S. Jasanoff et al. Sage.

Esser, Josef, Wolfgang Fach, and Kenneth Dyson. 1983. "Social Market" and Modernization Policy: West Germany. In *Industrial Crisis,* ed. K. Dyson and S. Wilks. Blackwell.

Evans, Peter, Dieter Rueschemeyer, and Theda Skocpol, eds. 1985. *Bringing the State Back In.* Cambridge University Press.

Evans, Steve, Keith Ewing, and Peter Nolan. 1992. Industrial Relations and the British Economy in the 1990s: Mrs. Thatcher's Legacy. *Journal of Management Studies* 29: 571–589.

Fenno, Richard. 1978. *Home Style: House Members in Their Districts.* Little, Brown.

Ferguson, James. 1994. *The Anti-Politics Machine: "Development," Depoliticization, and Bureaucratic Power in Lesotho.* University of Minnesota Press.

Ferguson, Thomas. 1989. Industrial Conflict and the Coming of the New Deal: The Triumph of Multinational Liberalism in America. In *The Rise and Fall of the New Deal Order, 1930–1980,* ed. S. Fraser and G. Gerstle. Princeton University Press.

Fischer, Frank. 1995. *Evaluating Public Policy.* Nelson-Hall.

Fischer, Frank. 1996. But Is It Scientific? Local Knowledge in Postpositivist Perspective. Manuscript, New York.

Fischer, Frank, and Jon Forester, eds. 1993. *The Argumentative Turn in Policy Analysis and Planning.* Duke University Press.

Fisher, Donald. 1990. Boundary Work and Science: The Relation Between Power and Knowledge. In *Theories of Science and Society,* ed. S. Cozzens and T. Gieryn. Indiana University Press.

Fleck, Roland. 1990. *Technologieförderung. Schwachstellen, europäische Perspektiven und neue Ansätze.* Deutscher Universitätsverlag.

Foucault, Michel. 1967. *Madness and Civilisation.* Tavistock.

Foucault, Michel. 1972. *The Archaeology of Knowledge and the Discourse on Language.* Pantheon.

Foucault, Michel. 1980. *The History of Sexuality,* volume I: *An Introduction.* Vintage.

Foucault, Michel. 1984. Nietsche, Genealogy, History. In *The Foucault Reader,* ed. P. Rabinow. Penguin.

Frank, Manfred. 1989. *What Is Neostructuralism?* University of Minnesota Press.

Frankland, E. Gene, and Donald Schoonmaker. 1992. *Between Protest and Power: The Green Party in Germany.* Westview.

Game, Ann. 1991. *Undoing the Social: Towards a Deconstructive Sociology.* Toronto University Press.

Gardener, Edward P. M., and Philip Molyneux. 1990. *Changes in West European Banking.* Unwin Hyman.

Gaudillière, Jean-Paul. 1991. Biologie moléculaire et biologists dans les année soixante: La naissance d'une discipline. Le cas français, Thèse, Université de Paris.

George, Jim. 1994. *Discourses of Global Politics: A Critical (Re)Introduction to International Relations.* Lynne Rienner.

George, Jim, and David Campell. 1990. Patterns of Dissent and the Celebration of Difference: Critical Social Theory and International Relations. *International Studies Quarterly* 34: 269–293.

Gerschenkron, Alexander. 1962. *Economic Backwardness in Historical Perspective.* Harvard University Press.

Gieryn, Thomas F. 1983. Boundary-Work and the Demarcation of Science from Non-Science: Strains and Interests in Professional Interests of Scientists. *American Sociological Review* 48: 781–795.

Gieryn, Thomas F. 1995. "Boundaries of Science. In *Handbook of Science and Technology Studies,* ed. S. Jasanoff et al. Sage.

Gill, Bernhard. 1991. *Gentechnik ohne Politik. Wie die Brisanz der Synthetischen Biologie von wissenschaftlichen Institutionen, Ethik- und anderen Kommissionen systematisch verdrängt wird.* Campus.

Gill, Stephen. 1991. Reflections on Global Order and Sociohistorical Time. *Alternatives* 16: 275–314.

Gilpin, R. 1968. *France in the Age of the Scientific State.* Princeton University Press.

Glaser, Vicki. 1992. Strong Growth in Biotechnology Market Sectors Predicted for 1992–2002. *Genetic Engineering News,* March 12.

Glémas, Patrick. 1989. BST. Une Hormone tres politique. *Agriculture Magazine,* February: 45–50.

Goffman, Erving. 1974. *Frame Analysis.* Harper.

Goldstein, Judith, and Robert O. Keohane, eds. 1993a. *Ideas and Foreign Policy: Beliefs, Institutions, and Political Change.* Cornell University Press.

Goldstein, Judith, and Robert O. Keohane. 1993b. Ideas and Foreign Policy: An Analytical Framework. In *Ideas and Foreign Policy,* ed. J. Goldstein and R. Keohane. Cornell University Press.

Gooding, Robert E., and Hans-Dieter Klingemann. 1996. Political Science: The Discipline. In *A New Handbook of Political Science,* ed. R. Gooding and H.-D. Klingemann. Oxford University Press.

Gordon, Colin. 1991. Government Rationality: An Introduction. In *The Foucault Effect,* ed. G. Burchell et al. University of Chicago Press.

Gottschling, Claudia. 1995. Forscher lernen feilschen. *Focus,* May 29.

Gottweis, Herbert. 1988. *Die Welt der Gesetzgebung: Rechtsalltag in Österreich.* Böhlau.

Graf, William D. 1992. Internationalization and Exoneration: Social Functions of the Transnationalizing West German Political Economy in the Post-War Era. In *The Internationalization of the German Political Economy,* ed. W. Graf. St. Martin's.

Grafstein, Robert. 1992. *Institutional Realism: Social and Political Constraints on Rational Actors.* Yale University Press.

Grant, Wyn. 1989. *Government and Industry: A Comparative Analysis of the US, Canada and the UK.* Edward Elgar.

Grant, Wyn. 1993. Pressure Groups and the European Community. An Overview. In *Lobbying in the European Community,* ed. S. Mazey and J. Richardson. Oxford University Press.

Grant, Wyn, William Paterson, and Colin Whitston. 1988. *Government and the Chemical Industry: A Comparative Study of Britain and West Germany.* Clarendon.

Greenwood, J., and K. Ronit. 1991. Organized Interests and the European Internal Market. *Environment and Planning C* 9: 467–484.

Groet, Suzanne S. 1991. Biotechnology and the US Government: The Pot at the End of the Rainbow? In *The Business of Biotechnology,* ed. R. Ono. Butterworth-Heinemann.

Gros, François François Jacob, and Pierre Royer. 1979. *Sciences de la Vie et Société. Rapport au Président de la République* (Paris: La Documentation Française.

Gummet, Philip. 1991. History, Development and Organisation of UK Science and Technology up to 1982. In *Science and Technology in the United Kingdom,* ed. R. Nicholson et al. Longman.

Gunnell, John G. 1975. *Philosophy, Science, and Political Inquiry.* General Learning Press.

Gunnell, John G. 1983. Political Theory: The Evolution of a Sub-Field. In *Political Science,* ed. A. Finifter. American Political Science Association.

Haas, Peter M. 1992. Introduction: Epistemic Communities and International Policy Coordination. *International Organization* 46: 1–36.

Hacking, Ian. 1984. *Representing and Intervening: Introductory Topics in the Philosophy of Natural Science.* Cambridge University Press.

Hagerdorn, Charles. 1989. Potential and Risk in Commercial Use of Microorganisms. *Forum for Applied Research and Public Policy* 4: 84–91.

Haila, Yrjö, and Lassi Heininen. 1995. Ecology. A New Discipline for Disciplining? *Social Text* 13: 153–171.

Hainsworth, Paul. 1990. Breaking the Mould: The Greens in the French Party System. In *French Political Parties in Transition,* ed. A. Cole. Dartmouth.

Hajer, Maarten A. 1994. Managing the Metaphors: Global Environmental Constructs and the Missing Public Domain. Paper prepared for workshop on Science Studies, International Relations and the Global Environment, Cornell University.

Hajer, Maarten A. 1995. *The Politics of Environmental Discourse: Ecological Modernization and the Policy Process.* Clarendon.

Hall, Peter A. 1986. *Governing the Economy: The Politics of State Intervention in Britain and France.* Oxford University Press.

Hall, Peter A., ed. 1989. *The Political Power of Economic Ideas: Keynesianism across Nations.* Princeton University Press.

Hall, Peter A., and Rosemary C. R. Taylor. 1994. Political Science and the Four New Institutionalisms. Paper given at annual meeting of the American Political Science Association, New York.

Hall, Stuart. 1989. The Toad in the Garden. Thatcherism among the Theorists. In *Marxism and the Interpretation of Culture,* ed. C. Nelson and L. Grossberg. Macmillan.

Halpern, Nina P. 1993. Creating Socialist Economies: Stalinist Political Economy and the Impact of Ideas. In *Ideas and Foreign Policy,* ed. J. Goldstein and R. Keohane. Cornell University Press.

Hampel, Frank. 1991. Politikberatung in der Bundesrepublik: Überlegungen am Beispiel von Enquete-Kommissionen. *Zeitschrift für Parlamentsfragen* 22: 110–133.

Harvey, David. 1989. *The Condition of Postmodernity: An Enquiry into the Origins of Cultural Change.* Blackwell.

Hassard, John, and Martin Parker. 1993. *Postmodernism and Organizations.* Sage.

Henriques, Julian, Wendy Holway, Cathy Urwin, Couze Venn, and Valerie Walkerdine. 1984. *Changing the Subject: Psychology, Social Regulation and Subjectivity.* Methuen.

Hértier, Adrienne. 1993. Einleitung. Policy-Analyse. Elemente der Kritik und Perspektiven der Neuorientierung. In *Policy-Analyse,* ed. A. Hértier. Westdeutscher Verlag.

Hértier, Adrienne, Susanne Mingers, Chrisoph Knill, and Martina Becka. 1994. *Die Veränderung von Staatlichkeit in Europa.* Leske + Budrich.

Herman, Ros. 1986. *The European Scientific Community.* Longman.

Hill, Julie, and Green Alliance. 1990. Influences on Specific Sectors: Environment. In *The Impact of New and Impending Regulations on UK Biotechnology,* ed. D. Bennett and B. Kirsop. Cambridge Biomedical Consultants.

Hirsch, Joachim. 1970. *Wissenschaftlich-technischer Fortschritt und politisches System. Organisation und Grundlagen administrativer Wissenschaftsförderung in der BRD.* Suhrkamp.

Hoffmann, Stanley. 1991. The Institutions of the Fifth Republic. In *Searching for the New France,* ed. J. Hollifield and G. Ross. Routledge.

Hollingworth, Rogers, Philippe Schmitter, and Wolfgang Streeck, eds. 1994. *Governing Capitalist Economies: Performance and Control of Economic Sectors.* Oxford University Press.

Holmes, Peter. 1993. Towards a Common Industrial Policy in the EC? *European Business Journal* 5: 25–35.

Horn, Ernst-Jürgen. 1987. Germany: A Market-Led Process. In *Managing Industrial Change in Western Europe,* ed. F. Duchêne and G. Sheperd. Frances Pinter.

Huber, Joseph. 1985. *Die Regenbogengesellschaft: Ökologie und Sozialpolitik.* Fischer.

Hucke, Jochen. 1990. Umweltpolitik: Die Entwicklung eines neuen Politikfelds. In *Politik in der Bundesrepublik Deutschland,* ed. K. von Beyme and M. Schmid. Westdeutscher Verlag.

Hunt, Alan. 1992. Foucault's Expulsion of Law: Toward a Retrieval. *Law and Social Inquiry* 17: 1–38.

Immergut, Ellen M. 1992. *Health Politics: Interests and Institutions in Western Europe.* Cambridge University Press.

Irvine, Julia. 1990. Risk-Averse Lenders in a High-Risk Area. *Accountancy* 106,134–135.

Jacob, François. 1988. *The Statue Within: An Autobiography.* Basic Books.

Jacobs, Francis, Richard Corbett, and Michael Shackleton. 1992. *The European Parliament.* Longman.

Japp, Klaus P. 1986. Neue soziale Bewegungen und die Kontinuität der Moderne. *Soziale Welt* Sonderband 4: 311–333.

Jasanoff, Sheila. 1985. Technological Innovation in a Corporatist State: The Case of Biotechnology in the Federal Republic of Germany. In: *Research Policy* 14: 23–38.

Jasanoff, Sheila. 1990. *The Fifth Branch: Science Adivsers as Policymakers.* Harvard University Press.

Jasanoff, Sheila, Gerald Markle, Trevor Pinch and James Petersen, eds. 1995. *Handbook of Science and Technology Studies.* Sage.

Jessop, Bob. 1988. Conservative Regimes and the Transition to Post-Fordism: The Case of Britain and West Germany. Papers in Politics and Government no. 47, University of Essex.

Jessop, Bob. 1990. *State Theory: Putting Capitalist States in Their Place.* Pennsylvania State University Press.

Jessop, Bob. 1991. Thatcherism and Flexibility: The White Heat of a Post-Fordist Revolution. In *The Politics of Flexibility,* ed. B. Jessop et al. Edward Elgar.

Jessop, Bob. 1992. From Social Democracy to Thatcherism: Twenty-Five Years of British Politics. In *Social Change in Contemporary Britain,* ed. N. Abercrombie and A. Warde. Polity Press.

Jobert, Bruno, and Pierre Muller. 1987. *L'Etat en Action. Politiques Publiques et Corporatismes.* Presses Universitaires de France.

Jordan, Grant. 1990. Policy Realism versus "New" Institutionalist Ambiguity. *Political Studies* 38: 470–484.

Jordan, Grant, and Jeremy Richardson. 1982. The British Policy Style or the Logic of Negotiation? In *Policy Styles in Western Europe,* ed. J. Richardson. Allen & Unwin.

Jordan, Grant, and Klaus Schubert. 1992. A Preliminary Ordering of Policy Network Labels. *European Journal of Political Research* 21: 7–27.

Katzenstein, Peter. 1987. *Policy and Politics in West Germany: The Growth of the Semisovereign State.* Temple University Press.

Kay, Lily E. 1993. *The Molecular Vision of Life: Caltech, the Rockefeller Foundation, and the Rise of the New Biology.* Oxford University Press.

Kay, Lily E. 1995. Who Wrote the Book of Life? Manuscript, Boston.

Keller, Evelyn Fox. 1992. Nature, Nurture, and the Human Genome Project. In *The Code of Codes,* ed. D. Kevles and L. Hood. Harvard University Press.

Kellner, Douglas. 1989. *Jean Baudrillard: From Marxism to Postmodernism and Beyond.* Stanford University Press.

Kenward, Michael. 1992. Little Ventured, Nothing Gained. *Director* 46: 35.

Kevles, Daniel J. 1985. *In the Name of Eugenics: Genetics and the Uses of Human Heredity.* Knopf.

Kitschelt, Herbert. 1989. *The Logics of Party Formation: Ecological Politics in Belgium and West Germany.* Cornell University Press.

Klingmüller, Walter, ed. 1980. *Genforschung im Widerstreit.* Wissenschaftliche Verlagsgesellschaft.

Kloepfer, Michael. 1989. *Umweltrecht.* C. H. Beck.

Kohler, Robert E. 1976. The Management of Science: The Experience of Warren Weaver and the Rockefeller Foundation Programme in Molecular Biology. *Minerva* 14: 249–293.

Kollek, Regine. 1989. Neue Kriterien für die Abschätzung des Risikos. In *Gentechnik—Wer kontrolliert die Industrie?* ed. M. Thurau. Fischer.

Krämer, Ludwig. 1990. *EEC Treaty and Environmental Protection.* Sweet & Maxwell.

Krimsky, Sheldon. 1982. *Genetic Alchemy: The Social History of the Recombinant DNA Controversy.* MIT Press.

Kuisel, Richard F. 1981. *Capitalism and the State in Modern France: Renovation and Economic Management in Twentieth Century.* Cambridge University Press.

Kusch, Martin. 1991. *Foucault's Strata and Fields: An Investigation into Archeological and Genealogical Science Studies.* Kluwer.

Laclau, Ernesto. 1990. *New Reflections on the Revolution of Our Time.* Verso.

Laclau, Ernesto, and Chantal Mouffe. 1985. *Hegemony and Socialist Strategy: Toward a Radical Democratic Politics.* Verso.

Latour, Bruno. 1987. *Science in Action: How to Follow Scientists and Engineers Through Society.* Harvard University Press.

Latour, Bruno. 1990. Drawing Things Together. In *Representation in Scientific Practice,* ed. M. Lynch and S. Woolgar. MIT Press.

Latour, Bruno. 1993. *We Have Never Been Modern.* Harvard University Press.

Leborgne, D., and A. Lipietz. 1988. New Technologies, New Modes of Regulation: Some Spatial Implications. *Environment and Planning D* 8: 263–280.

Lefebre, Henri. 1991. *The Production of Space.* Blackwell.

Leggewie, Claus, 1985. Propheten ohne Macht: Die neuen sozialen Bewegungen in Frankreich zwischen Resignation und Fremdbestimmung. In *Neue soziale Bewegungen in Westeuropa und den USA,* ed. K.-W. Brand. Campus.

Lenoir, Timothy. 1992. The Discipline of Nature and the Nature of Discipline. Manuscript, Stanford University.

Lenoir, Timothy. 1994. Was the Last Turn the Right Turn? The Semiotic Turn and A. J. Greimas. *Configurations* 1: 119–136.

Levidow, Les, and Joyce Tait. 1992. Release of Genetically Modified Organisms: Precautionary Legislation. *Project Appraisal* 7: 93–105.

Levin, Morris, and Harlee S. Strauss. 1991. *Risk Assessment in Genetic Engineering.* McGraw-Hill.

Lewin, Benjamin. 1990. *Genes IV.* Oxford University Press.

Lex, Maurice. 1990. The UK Science Base. In *The UK Biotechnology Handbook '90,* ed. A. Crafts-Lighty et al. Bioindustry Association.

Linstead, Stephen, and Robert Grafton-Small. 1992. On Reading Organizational Culture. *Organization Studies* 13: 331–355.

Linstead, Steve. 1993. Deconstruction in the Study of Organizations. In *Postmodernism and Organization,* ed. J. Hassard and M. Parker. Sage.

Lipietz, Alain. 1991. Governing the Economy in the Face of International Challenge: From National Developmentalism to National Crisis. In *Searching for the New France,* ed. J. Hollifield and G. Ross. Routledge.

Ludmerer, Kenneth L. 1972. *Eugenics and American Society: A Historical Survey.* Johns Hopkins University Press.

Lukes, Steven. 1974. *Power: A Radical View.* Macmillan.

Lukes, Rudolf, and Rupert Scholz, eds. 1986. *Rechtsfragen der Gentechnologie.* Carl Heymanns.

Lyotard, Jean-François. 1984. *The Post-Modern Condition: A Report on Knowledge.* Manchester University Press.

Malunat, Bernd M. 1994. Die Umweltpolitk der Bundesrepublik Deutschland. *Aus Politik und Zeitgeschichte. Beilage zur Wochenzeitung Das Parlament* B 49/94: 3–12.

Martin, Wallace. 1986. *Recent Theories of Narrative.* Cornell University Press.

Mayntz, Renate. 1994. Policy-Netzwerke und die Logik von Verhandlungssystemen. In *Policy-Analyse,* ed. A. Hértier. Westdeutscher Verlag.

Mazey, Sonia. 1986. Public Policy-Making in France: The Art of the Possible. *West European Politics* 9: 412–428.

McArthur, R. 1990. Replacing the Concept of High Technology: Towards a Diffusion-Based Approach. *Environment and Planning A* 22: 811–828.

McCormick, John. 1991. *British Politics and the Environment.* Earthscan.

McKay, David. 1993. Economic Difficulties and Government Response 1931–1993. In *Developing British Political System: the 1990s,* ed. I. Budge and D. McKay. Longman.

Melucci, Alberto. 1985. The Symbolic Challenge of Contemporary Movements. *Social Research* 52: 789–816.

Melucci, Alberto. 1988. Social Movements and the Democratization of Everyday Life. In *Civil Society and the State,* ed. J. Keane. Verso, 1988.

Messerlin, Patrick. 1987. France: The Ambitious State. In *Managing Industrial Change in Western Europe,* ed. F. Duchêne and G. Sheperd. Pinter.

Metha, Judith. 1993. Meaning in the Context of Bargaining Games: Narratives in Opposition. In *Economics and Language,* ed. W. Henderson et al. Routledge.

Middlemas, Keith. 1991. *Power, Competition and the State*, volume 3: *The End of the Postwar Era: Britain since 1974.* Macmillan.

Morgan, Joan, and W. J. Whelan. 1979. *Recombinant DNA and Genetic Experimentation.* Pergamon.

Morris, Aldon B., and Carol McClurg Mueller, eds. 1992. *Frontiers in Social Movement Theory.* Yale University Press.

Murray, Gordon C. 1992. A Challenging Marketplace for Venture Capital. *Long Range Planning* 25, no. 6: 79–86.

Narjes, Karl Heinz. 1988. Europe's Technological Challenge: a View from the European Commission. *Science and Public Policy* 15: 395–402.

Nelkin, Dorothy, and Susan Lindee. 1996. *The DNA Mystique: The Gene as Cultural Icon.* Freeman.

Newmark, Peter. 1989. One of the Better Models of European Cooperation. *Nature* 338: 724.

Nicolaides, Phedon. 1993. Industrial Policy: The Problem of Reconciling Definitions, Intentions and Effects. In *Industrial Policy in the European Community,* ed. P. Nicolaides. Martinus Nijhoff.

Nielsen, Klaus. 1991. Towards a Flexible Future—Theories and Politics. In *The Politics of Flexibility,* ed. B. Jessop et al. Edward Elgar.

Nordlinger, Eric. 1981. *On the Autonomy of the Democratic State.* Harvard University Press.

Oakey, Ray, Wendy Faulkner, Sarah Cooper, and Vivien Walsh. 1990. *New Firms in the Biotechnology Industry: Their Contribution to Innovation and Growth.* Pinter.

OECD. 1989. *Biotechnology. Economic and Wider Impacts.* OECD.

OECD, Alan T. Bull, Geoffrey Holt, and Malcolm D. Lilly. 1982. *Biotechnology. International Trends and Perspectives.* OECD.

Offe, Claus. 1985. New Social Movements: Challenging the Boundaries of Institutional Politics. *Social Research* 52: 817–868.

O'Neill, Patrick. 1994. *Fictions of Discourse: Reading Narrative Theory.* University of Toronto Press.

Orsenigo, Luigi. 1989. *The Emergence of Biotechnology: Institutions and Markets in Industrial Innovation.* St. Martin' s.

Overbeck, Henk. 1990. *Global Capitalism and National Decline: The Thatcher Decade in Perspective.* Unwin Hyman.

Patzelt, Werner J. 1993. *Abgeordnete und Repräsentation: Amtsverständnis und Wahlkreisarbeit.* Richard Rothe.

Patzelt, Werner J. 1987a. *Grundlagen der Ethnomethodologie. Theorie, Empirie und politikwissenschaftlicher Nutzen einer Soziologie des Alltags.* Wilhelm Fink.

Patzelt, Werner J. 1987b. *Theorie, Empirie und politikwissenschaftlicher Nutzen einer Soziologie des Alltags.* Wilhelm Fink.

Pauly, Philip J. 1993. Essay Review: The Eugenics Industry—Growth or Restructuring? *Journal of the History of Biology* 26: 131–145.

Peterson, John. 1991. Technology Policy in Europe: Explaining the Framework Porgramme and Eureka in Theory and Practice. *Journal of Common Market Studies* 29: 271–290.

Pinch, Trevor J., and Wiebe E. Bijker. 1987. The Social Construction of Facts and Artifacts: Or How the Sociology of Science and the Sociology of Technology Might Benefit Each Other. In *The Social Construction of Technological Systems,* ed. W. Bijker et al. MIT Press.

Pisano, Gary P. 1991. The Governance of Innovation: Vertical Integration and Collaborative Arrangements in the Biotechnology Industry. *Research Policy* 20: 237–249.

Pogunkte, Thomas. 1993. *Alternative Politics: The German Green Party.* Edinburgh University Press.

Pohlmann, Andreas. 1990. *Neuere Entwicklungen im Gentechnikrecht: Rechtliche Grundlagen und aktuelle Gesetzgebung für gentechnische Industrievorhaben.* Duncker & Humbold.

Prendville, Brendan. 1994. *Environmental Politics in France.* Westview.

Prier, Michel. 1984. *Droit de l'environment.* Dalloz.

Proctor, Robert N. 1992. Genomics and Eugenics: How Fair is the Comparison? In *Gene Mapping,* ed. G. Annas and S. Elias. Oxford University Press.

Pronier, Raymond, and Vincent Jacques le Seigneur. 1992. *Génération verte. Les écologistes en politique.* Presses de la Renaissance.

Raschke, Joachim. 1991. *Krise der Grünen Bilanz und Neubeginn.* Schüren.

Raugel, Pierre-Jean. 1990. Création des sociétés indépendent spécialisées en biotechnologies et en biologie en France. *Biofutur,* March: 95–104.

Raugel, Pierre-Jean. 1992. An Impressionist View of the French Biotechnology Industry. *BFE* 9: 206–211.

Reed, Michael, and Michael Hughes. 1992. *Rethinking Organization: New Directions in Organization Theory and Analysis.* Sage.

Regal, P. J. 1994. Scientific Principles for Ecologically Based Risk Assessment of Transgenic Organisms. *Molecular Ecology* 3: 5–13.

Reich, Simon. 1990. *The Fruits of Fascism: Postwar Prosperity in Historical Perspective.* Cornell University Press.

Rheinberger, Hans-Jörg. 1992. Experiment, Difference, and Writing, Part I and Tracing Protein Synthesis, Part II: The Laboratory Life of Transfer RNA. *Studies in History and Philosophy of Science* 23: 305–331 and 389–422.

Rheinberger, Hans-Jörg. 1993. Genetic Engineering and the Practice of Molecular Biology. Paper given at the Mellon Workshop on Genetic Engineering: Transformation in Science, Politics and Culture, Massachusetts Institute of Technology.

Rheinberger, Hans-Jörg. 1994. Representations: Essay Review. *Studies in History and Philosophy of Science* 25: 647–654.

Rheinberger, Hans-Jörg. 1995. Kurze Geschichte der Molekularbiologie. Preprint 24, Max-Planck-Institut für Wissenschaftsgeschichte, Berlin.

Rhodes, R. A. W. 1988. *Beyond Westminster and Whitehall.* Unwin Hyman.

Riessman, Catherine Kohler. 1993. *Narrative Analysis.* Sage.

Riker, William H., and Peter C. Ordeshook. 1973. *An Introduction to Positive Political Theory.* Prentice-Hall.

Robinson, Mike. 1992. *Greening of British Party Politics.* Manchester University Press.

Robson, Keith. 1993. Governing Science and Economic Growth at a Distance: Accounting Representation and the Management of Research and Development. *Economy and Society* 22: 461–481.

Ronayne, Jarlath. 1984. *Science in Government.* Arnold.

Ronge, Volker. 1986. Zur politischen Steuerbarkeit des technologischen Wandels. In *Forschungs- und Technologiepolitik in der Bundesrepublik Deutschland,* ed. W. Bruder. Westdeutscher Verlag.

Rorty, Richard. 1989. *Contingency, Irony, and Solidarity.* Cambridge University Press.

Rose, Nikolas. 1994. Expertise and the Government of Conduct. *Studies in Law, Politics and Society* 14 (1994), 359–397.

Rose, Nikolas. 1996. Governing "Advanced" Liberal Democracies. In *Foucault and Political Reason,* ed. A. Barry et al. University of Chicago Press.

Rose, Nikolas, and Peter Miller. 1992. Political Power Beyond the State: Problematics of Government. *British Journal of Sociology* 43: 172–205.

Rosenau, Pauline Marie. 1992. *Post-Modernism and the Social Sciences: Insights, Inroads, and Intrusions.* Princeton University Press.

Roth, Roland. 1985. Neue soziale Bewegungen in der politischen Kultur der Bundesrepublik—eine vorläufige Skizze. In *Neue soziale Bewegungen in Westeuropa und den USA,* ed. K.-W. Brand. Campus.

Roth, Roland. 1991. Abkehr vom Etatismus. In *Die Bundesrepublik in den achziger jahren,* ed. W. Süss. Leske+Budrich.

Roth, Roland, and Dieter Rucht. 1987. *Neue soziale Bewegungen in der Bundesrepublik Deutschland.* Campus.

Rothwell, Roy, and Mark Dodgson. 1992. European Technology Policy Evolution: Convergence Towards SMEs and Regional Technology Transfer. *Technovation* 12: 223–238.

Rouban, Luc. 1988. *L'etat et la science. la politique publique de la science et de la technologie.* Editions du CNRS.

Ryder, Maarten. 1994. Key Issues in the Deliberate Release of Genetically Manipulated Bacteria. *FEMS Microbiology Ecology* 15: 139–146.

Saalfeld, Thomas. 1990. The West German Bundestag after 40 Years: The Role of Parliament in a "Party Democracy." *West European Politics* 13: 68–89.

Salomon, Jean-Jacques. 1991. La Capacité d'Innovation. In *Entre "Etat et le Marché,"* ed. M. Lévy-Leboyer and J.-C. Casanove. Gallimard.

Salter, Brian and Ted Tapper. 1993. The Application of Science and Scientific Autonomy in Great Britain: A Case Study of the Science and Engineering Research Council. *Minerva* 31: 38–55.

Sandholtz, Wayne. 1992. *High-Tech Europe: The Politics of International Cooperation.* University of California Press.

Sarkar, Saral. 1993, 1994. *Green-Alternative Politics in West Germany* (two volumes). United Nations University Press.

Scharpf, Fritz W., and Volker Hauff. 1975. *Modernisierung der Volkswirtschaft.* Europäische Verlagsanstalt.

Scholz, Rudolf. 1986. Die Gentechnologie aus der Sicht des Rechts der Technik. *Deutsches Verwaltungsblatt* 37: 1221–1230.

Senker, Jacqueline. 1991. UK Biotechnology. Technology Transfer Involving Small and Medium-sized Firms. *Industry and Higher Education,* June: 108–113.

Senker, Jacqueline, and Margret Sharp. 1988. The Biotechnology Directorate of the SERC. Report and Evaluation of its Achievements 1981–1987. Report submitted to the Management Committee of the Biotechnology Directorate, Science Policy Unit, Sussex.

Shackley, Simon and Margaret Sharp. 1989. Environmental Release: Don't Trust the DoE. *International Industrial Biotechnology* 9: 26–28.

Shapiro, J., L. MacHattie, L. Eron, G. Ihler, K. Ippen, and J. Beckwith. 1969. The isolation of pure lac operon DNA. *Nature* 224: 768–774.

Sharp, Margret. 1989. European Technology—Does 1992 Matter? Papers in Science, Technology and Public Policy No. 19, Science Policy Research Unit, University of Essex.

Sharp, Margret. 1991a. Pharmaceuticals and Biotechnology: Perspectives for the European Industry. In *Technology and the Future of Europe,* ed. C. Freeman et al. Pinter.

Sharp, Margret 1991b. The Single Market and European Technology Policies. In *Technology and the Future of Europe,* ed. C. Freeman et al. Pinter.

Sharples, Frances E. 1987. Regulation of Products from Biotechnology. *Science* 235: 1329–1332.

Shattock, Michael L. 1991. Higher Education and the Research Councils. In *Science and Technology in the United Kingdom,* ed. R. Nicholson et al. Longman.

Sheperd, Geoffrey. 1987. United Kingdom: A Resistance to Change. In *Managing Industrial Change in Western Europ,* ed. F. Duchéne and G. Sheperd. Pinter.

Shepsle, Kenneth A. 1989. Studying Institutions: Some Lessons from the Rational Choice Approach. *Journal of Theoretical Politics* 1: 131–147.

Shepsle, Kenneth A., and Barry R. Weingast. 1994. Positive Theories of Congressional Institutions. *Legislative Studies Quarterly* 19: 149–179.

Simonis, Udo E. 1995. Ecological Modernization of Industrial Society: Three Strategic Elements. *International Social Science Journal* 41: 347–361.

Simons, Jon. 1995. *Foucault and the Political.* Routledge.

Smart, Barry. 1982. Foucault, Sociology, and the Problem of Human Agency. *Theory and Society* 11: 121–141.

Smart, Barry. 1985. *Michel Foucault.* Ellis Horwood.

Smart, Barry. 1992a. *Modern Conditions, Postmodern Controversies.* Routledge.

Smart, Barry. 1992b. The Politics of Truth. In *Foucault: A Critical Reader,* ed. D. Couzens Hoy. Blackwell.

Spaargaren, Gert, and Arthur P.J. Mol. 1992. Sociology, Environment, and Modernity: Ecological Modernization as a Theory of Social Change. *Society and Natural Ressources* 5: 323–344.

Star, Susan Leigh, and James R. Griesemer. 1989. Institutional Ecology, "Translations" and Boundary Objects: Amateurs and Professionals in Berkeley's Museum of Vertebrate Zoology, 1907–39. *Social Studies of Science* 19: 387–420.

Stent, Gunther S., and Richard Calendar. 1978. *Molecular Genetics: An Introductory Narrative,* second edition. Freeman.

Streeck, Wolfgang, and Philippe C. Schmitter, eds. 1985. *Private Interest Government: Beyond Market and State.* Sage.

Strohman. Richard C. 1993. Ancient Genomes, Wise Bodies, Unhealthy People: Limits of a Genetic Paradigm in Biology and Medicine. *Perspectives in Biology and Medicine* 37: 112–145.

Süss, Werner. 1991. Zukunft durch Modernisierungspolitik. Das Leitthema der 80er Jahre. In *Die Bundesrepublik in den achtziger Jahren,* ed. W. Süss. Lesk und Budrich.

Tait, Joyce, and Les Levidow. 1992. Proactive and Reactive Approaches to Risk Regulation. The Case of Biotechnology. *Futures* 24: 219–231.

Tarabusi, Claudio Casadio. 1993. Globalisation in the Pharmaceutical Industry: Technological Change and Competition in a Triad Perspective. *STI Review,* December: 123–161.

Tarrow, Sidney. 1988. National Politics and Collective Action: Recent Theory and Research in Western Europe and the United States. *Annual Review of Sociology* 14: 421–440.

Tauber, Alfred I., and Sahota Sarkar. 1992. The Human Genome Project: Has Blind Reductionism gone Too Far? *Perspectives in Biology and Medicine* 35: 222–235.

Teitelman, Robert. 1994. *Profits of Science: The American Marriage of Business and Technology.* Basic Books.

Tesier, Robert. 1993. Ethique Environmentale et Théorie du Fait moral chez Durkheim. *Social Compass* 40: 437–449.

Thelen, Kathleen, and Sven Steinmo. 1992. Historical Institutionalism in Comparative Perspective. In *Structuring Politics,* ed. S. Steinmo et al. Cambridge University Press.

Thiebaut, Carlos. 1992. The Complexity of the Subject, Narrative Identity and the Modernity of the South. *Philosophy and Social Criticism* 18: 313–331.

Thompson, Graham. 1990. *The Political Economy of the New Right.* Twayne.

Tiedje, James M., Robert K. Colwell, Yaffa L. Grossman, Robert E. Hodson, Richard E. Lenski, Richard N. Mack, and Philip Regal. 1989. The Planned Introduction of Genetically Engineered Organisms: Ecological Considerations and Recommendations. *Ecology* 70: 298–315.

Touraine, Alain. 1984. *Le retour de l'acteur.* Fayard.

Touraine, Alain. 1992. Is Sociology Still the Study of Society? In *Between Totalitarianism and Postmodernity,* ed. P. Beilharz et al. MIT Press.

Truman, David B. 1951. *The Governmental Process.* Knopf.

Tschannen, Olivier, and François Hainard. 1993. Sociologie et Environment: Tropismes Disciplinaires ou Nouveau Paradigme? *Schweizersiche Zeitschrift für Soziologie* 19: 421–443.

Van der Meer, Robert R. 1986. EC-Biotechnology: A European Challenge. *TIBTECH,* November: 277–279.

van Elsas, J. D., and J. T. Trevors. 1991. Environmental Risks and Fate of Genetically Engineered Microorganisms in Soil. *Journal of Environmental Science and Health* 26: 981–1001.

Väth, Werner. 1984. Konservative Modernisierungpolitik—ein Widerspruch in sich? Zur Neuausrichtung der Forschungs- und Technologiepolitik der Bundesrepublik. *Prokla* 56: 83–103.

Vig, Norman J. 1968. *Science and Technology in British Politics.* Pergamon.

Vogel, David. 1986. *National Styles of Regulation: Environmental Policy in Great Britain and the United States.* Cornell University Press.

von Alemann, Ulrich, Peter Jansen, Heiderose Kilper, and Leo Kissler. 1988. *Technologiepolitik: Grundlagen und Perspektiven in der Bundesrepublik Deutschland und in Frankreich.* Campus.

von Thienen, Volker. 1986. Künftig ein technikgestaltendes Parlament? Zu den Empfehlungen der Enquete-Kommission "Technologiefolgenabschätzung." *Zeitschrift für Parlamentsfragen* 17: 548–557.

Vowe, Gerhard. 1986. Wissen, Interesse und Macht. Zur Technikgestaltung durch Enquete-Kommissionen. *Zeitschrift für Parlamentsfragen* 17: 557–568.

Wagner, Cynthia K. 1992. International R&D is the Rule. *Bio/Technology* 10 (1992): 529–531.

Ward, Mike. 1993. Rhône-Poulenc: From Bioscience to Market. *Bio/Technology* 11: 799–801.

Warmuth, Ekkehard. 1991. Biotechnologie-Förderung in den neuen Bundesländern, *BioEngineering* 7: 6.

Warren, Mark E. 1996. What Should We Expect from More Democracy? Radically Democratic Responses to Politics. *Political Theory* 24: 241–270.

Weale, Albert, Timothy O'Riordan, and Louise Krammer. 1991. *Controlling Pollution in the Round: Change and Choice in Environmental Regulation in Britain and West Germany.* Anglo-German Foundation.

Weiner, Klaus-Peter. 1990. Between Political Regionalization and Economic Globalization. Problems and Prospects of European Integration. *International Journal of Political Economy* 19: 41–62.

Weingart, Peter, Jürgen Kroll, and Kurt Bayertz. 1988. *Rasse, Blut und Gene: Geschichte der Eugenik und Rassenhygiene in Deutschland.* Suhrkamp.

Wells, Susan. 1996. *Sweet Reason: Rhetoric and the Discourses of Modernity.* University of Chicago Press.

Welsch, Wolfgang. 1996. *Vernunft. Die zeitgenoessische Vernunftkritik und das Konzept der transversalen Vernunft.* Suhrkamp.

Wheale, Peter, and Ruth McNally. 1990. UK Government Control of the Release of Genetically Engineered Organisms into the Environment. A Critical Evaluation. In *European Workshop on Law and Genetic Engineering,* ed. D. Leskien and J. Spangenberg. BBU.

White, Hayden. 1981. The Value of Narrativity in the Representation of Reality. In *On Narrative,* ed. W. Mitchell. University of Chicago Press.

Wilkie, Tom. 1991. *British Science since 1945.* Blackwell.

Wilks, Steven, and Maurice Wright, eds. 1987. *Comparative Government-Industry Relations.* Clarendon.

Wilsford, David. 1988. Tactical Advantages versus Administrative Heterogeneity: The Strengths and the Limits of the French State. *Comparative Political Studies* 21: 126–168.

Wolstenholme, Gordon. 1963. *Man and His Future.* Ciba Foundation and Churchill.

Wright, Susan. 1994. *Molecular Politics: Developing American and British Regulatory Policy for Genetic Engineering, 1972–198.* University of Chicago Press.

Wurzel, Gabriele, and Ernst Merz. 1991. Gesetzliche Regelungen von Fragen der Gentechnik und Humangenetik, Gentechnikgesetz und Humangenetikgesetz. *Aus Politik und Zeitgeschichte. Beilage zur Wochenzeitung Das Parlament,* February 1: 12–24.

Yanow, Dvora. 1996. *How Does a Policy Mean? Interpreting Policy and Organizational Action.* Georgetown University Press.

Yoxen, Edward. 1978. *The Social Impact of Molecular Biology,* Ph.D. dissertation, Cambridge University.

Yoxen, Edward. 1981. Life as a Productive Force: Capitalising the Science and Technology of Molecular Biology. In *Studies in the Labour Process,* volume 1, ed. R. Young and L. Levidow. CSE Books.

Yoxen, Edward. 1982. Giving Life a New Meaning: The Rise of the Molecular Biology Establishment. In *Scientific Establishments and Hierarchies,* ed. N. Elias et al. Reidel.

Yoxen, Edward J. 1984. Assessing Progress with Biotechnology. In *Science and Technology Policy in the 1980s and Beyond,* ed. M. Gibbons et al. Longman.

Zysman, John. 1983. *Governments, Markets, and Growth: Financial Systems and the Politics of Industrial Change.* Cornell University Press.

Zarnitz, Marie Luise. 1968. *Molekulare und physikalische Biologie.* Vandenhoeck & Ruprecht.

Zeuner, Bodo. 1991. Die Partei der Grünen. Zwischen Bewegung und Staat. In *Die Bundesrepublik in den achtziger Jahren,* ed. W. Süss. Lesk und Budrich.

Index